Advances in Plant Breeding Strategies

Volume 9

Series Editors

Jameel M. Al-Khayri, Department of Agricultural Biotechnology, King Faisal University, Al-Hassa, Saudi Arabia

Shri Mohan Jain, Department Agricultural Sciences, University of Helsinki, Helsinki, Finland

This book series is intended to provide a platform for plant scientists to present the latest advances in both innovative and traditional methodologies applied in plant breeding for the development of new crop varieties under different environmental conditions leading to sustainable food production, food and nutrition security, food supply chain, germplasm conservation and utilization. This series is an excellent reference source for plant breeders, geneticists, biotechnologists, and molecular biologists using molecular tools, bioinformatics, artificial intelligence, and other emerging technologies. The subject matter of all volumes will relate to postgraduate students specializing in agriculture, biotechnology, and molecular breeding as well as policymakers, seed companies, researchers, and professionals interested in plant genetic improvement.

Jameel M. Al-Khayri · Shri Mohan Jain ·
Suprasanna Penna
Editors

Biodiversity and Genetic Improvement of Medicinal and Aromatic Plants I

Springer

Editors
Jameel M. Al-Khayri
Department of Agricultural Biotechnology
King Faisal University
Al-Ahsa, Saudi Arabia

Shri Mohan Jain
Department of Agricultural Sciences
University of Helsinki
Helsinki, Finland

Suprasanna Penna
Amity Center for Nuclear Biotechnology
Amity Institute of Biotechnology
Amity University of Maharashtra
Mumbai, Maharashtra, India

ISSN 3004-8737 ISSN 3004-8745 (electronic)
Advances in Plant Breeding Strategies
ISBN 978-3-031-81845-5 ISBN 978-3-031-81846-2 (eBook)
https://doi.org/10.1007/978-3-031-81846-2

This Springer imprint is published by the registered company Springer Nature Switzerland AG
The registered company address is: Gewerbestrasse 11, 6330 Cham, Switzerland

Preface of the Series

Contemporary plant breeders rely on conventional and emerging methodologies to provide sustainable food production worldwide. However, the human population is growing at an alarming rate, especially in developing countries, and food availability to feed new mouths could gradually become a serious problem. Agriculture production is negatively impacted by environmental pollution, rapid industrialization, water scarcity, soil erosion of fertile topsoil, limited areas of potentially arable land, lack of improvement of local plant types, erosion of genetic diversity, and dependence on just a few major crop species for the world's food supplies. Breeding programs of these crops have been very much dependent on the ready availability of genetic variation, whether spontaneous or induced. Plant breeders and geneticists are under constant pressure to sustain food production by using innovative breeding strategies and enhancing minor crops that are well adapted to the marginal land and source of nutrition as well as abiotic and biotic stresses. An inherent limitation in traditional breeding is the introgression of one or few genes in a cultivar, which is carried out via backcrossing for several generations requiring extended time.

With the development of new molecular tools, molecular marker-assisted breeding helped rapid introgression of transgene and reduction of linkage drag. The continued development and adaptation of plant biotechnology approaches including genetic engineering, genome editing, molecular markers, genomics, proteomics, bioinformatics, artificial intelligence, and nanotechnology have established new tools for the creation, analysis, and manipulation of genetic variation for the development of improved cultivars to achieve agricultural sustainability and food security. A multidisciplinary approach combining traditional plant breeding, mutation breeding, plant biotechnology, and molecular biology would be strategically ideal for developing improved crop varieties to feed the world, provide natural pharmaceutical sources, and supply fiber and ornamental crops for industry, esthetic and environmental suppose. However, breeders are challenged by the constraints of environmental stress, the limitation of arable land expansion, and erosion of genetic resources particularly under the impact of global climate change.

This book series is intended to provide a platform for plant scientists to present the latest advances in both innovative and traditional methodologies applied in plant breeding for the development of new crop varieties under different environmental conditions leading to sustainable food production, food and nutrition security, food supply chain, germplasm conservation, and plant sustainable utilization. This series is an excellent reference source for plant breeders, geneticists, biotechnologists, and molecular biologists using molecular tools, bioinformatics, artificial intelligence, and other emerging technologies. The subject matter relates to postgraduate students specializing in agriculture, biotechnology, and molecular breeding as well as policymakers, seed companies, researchers, and professionals interested in plant genetic improvement.

Al-Ahsa, Saudi Arabia Jameel M. Al-Khayri
Helsinki, Finland Shri Mohan Jain

Preface

Plant breeding approaches have always been the foundations of crop improvement to achieve productivity. However, due to several agronomic and genetic constraints, it has become necessary to explore newer techniques to understand and breed novel plant types for food, nutritional and health perspectives. Medicinal and aromatic plants (MAPs) are an integral part of the global biological heritage and, since ancient time, are cultivated for their application in traditional and modern healthcare. The wealth of information on genetic resources, breeding, conservation, propagation, cultivation, and biotechnological strategies is crucial for plant improvement.

Biotechnological and genomics advances have revolutionized opportunities for more efficient breeding of MAPs. However, successful application requires a concrete understanding of the science and technology. This book provides a comprehensive coverage of molecular and advanced methodologies that are integrated within plant breeding programs for the improvement of MAPs. The ten chapters cover research advances in conventional and modern breeding technologies of various medicinal and important aromatic plants individually, including Brahmi, Barberry, Betel, Blond plantain, Caper, Chinese woad, Common hop, Deadly nightshade (Atropa), Ginkgo, and Gymnema. Chapters comprehensively review the contemporary literature on the subject and reflect the experiences of the authors. Each chapter emphasizes introduction covering related backgrounds and provides in-depth discussion of the subject supported with high-quality color illustrations and relevant data, future research perspective, and a comprehensive list of pertinent references.

The book is an excellent reference source for plant biologists, breeders, biotechnologists and geneticists engaged in plant improvement using plant breeding programs. It is suitable for both advanced undergraduate postgraduate students and researchers specializing in agriculture, biotechnology, and molecular breeding as well as entrepreneurs and policymakers concerned with analysis improvement of plant varieties.

The editors extend their appreciation and gratitude to the contributing authors from 10 different countries for their efforts and diligence and to Springer team for the opportunity and constant support to publish this book.

Al-Ahsa, Saudi Arabia — Jameel M. Al-Khayri
Helsinki, Finland — Shri Mohan Jain
Mumbai, India — Suprasanna Penna

Acknowledgments The editors extend their appreciation to the support of the Deanship of Scientific Research, Vice Presidency for Graduate Studies and Scientific Research, King Faisal University, Saudi Arabia (Grant No. KFU251206).

Contents

Editors and Contributors

About the Editors

Professor Jameel M. Al-Khayri is affiliated with the Department of Agricultural Biotechnology, King Faisal University, Saudi Arabia. He received B.S. Biology in 1984 from the University of Toledo, M.S. in Agronomy in 1988, and Ph.D. in Plant Science in 1991 from the University of Arkansas. He is the National Correspondent of the International Association of Plant Tissue Culture and Biotechnology. Al-Khayri has authored 130 research articles, 70 chapters and edited several journal special issues. In addition, he edited 30 reference books on plant biotechnology, genetic resources, breeding, genomics and nanotechnology. In addition to teaching, advising and research, he held administrative responsibilities as the Assistant Director of Date Palm Research Center, Head of Department of Plant Biotechnology and Vice Dean for Development and Quality Assurance. Al-Khayri served as a Member of Majlis Ash-Shura (Saudi Legislative Council) for the 2009–2012 term. He is interested in agricultural innovations and sustainability.

Professor Shri Mohan Jain, (Adjunct) is currently associated with the Department of Agricultural Sciences, University of Helsinki, Helsinki, Finland. He received M.Phil. 1973 and Ph.D., 1978, Jawaharlal Nehru University, New Delhi, India. He served as a Technical Officer, Plant Breeding, and Genetics, International Atomic Energy Agency (IAEA), Vienna, Austria, 1999–2005 when the agency was awarded Nobel Peace Prize, 2005. He is on the Editorial Board of *Euphytica*, *In Vitro*, *Propagation of Ornamental Plants*; a reviewer in plant cell and organ culture, *Plant Cell Reports*, and *Plants*. His publications are more than 225 peer-reviewed journals, book chapters, and conference proceedings, and edited 72 books; invited speaker and acted as a Chairperson in several international conferences worldwide. He is a former consultant to the IAEA, the European Union, The Government of Grenada, the Iranian Private Company, and the Egyptian Government.

Professor Suprasanna Penna, Ph.D. (Genetics) is the Director of Amity Institute of Nuclear Biotechnology, Amity University of Maharashtra, Mumbai, India. He was formerly with Bhabha Atomic Research Centre, Mumbai, as the Head of Nuclear Agriculture and Biotechnology Division. He is also an IAEA Expert consultant. His significant contributions include plant biotechnology, mutation breeding, stress tolerance, and biopolymers for enhancing plant productivity in plants. His research on plant abiotic stress tolerance in crop plants and halophytes led to elucidating mechanisms of salinity tolerance. He is on the editorial board of several international journals and has published over 400 publications. He has edited several books on plant mutation breeding, salinity tolerance, plant genetic diversity, plant biotechnology and plant-metal interactions and was the Guest Editor of special issues of International journals.

Contributors

Mohamed A. Abdein Seeds Development Department, El-Nada Misr Scientific Research and Development Projects, Turrell, Mansoura, Egypt

Maryam Ahmadi Department of Plant and Animal Biology, Faculty of Biological Science and Technology, University of Isfahan, Isfahan, Iran

Suliman M. S. Alghanem Department of Biology, College of Science, Qassim University, Buraydah, Saudi Arabia

Haifa A. S. Alhaithloul Biology Department, College of Science, Jouf University, Sakaka, Saudi Arabia

Usmangani Abdul Attar Department of Botany, Devchand College, Arjunnagar, Kolhapur, Maharashtra, India

Rony Bhowal Department of Applied Biology, School of Biological Sciences, University of Science and Technology, Meghalaya, Baridua, India

Madhusmita Borthakur Department of Applied Biology, School of Biological Sciences, University of Science and Technology, Meghalaya, Baridua, India; Applied Biology, University of Science and Technology, Meghalaya, Ri Bhoi, India

Vijayalaxmi S. Dandin JSS RSH PU College, Vidyagiri, Dharwad, Karnataka, India

Maha M. Elshamy Botany Department, Faculty of Science, Mansoura University, Mansoura, Egypt

Amira A. Ibrahim Botany and Microbiology Department, Faculty of Science, Arish University, Al-Arish, Egypt

Arun Jerang Department of Applied Biology, School of Biological Sciences, University of Science and Technology, Meghalaya, Baridua, India; Applied Biology, University of Science and Technology, Meghalaya, Ri Bhoi, India

Kadanthottu Sebastian Joseph Department of Life Sciences, Christ University, Bengaluru, Karnataka, India

Sony Kumari Department of Applied Biology, School of Biological Sciences, University of Science and Technology, Meghalaya, Baridua, India; Applied Biology, University of Science and Technology, Meghalaya, Ri Bhoi, India

Jissa Theresa Kurian Department of Life Sciences, Christ University, Bengaluru, Karnataka, India

Ashwini Venkatrao Mohite Department of Botany, Yashvantrao Chavan Institute of Science, Satara, Maharashtra, India

Özlem Özbek Department of Molecular Biology and Genetics, Faculty of Arts and Science, Hitit University, Çorum, Türkiye

Mehdi Rezaei Horticulture Sciences Department, Agriculture Faculty, Shahrood University of Technology, Shahrood, Iran

Puja Roy Department of Applied Biology, School of Biological Sciences, University of Science and Technology, Meghalaya, Baridua, India;
Applied Biology, University of Science and Technology, Meghalaya, Ri Bhoi, India

Hojjatollah Saeidi Department of Plant and Animal Biology, Faculty of Biological Science and Technology, University of Isfahan, Isfahan, Iran

Khaled F. M. Salem Department of Biology, Scientific Departments, Shaqra University, Qwaieah, Saudi Arabia;
Department of Plant Biotechnology, Genetic Engineering and Biotechnology Research Institute (GEBRI), University of Sadat City, Sadat City, Egypt

Ali Sarkhosh Horticultural Sciences Department, University of Florida, Gainesville, FL, USA

Shilpa Susan Scaria Department of Life Sciences, Christ University, Bengaluru, Karnataka, India

Poojadevi Sharma Department of Biotechnology, Sir Parashurambhau College, Pune, Maharashtra, India

Neeta Shrivastava Shri B. V. Patel Education Trust, Ahmedabad, Gujarat, India

Anshu Srivastava Institute of Pharmacy, Nirma University, Ahmedabad, Gujarat, India

Mai M. Wahba Botany Department, Faculty of Science, Mansoura University, Mansoura, Egypt

Atif Khurshid Wani Department of Biotechnology, School of Bioengineering and Biosciences, Lovely Professional University, Jalandhar, Punjab, India

Prachi Saurabha Zimare Department of Botany, Comrade Godavari Shamrao Parulekar College of Arts, Commerce and Science, Talasari, Palghar, Maharashtra, India

Saurabha Bhimrao Zimare Department of Botany, Sir Parashurambhau College (Autonomous), Pune, Maharashtra, India

Chapter 1
Breeding of *Bacopa monnieri* (Brahmi): Traditional to Advanced Approaches and Future Outlook

Poojadevi Sharma, Anshu Srivastava, and Neeta Shrivastava

Abstract *Bacopa monnieri* (L.) Pennell, also called Brahmi, is a creeper used to enhance memory in humans with reported documentation around 6th Century A.D. The major chemical entity shown responsible for neuropharmacological effects of *B. monnieri* are Bacosides. Considering the high demand, *B. monnieri* breeding to obtain elite germplasm has become an active area to pursue. The desirable traits in an ideal *B. monnieri* germplasm, both from farmer and consumer viewpoint, include properties like high herbage yield, bigger and colorful flowers and high Bacoside content. Further, *B. monnieri* germplasms may serve to prevent its genetic erosion and help in conservation. This chapter focuses on traditional to advanced plant breeding approaches being used in *B. monnieri*. The most prominent breeding method used in plants is selection of desired germplasm from accessions showing morphological variation, genetic variation and phytochemical variation. Genotype X Environment interaction factor governing these variations is also evaluated. The approach by far is the most successful breeding approach in *B. monnieri* leading to high quality bred varieties like Subodhak, Pragyashakti, CIM Jagriti. This chapter highlights the approaches like hybridization, mutation and polyploidy, which can induce variation in the plant in combination with plant tissue culture techniques. *Agrobacterium* based genetic transformation technology developed for the plant is also discussed. Further, the chapter stretches out towards future applications of existing cutting edge technologies waiting to be utilized in the plant like data integration through omics technologies, metabolic engineering, genome editing and machine learning.

P. Sharma
Department of Biotechnology, Sir Parashurambhau College, Pune, Maharashtra, India
e-mail: poojadevi.sharma@spcollegepune.ac.in

A. Srivastava
Institute of Pharmacy, Nirma University, Ahmedabad, Gujarat, India
e-mail: anshu.srivastava@nirmauni.ac.in

N. Shrivastava (✉)
Shri B. V. Patel Education Trust, Ahmedabad, Gujarat, India
e-mail: deputydirector@bvpedutrust.com

J. M. Al-Khayri et al. (eds.), *Biodiversity and Genetic Improvement of Medicinal and Aromatic Plants I*, Advances in Plant Breeding Strategies 9,
https://doi.org/10.1007/978-3-031-81846-2_1

Keywords *Bacopa monnieri* · Bacoside · Breeding · Biotechnology · Genetic transformation · Omics

1.1 Introduction

The first descriptions of *Bacopa monnieri* as a memory enhancing drug is found as early as the 6th Century A.D., in ancient Sanskrit Ayurvedic texts including the Charaka Samhita and the Sushruta Samhita where the plant is mentioned as "Brahmi". These texts categorize *B. monnieri* as a Medhya Rasayana (Medhya meaning intellect and Rasayana meaning rejuvenation); a class of herbs known to improve memory and intelligence (Russo and Borrelli 2005). The underlying desire in humans to domesticate *B. monnieri* exists since then or maybe even before it. The quest for elite *B. monnieri* germplasm has been a long one and at present many high-quality bred varieties are readily available in the market. Some examples of these varieties are Subodhak, Pragyashakti, CIM Jagriti, and many more (Singh 2008).

Underlying such enthusiasm for *B. monnieri* breeding are three main reasons. Firstly, *B. monnieri* has high demand as a medicinal and ornamental plant (Karatas et al. 2013, Silpa et al. 2021). It is a 'herb of high volume trade' (2000–5000 MT annually) from India (Silpa et al. 2021). The growth of *B. monnieri* is expected to expand globally and the market for its extract is set to grow at a CAGR of 8.7% during the period of 2022–2030, owing to increasing demand for plant-based supplements and the rising prevalence of chronic diseases, such as diabetes and depression. The major chemical entity shown responsible for neuropharmacological effects of *B. monnieri* is Bacoside A (64.28%) and Bacoside B (27.11%); the latter differs only in optical rotation. The Bacoside A (bacogenins A1, A2, A3, and A4) derives from two triterpenoid saponins: pseudojujubogenin and jujubogenin on acid hydrolysis (Rajan et al. 2015). Some of the other major saponins present in the plant are Bacopasaponins A-G (Hou et al. 2002); Bacopaside I-VIII (Zhou et al. 2008); monnierasides I–III (Chakravarty et al. 2002) and bacosine (Bhandari et al. 2006). Alkaloids (brahmine, nicotine, herpestine), flavonoids (luteolin, apigenin) and phytosterols (β-sitosterol, stigmasterol) are some of the minor chemical constituents present in the plant (Bhandari et al. 2006). Bacoside A is used as a primary ingredient in many memory enhancing tonic and drug formulations. Secondly, due to its high demand, *B. monnieri* has come under threat and is listed as an endangered plant. To satisfy the high commercial demand, tons of raw plant material is used by herbal industries (Muthiah et al. 2013). Most of this raw material is collected from natural resources, which led to the rapid depletion of wild natural plants. Thirdly, the cultivation of *B. monnieri* is restricted to certain agroclimatic conditions in wetlands and muddy shores (Saran et al. 2022). The process of natural regeneration through seeds or cuttings is slow and requires good irrigation facilities. Thus, *B. monnieri* germplasms serve to prevent its genetic erosion and help in its conservation (Kindie 2021; Srivastava and Shrivastava 2008).

The chapter aims to discuss the need and types of plant breeding approaches used in *B. monnieri* till date. It critically evaluates the progress carried out in agronomic trait improvement using various strategies leading to the development of elite *B. monnieri* cultivars, addressing the lacunae and scope for improvement in the existing strategies, and possibilities of recruiting untapped cutting-edge advanced technologies for accelerating the process of *B. monnieri* breeding.

1.2 Plant Breeding Approaches in *B. monnieri*

Plant breeding is a technology used to enhance various characteristics of plants so that they become more desirable agronomically and economically. In essence, plant breeding implies crossing the best parents, and identifying and recovering progeny that outperform the parents. The efforts of plant breeding by humans can be traced back to 11,000–12,000 years ago when humans are believed to have started agriculture (Vergauwen and De Smet 2017). It started with the activity of bringing wild plant species under human management, termed as domestication, following which all other breeding methods can be applied to plant species (Doebley et al. 2006). Understandably, *B. monnieri* may also have been subjected to domestication through time. In many classical Ayurvedic texts, often "parayaya" meaning synonyms have been used for Brahmi describing its morphological characters and its growth habitat (Doshi et al. 2020). The relevant synonyms are Jalanimba (easily grows in water and have bitter test like neem plant), Jalodwaya (area with lot of water attracts the brahmi to grow), Kapotvanka (leaves of plant resemble the wings of pigeon), Matsyakshi (shape of flowers resemble the eyes of fish), Soumya (pleasant looking creeper), Jalabrahmi (easily grows in water places) and Somavallari (creeper which prefers marshy places for its growth). These synonyms have acted as indicators to identify and collect Brahmi from the wild by Ayurvedic practitioners for many centuries.

A successful plant variety, developed through plant breeding, is one with a total balance of traits that makes it more profitable for growers and which may include both farmer-preferred traits (yield potential) and consumer-preferred traits (superior quality) (Bradshaw 2017; Swarup et al. 2021). In present bioeconomy, one may consider high herbage yield, bigger and colorful flowers and high Bacoside A content as desirable traits for *B. monnieri* (Escandón et al. 2006; Saran et al. 2022). Various plant breeding techniques applied to *B. monnieri* with the focus of developing successful plant variety are discussed subsequently.

1.2.1 Selection

The presence of genetic variation in plant species and utilization of this variation through selection is a basic approach for plant breeding (Nalina et al. 2022). Genotype variation leads to differences in the biochemical makeup of cells, which in turn together with the environment influences organ formation, plant growth, and eventually desirable traits (Dalziel et al. 2009). Thus, genotype variation is linked to phenotypic variation and phytochemical variation. The latest botanical classification and global map distribution of *B. monnieri* is covered in review by Sanyal et al. (2022). Screening of these variations among plant populations is a must so that superior genotypes can be selected (Varma et al. 2011). Many studies have reported collection of *B. monnieri* accessions from semi-temperate, subtropical and tropical environments at geographically distinct locations in India. After assembling germplasm, these accessions were subjected to analysis of phenotypic, genetic and phytochemical variability which led to screening and selection of superior ecotypes. The selection approaches explored for the plant are as follows:

1.2.1.1 *Phenotype Diversity Analysis and Selection*

Characterization of herbage morphology is a classical approach which is well suited for analysis of phenotype variation in a plant species. Morphological markers are based on visually accessible traits such as flower color, leaf size/shape, growth habits, pigmentation, etc. (Kindie 2021; Lamichhane and Thapa 2022). In modern taxonomy, Brahmi is described as a small creeping herb with numerous branches. The leaves of the plant are oblanceolate, succulent, relatively thick, and are arranged oppositely on the stem which is a distinctive phenotype feature for the plant as shown in Fig. 1.1. It is found throughout the Indian subcontinent in wet, damp and marshy areas. Its flowers are small and actinomorphic which grow in leaf axils. The color may vary from whitish, blue to medium purple in color with various sizes of flower-stalk length (Saran et al. 2022).

Further, it is known to become dormant during the winter months, except when grown in the vicinity of running water. Also in a field trial study, it was found that, *B. monnieri* can show loss of shoot biomass at low temperatures in field conditions (Mathur et al. 2002). Thus, in South and Central India, it can be grown throughout the year as the winters are pleasant, summers are hot, and the monsoon shows abundant

Fig. 1.1 *Bacopa monnieri* creeper herb growing in wild (photo by the authors)

rainfall which results in an ideal growth environment for the plant. In North India, where winters are very harsh, it is mainly grown in the summer-rainy season.

In one study, diversity of different *B. monnieri* accessions collected from different parts of India was evaluated for horticultural traits as well as Bacoside A content. When evaluating general known morphological characteristics of the plant in these accessions, few remarkable trait variations were observed, including larger leaf size or leaf area, very tiny leaf size, high stolon length, high flower pedicel length, and high pedicle diameter. Flower color was purple to purplish white in most of the genotypes, although whitish purple flower color was also observed in some cases. Among these accessions, the study finally selected two accessions, called DBM-4 and DBM-5, as new selections for further crop improvement and possible commercial cultivation. Elite DBM-4 selection exhibited high maximum average Bacoside-A3 and Bacopaside-II content, with maximum pedicel length trait associated with it. The DBM-5 selection provided the highest fresh herbage yield, with traits like large leaf length and stolon length associated with it (Saran et al. 2022).

1.2.1.2 Genetic Diversity Analysis and Selection

Genetic diversity can be analyzed by using genetic or DNA based markers such as RFLP (restriction fragment length polymorphism), RAPD (random amplified polymorphic DNA), SSR (simple sequence repeats), and AFLP (amplified fragment length polymorphism) (Agarwal et al. 2008).

In *B. monnieri*, different molecular markers have been reported to determine genetic diversity, and most of the work has been reported from India. RAPD molecular markers have been mostly used in *B. monnieri,* where some reports have emerged regarding the presence of polymorphism in *B. monnieri* collected from different geographical locations. In one study, 25 accessions of *B. monnieri* from different agroclimatic zones in India were analyzed, revealing medium polymorphism level among the accessions (Darokar et al. 2001). In another study, 25 accessions of *B. monnieri* were collected from four Southern Indian states, and cluster analysis done on the basis of similarity coefficient generated from RAPD profiles indicated that all the accessions were divided into two subgroups (Karthikeyan et al. 2011). Similar study in 18 accessions of *B. monnieri* collected from the Southern part of India revealed medium polymorphism after UPGMA clustering (Kumar et al. 2013). A study of 15 accessions of *B. monnieri* collected from Western Ghat region of South India revealed high levels of polymorphism in the samples (Anu et al. 2017). A combination of two different markers, i.e. RAPD and ISSR has also been used for genetic diversity analysis of 15 accessions collected from Central Indian States (Tripathi et al. 2012). Both markers revealed medium to high polymorphism in plant samples.

When considering potential yield of plants, it is a well-established consensus that direct and indirect interactions between both environmental niches and plant genotypes significantly affect potential yield of plants (Nemeth-Zamborine et al. 2019; Varma et al. 2011). Choosing a genotype for a target trait becomes a complex and difficult task as different genotypes respond differently to varied environmental conditions. Genotype X environment interaction (GXE) is an important factor that can cause variations in yield and phenotypic traits of a specific genotype. The presence of GXE has potential to reduce advantages from selection and, therefore, needs to be studied for genotypes developed through plant breeding (Mishra et al. 2018). GXE study provides ideas regarding stability of genotypes across environments, the identification of superior genotypes and the location that best represents the target environment for production (Niazian and Niedbała 2020).

In this direction, we made an attempt to study the effect of complex interaction between genotype and geographical conditions on Bacoside A content using principal coordinate analysis to analyze large scale data. In our study, Bacoside A content analysis of 75 accessions of *B. monnieri* from different geographical locations was carried out and significant variation in Bacoside A content was obtained. The accessions were divided into high- and low-yielding accessions based on phytochemical analysis and then subjected to AFLP profiling for genetic diversity analysis. However, no clustering was observed based on AFLP studies indicating predominance of agro-climatic conditions on differential Bacoside A content in *B. monnieri* (Srivastava et al. 2016). This study indicated that during the plant breeding programme of *B. monnieri*, growth habitat optimization can play a major role in obtaining desirable traits.

Another aspect of using molecular markers for plant breeding is discovery of quantitative trait loci (QTL) by joint study of segregation of marker genotypes and of phenotypic values of individuals or lines (Asins et al. 2010). A QTL is a polymorphic locus that contains alleles that differentially affect the expression of a continuously distributed phenotypic trait. QTL mapping helps in analysis of candidate genes, which could be protein-coding genes or regulatory genes. Once the marker linked to QTL is identified, it can be used in the plant selection program. Marker assisted selection (MAS) encompasses selection of desired individuals by using molecular markers linked with QTL governing desired phenotypic trait (Hasan et al. 2021). However, none of the molecular marker studies mentioned above have looked into QTL analysis for Bacoside A content in the plant.

1.2.1.3 *Chemodiversity Analysis and Selection*

Bacoside A content analysis is an important criterion while screening *B. monnieri* genotypes for selection. Bacoside A content can be determined in *B. monnieri* using several analytical approaches, such as HPLC and HPTLC. Chemometric approaches such as hierarchical cluster analysis, principal component analysis, and partial least square discriminant analysis can be used to distinguish and classify the low and high Bacoside A yielding *B. monnieri* varieties in accordance with the geographical origin (Srivastava et al. 2016).

Screening of different *B. monnieri* chemotypes could also be useful in its breeding. In a recent study, search for elite lines of *B. monnieri* from various geographic locations in South India was carried out. Sixty *B. monnieri* field accessions were collected to assess the contents of bacoside A and bacopaside I using HPTLC-densitometry and eight elite accessions were screened out. Additionally, an assessment was conducted on the impact of edaphic and ecological conditions on the biomass and bacoside content of these elite lines. The study concluded that *B. monnieri* showed preferences for low altitude wet, muddy, and marshy habitats (Kunjumon et al. 2023).

Another study reported simultaneous separation and determination of multiple bioactive components in *B. monnieri*. Twenty-four accessions of *B. monnieri* samples were studied from two provinces of eastern India. Two different types of chemotypes were identified: one chemotype in Odisha was Bacopasaponin C rich and another in West Bengal was Bacopaside II rich (Padhiari et al. 2021).

Expanding the study further, Padhiari et al. (2022) ventured into a computer based machine learning approach and evaluated whether Artificial Neural Network (ANN) technique can be used to predict ideal environmental niche that may bring about maximum Bacoside A production in *B. monnieri*. ANN has the ability to model complex problems, which enables discovery of relationships among multiple dependent and independent variables (Azevedo et al. 2015). The recruitment of ANN models for yield component analysis and indirect selection studies in plant breeding is a relatively recent approach and still in its infancy. The adaptation of the technology is not wide scale. In this regard, it is worthy to discuss a remarkable pilot field study, carried out recently by Niedbala et al. (2020) to develop ANN models that would examine the impact of 23 winter wheat germ genotypes and weather conditions on the concentration of ferulic acid, and link its content with the concentration of deoxynivalenol and nivalenol. The study revealed that genotype was an important factor significantly affecting the secondary metabolite content of winter wheat seeds. This confirms the significant role of growing new varieties.

In the study by Padhiari et al. (2022), plant samples were collected from 81 different geographical locations in India (Odisha and West Bengal provinces). There were 13 input variables i.e. precipitation, altitude, maximum temperature, average annual temperature, minimum temperature, average relative humidity (ARH), ultra-violet index, pH, electrical conductivity (EC), organic carbon (OC), N, P and K as the input layers, and one dependent variable (Bacoside A) as the output layer. The study determined that variation in Bacoside A content among different wild accessions was mainly due to varied soil nutrient levels and climatic factors. The ANN analysis specifically implicated nitrogen levels and altitude having the highest impact on Bacoside A content. However, these chemo diversity studies did not offer any insight regarding genotype profile of Bacopa wild accessions.

From the above discussion, we suggest that ANN modeling approach should be designed for GXE study correlating with Bacoside A content in *B. monnieri* to harness its predictive power for selection of desirable genotypes that may show stable yields at different environmental niches.

1.2.2 Hybridization

The mating or crossing of two plants or lines of dissimilar genotypes is known as hybridization. Individuals produced as a result of cross between two genetically different parents are known as hybrids. In plants, crossing is done by placing pollen grains from one genotype, called a male parent, onto the stigma of flowers of the female plant (Goulet et al. 2017). The seeds as well as progeny resulting from hybridization are known as hybrids or F1.The aim of hybridization may be transfer and or improvement of one or few qualitative characters in F1 hybrid. The progeny of F1, obtained by integrating F1 plants, and the subsequent generations are termed segregating generations. Segregation of and recombination among the heterozygous genes would produce many new gene combinations in F2 and the later generations i.e. the segregating generation. The degree of genetic variation produced in segregating generations would, therefore, depend on the number of genes that differ in the parents of F1.

With regards to *B. monnieri*, there is a dearth of reports concerning hybridization studies. This may be partly attributed to the fact that the plant has a small flower and a high number of chromosomes ($2n = 2x = 64$) as indicated by reproductive biology and karyotyping studies carried out on the plant. Further, propagation through seeds is limited due to low availability and viability of seeds (Samaddar et al. 2012). Therefore, development of new varieties of Brahmi by hybridization may be time-consuming and difficult.

1.2.3 Mutation

Treating a biological material with a mutagen in order to induce mutation is known as mutagenesis. Mutation breeding refers to the process of inducing and isolating mutants with improved traits for plant improvement. Mutation breeding is an important and commonly practiced technique to create genetic variation among existing plant gene pools, which may help in the selection process in a given environment (Ibrahim et al. 2018). Mutation carried out using X-rays, gamma rays, and ion beam radiation is a routine practice to create new varieties (Ibrahim et al. 2018).

An interesting study involving γ-radiation was carried out on two selections of *B. monnieri* i.e. Pragyashakthi and Calcutta Local cultivar. After γ-radiation treatment, nodal segment explant of Pragyashakthi gave enhanced Bacoside A contents compared to calli and control, whereas Bacoside A content was high in γ-radiation irradiated calli compared to control and nodal segment from Calcutta Local cultivar (Varghese and Satyanarayana 2007).

1.2.4 Polyploidy

When chromosome numbers are exact multiples of the basic or genomic number of a species, then these variations are called polyploids. Along with its medicinal value, *B. monnieri* also has ornamental value. It is used as an ornamental plant in aquatic gardens (Karatas et al. 2013). *B. monnieri* has very small sized flowers. Bigger flowers are a desired trait that could improve ornamental value of the species (Escandón et al. 2006). Polyploidy agents such as colchicine and oryzalin can be used to modify the plant shape and obtain flowers with enhanced size. These chemicals are reported to induce polyploidy and give rise to new genotypes with variants in size and/or colour of flowers and leaves. Chromosome doubling under in vitro conditions is a widely used approach for domestication of wild germplasm to be used as ornamentals (Trojak-Goluch et al. 2021). These improved ornamental products could be either included in a breeding program or transferred to potential users together with the technology developed. Chromosome doubling may also induce differential gene expression between diploid and synthesized tetraploid, therefore affecting the various phenotypic traits.

Enlargement of organs (flowers and leaves), intensification of colours, harder and more robust plants, thicker and more rigid foliage, a discernible increase in the tolerance to different stresses, and the resistance to diseases and pests are outcomes of polyploidy. Thus, chromosome doubling is accepted as a source of evolution in flowering plants, and breeders benefit from it for the domestication of certain genotypes.

Studies have also been reported on treatment of *B. monnieri* with colchicine, a chemical mutagen. In one study, a tetraploid *B. monnieri* plant obtained via colchicine treatment resulted in development of a new variety named Ali INTA-JICA with enlarged leaves and intense flower color (Escandón et al. 2006). In another study, colchicine induced autotetraploid plants were superior in agricultural characteristics (such as fresh weight, dry weight and total Bacoside content) to the original parental diploid plants (Inthima and Sujipuli 2019). *B. monnieri* plants regenerated with colchicine showed morphological variations in leaf shape, number of leaves per node, leaf arrangement, etc. and enhanced the Bacoside content (Kharde et al. 2017).

1.2.5 In Vitro Cultivation of Plant Cells and Regeneration of Plants from Cultured Cells

In field conditions, *B. monnieri* plant can be easily cultivated in damp areas on a variety of soils. It requires wet and marshy soil and can be propagated by seeds and also vegetatively by cuttings. It can grow in a wide range of temperatures (15–40 °C) and soil pH (5–7.5). The plant grows best in high moisture locations, which offer partial shade during a part of the day. The best planting material for cultivation of plants are freshly collected propagules of 10–15 cm length with internodes and rootlets (Jat et al. 2016).

Plant tissue culture (PTC) is a technique involving cultivation of plant organs, tissues or cells in vessels using artificial media which can also be termed as in vitro cultivation. These in vitro grown cultures have potential to act as a continuous source of plant biomass and bioactive compounds including secondary metabolites. Furthermore, plant tissue cultures generated in vitro are easily amenable to induction techniques such precursor feeding, elicitation, etc., to temporarily boost the synthesis of bioactives and obtain higher content (Narayani and Srivastava 2017; Shukla et al. 2023). Thus, PTC allows renewable and sustainable productivity independent of seasonal variation from field conditions (Lakshmaiah et al. 2022).

The successful application of plant tissue culture in plant improvement depends on regeneration of organs or whole plants directly from plant parts or indirect regeneration through callus and suspension cultures. Somatic embryo cultures, meristem cultures, anther or pollen cultures, somatic hybrids and cell culture are different types of plant tissue cultures recruited routinely in plant breeding for different purposes (Bridgen et al. 2018; Espinosa-Leal et al. 2018). Meristem culture is used to obtain virus free plants. Haploids obtained through pollen culture are used to obtain disease resistant plants and superior diploid lines. Conventional plant breeding methods like hybridization, induced mutation, induced polyploidy, etc. are supplemented with plant tissue culture techniques either to increase their efficiency or obtain faster results (Bridgen et al. 2018).

As already explained, genetic variation is a fundamental requirement to successful plant breeding. Occurrence of high genetic variation in in vitro cultures can be attributed to somaclonal variation. The cells of the plants regenerated from cell cultures are known to display heritable variation for both quantitative and qualitative traits; such a variation is known as somaclonal variation (Bairu et al. 2011). The occurrence of somaclonal variants may be due to chromosome structural changes, gene mutations, plasmagene mutations, gene duplications, epigenetic changes in gene expressions, mitotic crossing over, transposons, etc. (Bairu et al. 2011). Isolation of somaclonal involves a screening phase where a large number of cells or regenerated plants are observed for variation in morphological, molecular and phytochemical characteristics. In general, progeny R1 are scored for identification of variant plants and R2 progeny lines are evaluated for confirmation (Krishna et al. 2016). In comparison to mutation breeding, somaclonal variation is reported to be a faster approach to obtain a new variety. Further, somaclonal variation being a characteristic feature of plant tissue culture regeneration, highly effective and large scale screening and selection can be carried out in vitro. Thus somaclonal variants have huge applications in plant breeding for plant improvement (Krishna et al. 2016).

PTC is an alternative and fast propagation method for mass scale production of *B. monnieri* plant helping to meet its market demand as well help in its conservation as well (Srivastava and Shrivastava 2008). Several protocols have been reported for cell, suspension cultures, shoot cultures and micropropagation. *B. monnieri* has a very high regeneration capacity and responds readily to treatment with plant growth regulators (Murthy 2022). General requirements to develop plant tissue cultures of

B. monnieri along with approaches for in vitro bacoside induction is summarized in Table 1.1. The PTC technology for *B. monnieri* can emerge as a key driver to accelerate plant breeding in the plant.

Molecular marker based screening of somaclonal variants at DNA level has been attempted for in vitro regenerated *B. monnieri* plants. In one study, RAPD was used to confirm occurrence of polymorphism among the in vivo and in vitro plants of *B. monnieri* at DNA level (Pathak et al. 2013). In another study, RAPD and ISSR study was used to screen for genetic variation where they detected 10% polymorphism in in vitro regenerated plants when compared to mother plants. However, no studies have been reported, till now, where somaclonal variants were isolated for high Bacoside content. Selection of high Bacoside producing somaclonal variants can emerge as an effective plant breeding strategy to obtain elite varieties.

Table 1.1 Overview of requirements for plant tissue culture (PTC) of *Bacopa monnieri* and in vitro Bacoside induction

Conditions	Requirements	References
Culture types	*Growth Hormones*	Asha et al. (2013), Begum and Mathur (2014), Behera et al. (2015), Dey et al. (2019), Escandón et al. (2006), Gurnani et al. (2012), Haque et al. (2017), Hegazi et al. (2017), Jain et al. (2013), Jain et al. (2014), Kashyap et al. (2017), Kaur et al. (2013), Koul et al. (2014), Kumari et al. (2014), Mishra et al. (2015), Mohanta and Sahoo (2014), Mohapatra and Rath (2005), Nagarajan et al. (2015), Naik et al. (2014), Nandhini et al. (2015), Pandiyan and Selvaraj (2012), Rahman et al. (2002), Ranjan and Kumar (2018), Rao et al. (2012), Rency et al. (2016), Rout et al. (2011), Sharath et al. (2007), Sharma et al. (2013a, b), Sharma et al. 2013a; Showkat et al. (2010), Shrivastava and Rajani (1999), Soundararajan and Karrunakaran (2011), Srivastava and Shrivastava (2008), Srivastava et al. (2017), Subashri and Pillai (2014), Taha (2017), Tiwari et al. (1998), Tiwari et al. (2001), Tiwari et al. (2006), Umesh et al. (2014), Vijayakumar et al. (2010), Wangdi and Sarethy (2016), Yusuf et al. (2011), and Zote et al. (2018)
Callus cultures	KIN with NAA 2,4-D with NAA IAA with KIN and BAP 2,4-D and BAP	
Suspension cultures	2,4-D and BAP	
Somatic embryos	2,4-D and KIN	
Shoot cultures	*Only cytokinins* BAP, TDZ BAP and TDZ *Cytokinin with auxin* BAP with IAA or NAA or IBA IBA with NAA and KIN	
Whole plant cultures	*Direct rooting* No PGRs required *Root induction after shooting* with auxins IBA, IAA	
Bacoside induction	*Induction agents*	
In vitro cultures	*Precursor feeding* Pyruvic acid Mevalonic acid *Abiotic elicitors* Methyl jasmonate Chitosan DMSO *Biotic elicitors* *Saccharomyces cerevisiae* *Chitiniphilus* sp. MTN22 *Streptomyces* sp. MTN14	

Another approach to introduce genetic variation for trait improvement in in vitro cultures is through induced mutation and polyploidy. In this regard, in vitro cultures of *B. monnieri* have been subjected to colchicine treatment and strategy has been successful in regeneration of superior plants and varieties (Escandón et al. 2006; Inthima and Sujipuli 2019; Kharde et al. 2017).

1.3 Genetic Transformation in *Bacopa monnieri*

The approaches discussed till now for plant breeding deal with generation of non-targeted genetic variation. The most potent biotechnological approach for creating desired genetic variation is transfer of specifically constructed gene assemblies through genetic transformation (Ahmar et al. 2020). The DNA segment which is introduced in this process contains the gene of interest and a cassette containing additional genetic material including a promoter (determines the site and timing of expression of the introduced gene), a terminator (identify the end of transcription) and a marker gene (allows selection of plants having the introduced gene) (Anami et al. 2013). Then insertion of cassette into the plant expression vector is carried out through restriction digestion and ligation or site specific recombination through end site specific homology sequences (Anami et al. 2013). The cells that receive the new gene are called transgenic and are selected from those that are not transgenic. The best transgenic lines are grown to maturity in greenhouses. When the focus is plant breeding, the specificity and stability of the transgenic line for transgene expression is evaluated for a few more generations in field conditions to decide whether it can be considered an "elite" line and used for further crosses (De Block 1993). *Agrobacterium*-mediated genetic transformation is the most widely used genetic transformation method for plants (Desai et al. 2010; Guo et al. 2019). Both *Agrobacterium tumefaciens* and *Agrobacterium rhizogenes* strains are used for plant genetic transformation (Sharma et al. 2013a, b).

B. monnieri has been revealed to be amenable to *Agrobacterium*-mediated genetic transformation (Sanyal et al. 2022). Majority of studies reported till now are limited to standardization and optimization of *Agrobacterium*-mediated transformation of in vitro regenerated whole plant culture of *B. monnieri*. These studies are summarized in Table 1.2. Researchers used different *Agrobacterium* strains, different plant expression vectors, selection medium, explant type, and different techniques for confirmation of transgenes (Aasim et al. 2019; Sanyal et al. 2022). After thorough literature survey, we realized that somatic embryos or embryogenic cells should also be used for genetic transformation in *B. monnieri,* which has remained unexplored so far. Somatic embryos are easily amenable to genetic transformation, and the transformed cells could be selected and directly regenerated into plants without altering any hormone combination or requirements.

Further, very few studies have been reported for trait modification in *B. monnieri* using genetic engineering. In a study, heterologous transfer genes of terpene indole alkaloid pathway in *Catharanthus roseus* (L.) G. Don. was carried out in *B. monnieri.*

Table 1.2 General overview of *Agrobacterium*-mediated genetic transformation scheme in *Bacopa monnieri*

Parameters	Types	References
Explants	Leaf explant, nodes, internodes, thin cell layers	Aggarwal et al. (2013), Bansal et al. (2014), Croom et al. (2016), Kumari et al. (2015), Largia et al. (2016), Nisha et al. (2003), Paul et al. (2015), Ramesh et al. (2011), Sharma et al. (2017), and Yadav et al. (2014)
Agrobacterium species	*Agrobacterium tumefaciens strains* (for in vitro regenerated shoot cultures/whole plants) LBA4404, GV3101, EHA2015, LBA119 *Agrobacterium rhizogenes strains* (for hairy root cultures) , A4, R1000, SA79, MTCC 532, MTCC 2364	
Plant transformation vectors	*Agrobacterium tumefaciens* pBE2113, pCAMBIA 1301, pCAMBIA 1302, pMOG22 *Agrobacterium rhizogenes* Ri plasmid	
Reporter gene	GUS: Beta-glucuronidase enzyme GFP: Green fluorescent protein	
Selectable marker gene	Neomycin phosphotransferase (*npt*II) Hygromycin phosphotransferase (*hpt*)	
Confirmation and screening of in vitro cultured transformants	*At genetic level* Polymerase chain reaction (PCR) analysis, southern blot hybridization *At mRNA level* Reverse transcription polymerase chain reaction (RT-PCR) *At protein level* GUS (β-glucuronidase) activity *At metabolite level* HPLC analysis for secondary metabolite analysis	

The target genes of study were tryptophan decarboxylase (tdc) or strictosidine synthase (str) genes. Enhancement in tryptophan and tryptamine in transgenic *B. monnieri* was confirmed using HPLC, which indicated overexpression of the *tdc* gene in the plant (Sharma et al 2017).

However, surprisingly, no study has been reported till now for genetic manipulation of any ornamental trait or precise metabolic engineering of Bacoside biosynthetic pathway in the plant. There is a vast open field offering scope for generating improved varieties of *B. monnieri* via gene transfer for plant breeding. It is indeed high time to explore DNA manipulation strategies like genetic engineering, metabolic engineering, and genome editing for this plant.

1.4 Conclusion and Future Perspectives

Emergence of high throughput omics technologies in the past decade has made it possible to design effective knowledge-based plant breeding techniques for trait improvement (Priyadarshi et al. 2021; Thulasiram et al. 2022). Different datasets generated from genomics, transcriptomics, proteomics, or metabolomics studies can be analyzed individually, or together and desired trait associated genetic sequences (molecular markers, QTL, etc.), genes, metabolite markers, gene networks or expression patterns can be elucidated (Thulasiram et al. 2023; Yang et al. 2021). The knowledge generated can be used to enhance the tolerance to abiotic and biotic stress in plant varieties by various breeding approaches and create elite lines to improve the germplasm. Further, identification of agronomically important genes combined with genetic modification can ensure precision in trait optimization in terms of yield, nutritional value, and plant fitness (Yang et al. 2021).

With regards to omics studies in *Bacopa monnieri* shoot and root transcriptome resource of the plant has been generated providing insight into genes involved in the secondary metabolite biosynthetic pathway in the plant (Jeena et al. 2017). Functional characterization of these genes and identification of putative terpene synthases can elucidate Bacoside biosynthetic genes. For example, recently, in a study many novel MicroRNAs (miRNA) were screened out in *B. monnieri*. miRNAs are small noncoding, endogenous RNAs that riboregulate the expression of target genes involved in various plant processes. Three miRNA targets, including asparagine synthetase, cycloartenol synthase and ferulate 5 hydroxylase (F5H) were validated by confirming presence of miRNA cleavage site in their sequence (Jeena et al. 2021). Also, overexpression of one miRNA i.e. Bm-miR172c-5p in *B. monnieri* by *Agrobacterium*-mediated genetic transformation resulted in reduced lignification and secondary xylem thickness responsible for the drought-sensitive phenotype in the plant via suppression of F5H gene (Jeena et al. 2021). However, mutation in Bm-miR172c-5p showed increased lignifications, increased secondary xylem thickness, and drought-tolerant phenotype in the plant. It was concluded that Bm-miRNA172c-5p might be a key player in maintaining the native phenotype of *B. monnieri* under different environmental conditions. Targeted expression of miRNAs for desired trait modification may be an effective strategy that can be explored for *B. monnieri* breeding.

More recently, a study reported functional characterization of a UDP-glycosyltransferase. UGT79A18, using gene overexpression and knockdown lines of UGT79A18 in *B. monnieri* and found it to be involved in the glycosylation of pseudojujubogenin in Bacoside biosynthetic pathway (Kumar et al. 2024). Suitable metabolic engineering strategies using these genes can be devised with the focus of trait improvement in the plant.

Additionally, it might be interesting to evaluate *B. monnieri* flower buds and flower tissue in order to generate genomic and transcriptomic resources. This will allow for deciphering of developmental genes and regulatory miRNAs that govern desired ornamental features like, enhanced flowering size, enhanced leaf biomass, etc., for the purpose of ornamental trait modification in the plant which has not yet been investigated until now.

However, a general issue with using gene transfer approach for plant breeding is that transgenic plants contain foreign DNA fragments with random insertions in the genome, the effects of which cannot be governed. To circumvent this issue, many modifications have been reported in gene transfer approach including cisgenesis and intragenesis, grafting, agroinfiltration, reverse breeding, RNA-directed DNA methylation, genome elimination, oligonucleotide-directed mutagenesis (ODM), and site-directed nucleases (SDNs). For detailed understanding of these concepts, readers are suggested to read an excellent review by Van Vu et al. (2022) to gain deeper insight in the issue. At present, public and government opinions do not side with plants developed through transgenic technology due to a variety of concerns ranging from ethical issues, health issues and posing a threat to ecosystems and biodiversity (Sharma et al. 2014). Therefore, release of plant varieties developed through plant breeding requires clearance through many regulatory hurdles.

Genome editing is another genetic engineering concept involving deletion or substituting a foreign gene in the organism's DNA (Ahmar et al. 2020). Advanced technique like "Clustered regularly interspaced short palindromic repeats" CRISPR/CRISPR-associated protein Cas9 is used for genome editing in plants. It has successfully created many elite traits in a short span of time as compared to traditional breeding approaches (Rahman et al. 2022; Van Vu et al. 2022). This technique can circumvent regulatory limitations associated with transgenic breeding techniques, as the end product acquires no foreign genes through genome editing (Schaart et al. 2016). With regards to *B. monnieri*, CRISPR/Cas9 should be explored for functional gene characterization as well as alteration of flower size and color, epigenetic modifications to alter regulation of developmental genes, and metabolic engineering of Bacoside biosynthetic pathways.

Further, large-scale complex data flow generated from population screening studies, GXE, plant tissue culture, omics, etc. studies can be effectively analyzed and utilized by machine learning for efficient interpretation of results making it a data-rich, evidence-based, and team-oriented process (Cobb et al. 2019). ANN modeling is suggested to enable breeders to evaluate the GXE and genetic stability of a large number of genotypes faster and more precisely (Niedbała et al. 2020). Thus, machine learning can aid in selection of elite *B. monnieri* varieties showing desired performance.

As the saying goes "All roads lead to Rome", it is easy to comprehend that use of omics technologies, gene transfer, metabolic engineering, genome editing, machine learning and will ultimately converge into advanced field of synthetic biology (Syn Bio) (Wurtzel et al. 2019). Synthetic biology is an advanced concept which aims to combine the disciplines of biology and engineering to design and create new biological systems to perform functions that are not found in nature (Xin et al. 2019). Synthetic biology is proposed to play a major role in development of novel agronomic traits, and it is conceivable that it may find its way in *B. monnieri* breeding ventures in near future.

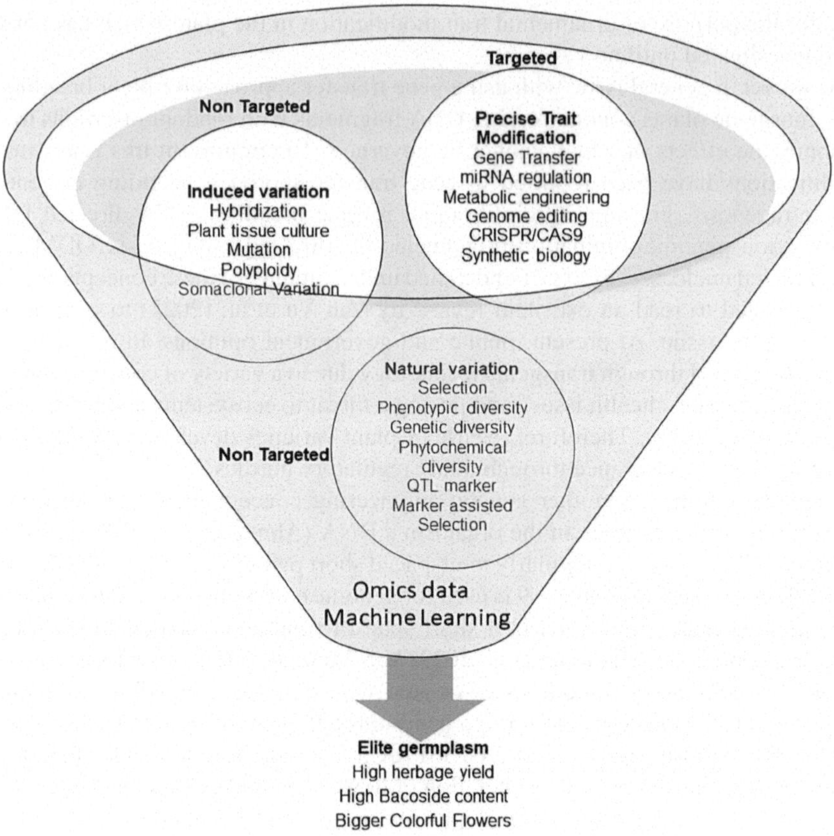

Fig. 1.2 *Bacopa monnieri* breeding technology implementation: a perspective

The present discussion surmises that *B. monnieri* breeding is highly underexplored with a lot of advanced untapped technologies waiting to be recruited for the purpose. A brief outline of use of these technologies in context with *B. monnieri* breeding is presented in Fig. 1.2. Strategic efforts are required by the scientific community armed with the latest technology to accelerate plant breeding in *B. monnieri.*

References

Aasim M, Sameeullah M, Karataş M et al (2019) An insight into biotechnological approaches used for the improvement of secondary metabolites from the medicinal aquatic plant, Water Hyssop (*Bacopa monnieri* L.). In: Akhtar MS, Swamy MK (eds) Natural bio-active compounds, vol 3. Biotechnology, bioengineering, and molecular approaches. Springer, Singapore, pp 123–152

Agarwal M, Shrivastava N, Padh H (2008) Advances in molecular marker techniques and their applications in plant sciences. Plant Cell Rep 27:617–631

Aggarwal D, Jaiswal N, Kumar A et al (2013) Factors affecting genetic transformation and shoot organogenesis of *Bacopa monnieri* (L.) Wettst. J Plant Biochem Biotechnol 22:382–391
Ahmar S, Gill RA, Jung KH et al (2020) Conventional and molecular techniques from simple breeding to speed breeding in crop plants: recent advances and future outlook. Int J Mol Sci 21:2590
Anami S, Njuguna E, Coussens G et al (2013) Higher plant transformation: principles and molecular tools. Int J Dev Biol 57:483–494
Anu S, Remmies R, Jose C (2017) Phylogenetic polymorphism in Neera-Brahmi, *Bacopa monnieri* (L.) using molecular markers. Int J Adv Sci Eng Technol 5:68–70
Asha KI, Devi AI, Dwivedi NK et al (2013) In vitro regeneration of brahmi (*Bacopa monnieri* (Linn.) Pennell)—an important medicinal herb through nodal segment culture. Res Plant Biol 3:1–7
Asins MJ, Bernet GP, Villalta I et al (2010) QTL analysis in plant breeding. In: Jain SM, Brar DS (eds) Molecular techniques in crop improvement, 2nd edn. Springer, Dordrecht, pp 3–21
Azevedo AM, Andrade Júnior VD, Pedrosa CE et al (2015) Application of artificial neural networks in indirect selection: a case study on the breeding of lettuce. Bragantia 74:387–393
Bairu MW, Aremu AO, Van Staden J (2011) Somaclonal variation in plants: causes and detection methods. Plant Growth Regul 63:147–173
Bansal M, Kumar A, Sudhakara RM (2014) Influence of *Agrobacterium rhizogenes* strains on hairy root induction and 'Bacoside A' production from *Bacopa monnieri* (L.) Wettst. Acta Physiol Plant 36:2793–2801
Begum T, Mathur M (2014) In vitro regeneration of *Catharanthus roseus* and *Bacopa monnieri* and their survey around Jaipur district. Int J Pure Appl Biosci 2:210–221
Behera S, Nayak N, Shasmita et al (2015) An efficient micropropagation protocol of *Bacopa monnieri* (L.) Pennell through two-stage culture of nodal segments and ex vitro acclimatization. J Appl Biol Biotechnol 3:16–21
Bhandari P, Kumar N, Singh B et al (2006) Bacosterol glycoside, a new 13,14-seco-steroid glycoside from *Bacopa monnieri*. Chem Pharm Bull 54:240–241
Bradshaw JE (2017) Plant breeding: past, present and future. Euphytica 213:1–12
Bridgen MP, Houtven WV, Eeckhaut T (2018) Plant tissue culture techniques for breeding. In: Huylenbroeck JV (ed) Ornamental crops. Springer, Cham, pp 127–144
Chakravarty AK, Sarkar T, Nakane T et al (2002) New phenylethanoid glycosides from *Bacopa monnieri*. Chem Pharm Bull 50:1616–1618
Cobb JN, Juma RU, Biswas PS et al (2019) Enhancing the rate of genetic gain in public-sector plant breeding programs: lessons from the breeder's equation. Theor Appl Genet 132:627–645
Croom LA, Jackson CL, Vaidya BN et al (2016) Thin cell layer (TCL) culture system for herbal biomass production and genetic transformation of *Bacopa monnieri* L. Wettst. Am J Plant Sci 7:1232–1245
Dalziel AC, Rogers SM, Schulte PM (2009) Linking genotypes to phenotypes and fitness: how mechanistic biology can inform molecular ecology. Mol Ecol 18:4997–5017
Darokar MP, Khanuja SS, Shasany AK et al (2001) Low levels of genetic diversity detected by RAPD analysis in geographically distinct accessions of *Bacopa monnieri*. Genet Resour Crop Evol 48:555–558
De Block M (1993) The cell biology of plant transformation: current state, problems, prospects and the implications for the plant breeding. Euphytica 71:1–14
Desai P, Shrivastava N, Padh H (2010) Production of heterologous proteins in plants: strategies for optimal expression. Biotechnol Adv 28:427–435
Dey A, Hazra AK, Nongdam P et al (2019) Enhanced bacoside content in polyamine treated in-vitro raised *Bacopa monnieri* (L.) Wettst. S Afr J Bot 123:259–269
Doebley JF, Gaut BS, Smith BD (2006) The molecular genetics of crop domestication. Cell 127:1309–1321
Doshi KA, Patel MH, Virpariya PA (2020) Brahmi (*Bacopa monnieri linn.*)" Tiny plant with tremendous properties—a comprehensive review World J Pharm Life Sci 6:116–121

Escandón AS, Hagiwara JC, Alderete LM (2006) A new variety of *Bacopa monnieri* obtained by in vitro polyploidization. Electron J Biotechnol 9

Espinosa-Leal CA, Puente-Garza CA, García-Lara S (2018) In vitro plant tissue culture: means for production of biological active compounds. Planta 248:1–18

Goulet B, Roda F, Hopkins R (2017) Hybridization in plants: old ideas, new techniques. Plant Physiol 173:65–78

Guo M, Ye J, Gao D et al (2019) *Agrobacterium*-mediated horizontal gene transfer: mechanism, biotechnological application, potential risk and forestalling strategy. Biotechnol Adv 37:259–270

Gurnani C, Kumar V, Mukhija S et al (2012) In vitro regeneration of brahmi (*Bacopa monnieri* (l.) Penn.)—a threatened medicinal plant. Kathmandu Univ J Sci Eng Technol 8:97–99

Haque SM, Chakraborty A, Dey D et al (2017) Improved micropropagation of *Bacopa monnieri* (L.) Wettst. (Plantaginaceae) and antimicrobial activity of in vitro and ex vitro raised plants against multidrug-resistant clinical isolates of urinary tract infecting (UTI) and respiratory tract infecting (RTI) bacteria. Clin Phytoscience 3:17

Hasan N, Choudhary S, Naaz N et al (2021) Recent advancements in molecular marker-assisted selection and applications in plant breeding programmes. J Genet Eng Biotechnol 19:128

Hegazi GAE-M, Taha HS, Sharaf AMM et al (2017) Enhancing in vitro production of Bacoside A from *Bacopa monnieri* using precursor and elicitors feeding. J Basic Appl Sci Res 7:27–35

Hou CC, Lin SJ, Cheng JT et al (2002) Bacopaside III, Bacopasaponin G, and Bacopasides A, B, and C from *Bacopa monniera*. J Nat Prod 65:1759–1763

Ibrahim R, Ahmad Z, Salleh S et al (2018) Mutation breeding in ornamentals. In: Huylenbroeck JV (ed) Ornamental crops. Springer, Cham, pp 175–211

Inthima P, Sujipuli K (2019) Improvement of growth and bacoside production in *Bacopa monnieri* through induced autotetraploidy with colchicine. PeerJ 7:e7966

Jat BL, Panwar R, Gena D et al (2016) In vitro production of bacosides in tissue cultures of *Bacopa monneri*. World J Pharm Res 5:1087–1107

Jain A, Pandey K, Benjamin D et al (2014) In vitro approach of medicinal herb: *Bacopa monnieri*. Int J Innov Res Sci Eng Technol 3:12088–12093

Jain R, Prasad B, Jain M (2013) In-vitro regeneration of *Bacopa monnieri* (L.): a highly valuable medicinal plant. Int J Curr Microbiol Appl Sci 2:198–205

Jeena GS, Fatima S, Tripathi P et al (2017) Comparative transcriptome analysis of shoot and root tissue of *Bacopa monnieri* identifies potential genes related to triterpenoid saponin biosynthesis. BMC Genom 18:1–15

Jeena GS, Joshi A, Shukla RK (2021) Bm-miR172c-5p regulates lignin biosynthesis and secondary xylem thickness by altering the Ferulate 5 hydroxylase gene in *Bacopa monnieri*. Plant and Cell Physiol 62:894–912

Karatas M, Aasim M, Dogan M et al (2013) Adventitious shoot regeneration of the medicinal aquatic plant water hyssop (*Bacopa monnieri* (L.) Pennell) using different internodes. Arch Biol Sci 65:297–303

Karthikeyan A, Madhanraj A, Pandian SK et al (2011) Genetic variation among highly endangered *Bacopa monnieri* (L.) Pennell from Southern India as detected using RAPD analysis. Genet Resour Crop Evol 58:769–782

Kashyap S, Kapoor N, Kale RD (2017) Micropropagation of *B. monnieri* using humin media in plant tissue culture. Ann Plant Sci 6:1625–1629

Kaur J, Nautiyal K, Pant M (2013) In vitro propagation of *Bacopa monnieri* (L.) Wettst—a medicinally priced herb. Int J Curr Microbiol App Sci 2:131–138

Khardc AV, Chavan NS, Chandre MA et al (2017) In vitro enhancement of bacoside in Brahmi (*Bacopa monnieri*) using colchicine. J Plant Biochem Physiol 5:2–6

Kindie B (2021) Assess molecular marker applications for genetic variety analysis in biodiversity conservation status. Mol Biol 10:287

Koul A, Sharma A, Gupta S et al (2014) Cost effective protocol for micropropagation of *Bacopa monnieri* using leaf explants. Int J Sci Res 3:210–212

Krishna H, Alizadeh M, Singh D et al (2016) Somaclonal variations and their applications in horticultural crops improvement. 3 Biotech 6:1–18

Kumar MM, Gopi R, Lakshmanan GMA et al (2013) Study on genetic diversity of *Bacopa monnieri* (L.) Pennell ecotype variants from Tamil Nadu by RAPD markers. Rom J Biol Plant Biol 58:9–17

Kumar S, Singh N, Lahane V et al (2024) Methyl jasmonate inducible UGT79A18 is a novel glycosyltransferase involved in the bacoside biosynthetic pathway in *Bacopa monnieri*. Physiol Planta 176:14260

Kumari R, Priyadarshni M, Kumari A et al (2014) In vitro mass multiplication of *Bacopa monnieri* (L.) an endangered and valuable medicinal herb. Indian J Sci Res 7:1248–1253

Kumari U, Vishwakarma R, Gupta N et al (2015) Efficient shoots regeneration and genetic transformation of *Bacopa monniera*. Physiol Mol Biol Plants 21:261–267

Kunjumon R, Johnson AJ, Remadevi RS et al (2023) Dammarane-type triterpenoid saponin-rich elite genotypes of *Bacopa monnieri* from south India and their edaphic-ecological preferences. Ind Crops Prod 202:117075

Lakshmaiah VV, Banadka A, Thirumoorthy G et al (2022) In vitro production of bioactive compounds from plant cell culture. In: Belwal T, Georgiev MI, Al-Khayri JM (eds) Nutraceuticals production from plant cell factory. Springer, Singapore, pp 29–67

Lamichhane S, Thapa S (2022) Advances from conventional to modern plant breeding methodologies. Plant Breed Biotechnol 10:1–14

Largia MV, Satish L, Johnsi R et al (2016) Analysis of propagation of *Bacopa monnieri* (L.) from hairy roots, elicitation and Bacoside A contents of Ri transformed plants. World J Microbiol Biotechnol 32:131

Mathur S, Gupta MM, Ram M et al (2002) Herb yield and bacoside-A content of field-grown *Bacopa monnieri* accessions. J Herbs Spices Med Plants 9:11–18

Mishra CN, Kumar S, Venkatesh K et al (2018) Application of genotype and environmental interactions in plant breeding. In: Bharadwaj DN (ed) Advanced molecular plant breeding: meeting the challenge of food security, 1st edn. Apple Academic Press, New York, pp 31–48

Mishra SK, Tiwari KN, Shivna PL et al (2015) Micropropagation and comparative phytochemical, antioxidant study of *Bacopa monnieri* (L.) Pennell. Res J Pharm Biol Chem Sci 6:902–912

Mohanta YK, Sahoo S (2014) In vitro culture of highly valuable medicinal plant *Bacopa monnieri* (L.) Penn. for rapid and mass multiplication. Int J Pharm Sci Invent 3:41–45

Mohapatra HP, Rath SP (2005) In vitro studies of *Bacopa monnieri*—an important medicinal plant with reference to its biochemical variations. Indian J Exp Biol 43:373–376

Murthy HN (2022) Biotechnological production of bacosides from cell and organ cultures of *Bacopa monnieri*. Appl Microbiol Biotechnol 106:1799–1811

Muthiah JL, Shunmugiah KP, Manikadan R (2013) Genetic fidelity assessment of encapsulated in vitro tissues of *Bacopa monnieri* after 6 months of storage by using ISSR and RAPD markers. Turk J Botany 37:1–10

Nagarajan T, Alagumanian S, Jahirhussain G et al (2015) In vitro mass propagation of *Bacopa monnieri* (Linn.) Wettst from nodal explant-a multipurpose medicinal plant. World J Pharm Res 4:1970–1982

Naik PM, Patil BR, Kotagi KS et al (2014) Rapid one step protocol for in vitro regeneration of *Bacopa monnieri* (L.). J Cell Tissue Res 14:4293–4296

Nalina L, Rajamani K, Shanmugasundaram KA et al (2022) Breeding and conservation of medicinal plants in India. In: Máthé Á, Khan IA (eds) Medicinal and aromatic plants of India, vol 1. Springer, Cham, pp 201–236

Nandhini S, Bayyapureddy A, Varghese RJ (2015) An enhanced in-vitro production of saponins and other bioactives from *Bacopa monnieri* L. Penn. Res J Pharm Biol Chem Sci 6:446–451

Narayani M, Srivastava S (2017) Elicitation: A stimulation of stress in in vitro plant cell/tissue cultures for enhancement of secondary metabolite production. Phytochem Rev 16:1227–1252

Nemeth-Zamborine E, Rajhart P, Seidler-Łożykowska K et al (2019) Intraspecific responses of medicinal plants: genotype-environment interaction may alter drug quality of aromatic plants. Biochem Syst Ecol 86:103914

Niazian M, Niedbała G (2020) Machine learning for plant breeding and biotechnology. Agriculture 10:436

Niedbała G, Kurasiak-Popowska D, Stuper-Szablewska K et al (2020) Application of artificial neural networks to analyze the concentration of ferulic acid, deoxynivalenol, and nivalenol in winter wheat grain. Agriculture 10:127

Nisha KK, Seetha K, Rajmohan K et al (2003) *Agrobacterium tumefaciens*-mediated transformation of Brahmi [*Bacopa monniera* (L.) Wettst.], a popular medicinal herb of India. Curr Sci 85:85–89

Padhiari BM, Ray A, Champati BB et al (2022) Artificial neural network (ANN) model for prediction and optimization of bacoside A content in *Bacopa monnieri*: a statistical approach and experimental validation. Plant Biosyst 156:1346–1357

Padhiari BM, Ray A, Halder T et al (2021) Quality control of *Bacopa monnieri* by high-performance liquid chromatography fingerprinting combined with chemometric methods. Pharmacogn Mag 1:7872–7881

Pandiyan P, Selvaraj T (2012) In vitro multiplication of *Bacopa monnieri* (L.) Pennell from shoot tip and nodal explants. J Agric Technol 8:1099–1108

Pathak A, Dwivedi M, Laddha N et al (2013) Detection of somaclonal variants using RAPD marker in *Bacopa monnieri* and *Tylophora indica*. J Agric Technol 9:1253–1260

Paul P, Sarkar S, Jha S (2015) Effects associated with insertion of cryptogein gene utilizing Ri and Ti plasmids on morphology and secondary metabolites are stable in *Bacopa monnieri* transformed plants grown in vitro and ex vitro. Plant Biotechnol Rep 9:231–245

Priyadarshi R, Sinha P, Dasari A et al (2021) Application of "Omics" technologies in crop breeding. In: Kumar A, Kumar R, Shukla P, Patel HK (eds) Omics technologies for sustainable agriculture and global food security. Springer, Singapore, pp 25–45

Rahman LU, Verma PC, Singh D et al (2002) Bacoside production by suspension cultures of *Bacopa monnieri* (L.) Pennell. Biotechnol Lett 24:1427–1429

Rahman SU, McCoy E, Raza G et al (2022) Improvement of soybean; a way forward transition from genetic engineering to new plant breeding technologies. Mol Biotechnol 65:162–180

Rajan KE, Preethi J, Singh HK (2015) Molecular and functional characterization of *Bacopa monniera*: a retrospective review. Evid Based Complementary Altern Med 945217

Ramesh M, Karthikeyan A, Vijayakumar K et al (2011) *Agrobacterium*-mediated transformation of pharmaceutically important Indian medicinal herb *Bacopa monnieri* (L.). J Med Plant Res 5:2316–2321

Ranjan R, Kumar S (2018) A rapid in vitro propagation protocol of local germplasm of *Bacopa monnieri* (L.) induced through direct organogenesis from nodal explants. J Pharmacogn Phytochem 7:2515–2518

Rao S, Rajkumar P, Kaviraj C et al (2012) Efficient plant regeneration from leaf explants of *Bacopa monniera* (L.) Wettst.: a threatened medicinal herb. Ann Phytomed 1:110–117

Rency AS, Satish L, Pandian S et al (2016) In vitro propagation and genetic fidelity analysis of alginate-encapsulated *Bacopa monnieri* shoot tips using *Gracilaria salicornia* extracts. J Appl Phycol 29:481–489

Rout JR, Sahoo SL, Ray SS et al (2011) Standardization of an efficient protocol for in vitro clonal propagation of *Bacopa monnieri* L.—an important medicinal plant. J Agric Technol 7:289–299

Russo A, Borrelli F (2005) *Bacopa monnieri*, a reputed nootropic plant: an overview. Phytomedicine 12:305–317

Samaddar T, Nath S, Halder M et al (2012) Karyotype analysis of three important traditional Indian medicinal plants, *Bacopa monnieri*, *Tylophora indica* and *Withania somnifera*. Nucleus 55:17–20

Sanyal R, Nandi S, Pandey S et al (2022) Biotechnology for propagation and secondary metabolite production in *Bacopa monnieri*. Appl Microbiol Biotechnol 106:1837–1854

Saran PL, Damor HI, Lodaya DH et al (2022) Identification of potential accessions of *Bacopa monnieri* L. for herbage yield and bacosides A content. Ind Crops Prod 176:114348

Schaart JG, van de Wiel CCM, Lotz LAP et al (2016) Opportunities for products of new plant breeding techniques. Trends Plant Sci 21:438–449

Sharath R, Krishna V, Sathyanarayana BN et al (2007) High frequency regeneration through somatic embryogenesis in *Bacopa monnieri* (L.) Wettest, an important medicinal plant. Med Arom Plant Sci Biotechnol 1:138–141

Sharma A, Verma N, Verma P et al (2017) Optimization of a *Bacopa monnieri*-based genetic transformation model for testing the expression efficiency of pathway gene constructs of medicinal crops. In Vitro Cell Dev Biol Plant 53:22–32

Sharma P, Padh H, Shrivastava N (2013a) Hairy root cultures: a suitable biological system for studying secondary metabolic pathways in plants. Eng Life Sci 13:62–75

Sharma P, Sharma S, Yadav S et al (2014) Plant derived bioactive molecules: culture vessels to bioreactors. In: Paek KY, Murthy HN, Zhong JJ (eds) Production of biomass and bioactive compounds using bioreactor technology. Springer, Dordrecht, pp 47–60

Sharma P, Yadav S, Srivastava A et al (2013b) Methyl jasmonate mediates upregulation of Bacoside A production in shoot cultures of *Bacopa monnieri*. Biotechnol Lett 35:1121–1125

Showkat P, Zaidi Y, Asghar S et al (2010) In vitro propagation and callus formation of *Bacopa monnieri* (L.) Penn. Plant Tissue Cult Biotechnol 20:119–125

Shrivastava N, Rajani M (1999) Multiple shoot regeneration and tissue culture studies on *Bacopa monnieri* (L.) Pennell. Plant Cell Rep 18:919–923

Shukla S, Bhutani R, Das S et al (2023) Secondary metabolite enhancement via in vitro techniques and its industrial prospects. In: Singh RS, Kumar N (eds) Genetic manipulation of secondary metabolites in medicinal plant. Springer, Singapore, pp 307–337

Silpa SG, Smitha GR, Ranjitha K (2021) Drying and packaging methods impact the bacoside profile and microbiological quality of Brahmi herb (*Bacopa monnieri* L.) during storage. Ind Crops Prod 159:113064

Singh A (2008) A note on variation of active principles in Indian medicinal plants and TIM formulations. Ethnobot Leaflets 12:603–606

Soundararajan T, Karrunakaran CM (2011) Micropropagation of *Bacopa monnieri* through protoplast. Asian J Biotechnol 3:135–152

Srivastava A, Garg G, Sharma P et al (2016) Genetic diversity in chemically diverse accessions of *Bacopa monnieri*. J Planar Chromat 29:203–208

Srivastava A, Shrivastava N (2008) Phytochemical and preclinical screening of aseptically produced herbal raw material *Bacopa monnieri*. Med Aromat Plant Sci Biotechnol 2:123–127

Srivastava P, Tiwari KN, Srivastava G (2017) Effect of different carbon sources on in vitro regeneration of Brahmi *Bacopa monnieri* (L.) An important memory vitalizer. J Med Plant Stud 5:202–208

Subashri B, Pillai YK (2014) In vitro regeneration of *Bacopa monnieri* (L.) Pennell.—a multipurpose medicinal plant. Int J Pharm Pharm Sci 6:559–563

Swarup S, Cargill EJ, Crosby K et al (2021) Genetic diversity is indispensable for plant breeding to improve crops. Crop Sci 61:839–852

Taha HS (2017) Micropropagation of *Bacopa monnieri* and enhancing Bacoside A production in shoot cultures. J Appl Environ Biol Sci 7:115–126

Thulasiram HV, Warrier RR, Nataraja KN (2022) Omics in sandalwood. In: Arunkumar AN, Joshi G, Warrier RR, Karaba NN (eds) Indian Sandalwood. Springer, Singapore, pp 363–384

Thulasiram HV, Karegaonkar SJ, Sharma P et al (2023) Targeted metabolite profiling and *de novo* transcriptome sequencing reveal the key terpene synthase genes in medicinally important plant, *Couroupita guianensis* Aubl. Plant Genet Resour 21:558–570

Tiwari V, Singh BD, Tiwari KN (1998) Shoot regeneration and somatic embryogenesis from different explants of Brahmi [*Bacopa monniera* (L.) Wettst.]. Plant Cell Rep 17:538–543

Tiwari V, Tiwari KN, Singh BD (2001) Comparative studies of cytokinins on in vitro propagation of *Bacopa monnieri*. Plant Cell Tiss Organ Cult 66:9–16

Tiwari V, Tiwari KN, Singh BD (2006) Shoot bud regeneration from different explants of *Bacopa monniera* (L.) Wettst. by trimethoprim and bavistin. Plant Cell Rep 25:629–635

Tripathi N, Chouhan DS, Saini N et al (2012) Assessment of genetic variations among highly endangered medicinal plant *Bacopa monnieri* (L.) from Central India using RAPD and ISSR analysis. 3 Biotech 2:327–336

Trojak-Goluch A, Kawka-Lipińska M, Wielgusz K et al (2021) Polyploidy in industrial crops: applications and perspectives in plant breeding. Agronomy 11:2574

Umesh TG, Sharma A, Rao NN (2014) Regeneration potential and major metabolite analysis in nootropic plant—*Bacopa monnieri* (L.) Pennell. Asian J Pharm Clin Res 7:134–136

Van Vu T, Das S, Hensel G et al (2022) Genome editing and beyond: what does it mean for the future of plant breeding? Planta 255:130

Varghese DB, Satyanarayana BN (2007) Induction of variations in two cultivars of *Bacopa monnieri* by gamma irradiation of in vitro cultures. Med Aromat Plant Sci Biotechnol 1:86–89

Varma A, Padh H, Shrivastava N (2011) Ecogeographical phytochemistry of *Adhatoda vasica* Nees in relation to quantitative variations of alkaloids. J Planar Chromat 24:406–411

Vergauwen D, De Smet I (2017) From early farmers to Norman Borlaug—the making of modern wheat. Curr Biol 27:R858–R862

Vijayakumar M, Vijayakumar R, Stephen R (2010) In vitro propagation of *Bacopa monnieri* L.—a multipurpose medicinal plant. Indian J Sci Technol 3:781–786

Wangdi K, Sarethy IP (2016) Evaluation of micropropagation system of *Bacopa monnieri* L. in liquid culture and its effect on antioxidant properties. J Herbs Spices Med Plants 22:69–80

Wurtzel ET, Vickers CE, Hanson AD et al (2019) Revolutionizing agriculture with synthetic biology. Nat Plants 5:1207–1210

Xin F, Dong W, Dai Z et al (2019) Biosynthetic technology and bioprocess engineering. In: Singh SP, Pandey A, Du G, Kumar S (eds) Current developments in biotechnology and bioengineering: synthetic biology, cell engineering and bioprocessing technologies, 1st edn. Elsevier, pp 207–232

Yadav S, Sharma P, Srivastava A et al (2014) Strain specific *Agrobacterium*-mediated genetic transformation of *Bacopa monnieri*. J Genet Eng Biotechnol 12:89–94

Yang Y, Mumtaz AS, Liyun H et al (2021) Applications of multi-omics technologies for crop improvement. Front Plant Sci 12:563953

Yusuf A, Kumar TR, Nikhilesh S et al (2011) Effects of antioxidants and gelling agents on regeneration, in vitro conservation and genetic stability of *Bacopa monnieri* (L.) Pennell. Int J Ayurved Herbal Med 1:51–67

Zhou H, Xie X, Tang Y (2008) Engineering natural products using combinatorial biosynthesis and biocatalysis. Curr Opin Biotechnol 19:590–596

Zote RK, Patil YK, Londhe SS et al (2018) In vitro regeneration of *Bacopa monnieri* (L.) from leaf and stem explants. Int J Chem Stud 6:1577–1580

Chapter 2
Barberry (*Berberis* spp.) Domestication and Breeding

Mehdi Rezaei and Ali Sarkhosh

Abstract Barberry, a shrub widely spread across temperate and subtropical regions, holds a significant place in the traditional medicine practices of ancient civilizations. The genus *Berberis*, particularly notable for its alkaloids like berberine, along with a wealth of antioxidant compounds such as phenolic acids and anthocyanins, is gaining increasing attention. Extensive research has delved into the pharmacological potential of barberry, exploring its efficacy in treating heart disease, its anti-cancer properties, and its antimicrobial effects. A significant amount of research has been conducted on the genetic diversity of barberry intra- and inter-species in morphological, biochemical and molecular markers. However, there are a few breeding programs focused on the genetic improvement of barberry for medicinal proposes. Furthermore, molecular research has focused on botany and species classification, chiefly between *Mahonia* and *Berberis*. There are also concerns related to susceptible *Berberis* spp. roles as a wheat-black-rust alternate host and *Berberis* as an invasive species. Some species of *Berberis* are resistant to wheat-black-rusts and there are reports on resistant black-rust gene map loci in *Berberis*. Additionally, assessments of low-risk areas for the spread of black rust show great promise for the future. Except for ornamental barberries cultivars, only a Iranian barberry cultivar '*Zereshk Bidaneh*' is domesticated and cultivated on a commercial scale for the consumption of its fruit. This cultivar has seedless berries and is very productive, resistant to adverse environmental conditions, and is mostly used as a food seasoning. There are many concerns for barberry cultivation and, the future depends on the development of wheat-black-rust resistant cultivars, and expanded commercial usage of their alkaloids and bioactive components.

Keywords Bioactive properties · Berberine · *Berberis* · Antioxidants

M. Rezaei (✉)
Horticulture Sciences Department, Agriculture Faculty, Shahrood University of Technology, Shahrood, Iran
e-mail: mhrezaei@shahroodut.ac.ir

A. Sarkhosh
Horticultural Sciences Department, University of Florida, Gainesville, FL 32611, USA
e-mail: sarkhosha@ufl.edu

J. M. Al-Khayri et al. (eds.), *Biodiversity and Genetic Improvement of Medicinal and Aromatic Plants I*, Advances in Plant Breeding Strategies 9,
https://doi.org/10.1007/978-3-031-81846-2_2

2.1 Introduction

Barberries, belonging to the genus *Berberis* in the family Berberidaceae, are ubiquitous shrubs found in temperate and subtropical regions across the world. Taxonomically, barberries are categorized within the tribe Ranales, with the genus *Berberis* being the largest within the Berberidaceae family (Ahrendt 1961). The diversity of barberry species is particularly pronounced in Asia and South America, with native species also present in Europe, Africa, and North America. However, significant differences in barberry taxonomy exist between Old and New World species, likely attributable to geographical isolation and hybridization (Ahrendt 1961). The classification of the genus *Berberis* has been subject to historical debate, especially regarding its relationship with closely related species, *Mahonia* (Ahrendt 1961; Fedde 1902; Hsieh et al. 2021; Yu and Chung 2017). Currently, four genera were introduced in the Berberideae subfamily including: *Mahonia*, *AllBerberis*, *Moranothamnus*, and *Berberis* (Yu and Chung 2017. The genus *Moranothamnus* was formerly referred to endemic *Berberis* species of California, USA, with the scientific name of *B. claireae* (Moran 1982). *AllBerberis* was also listed as a compound within barberry in the study done by Kim et al. (2004). Nevertheless, the inter-species relationship between the genera *Berberis* and *Mahonia* has not yet been resolved due to their wider geographical distributions, morphological ductility, and the ease of hybridization between the two species.

The genus *Berberis* comprises a vast array of species, with notable diversity across various regions. However, only one cultivar, '*Zereshk Bidaneh*', known for its seedless red berries, is commercially cultivated on a large scale only in eastern area of Iran (Alemardan et al. 2013). These dried berries are primarily utilized as natural colorants and seasonings in local cuisine (Alemardan et al. 2013). Additionally, the raw cut or powdered forms of the roots, stems, and fruits of *B. asiatica* (known as turmeric tree), *B. aristata*, and *B. lycium*, referred to locally as 'Kingor', 'Daruharidra', and 'Kashmal', are utilized as herbal remedies in Nepal, India, and Pakistan (Choudhary et al. 2021). Berries of *Berberis* spp. also are predominantly gathered from their natural habitats in mountainous regions or planting in restricted areas (Choudhary et al. 2021; Pradhan and Saha 2016). In countries such as Turkey, Turkmenistan, Azerbaijan and Iran, the berries of wild barberry species such as *B. cratagenia*, *B. orthobotryis* (with black fruits) and *B. vulgaris* and *B. integerrima* (with red fruits) are used to prepare local sour drinks with high amounts of antioxidants (Yahya 2017; Yilmaz et al. 2017). The *B. microphylla* and *B. buxifolia* species are also used in Argentina and Chile for their fruit and as natural medical remedies (Ruiz et al. 2014; Salehi et al. 2019). They are also grown as a garden plant or bonsai (Rodoni et al. 2014). Until the 1900s, the fruits of the common barberry, *B. vulgaris*, were often used throughout Europe and North America in herbal remedies, beverages, local food seasoning, and coloring (Barnes et al. 2020). However, due to the role of this species on invasion of the cereal black stem rust, programs were implemented to eliminate this species from all over Europe and North America. Today, there is very limited medicinal or culinary use of barberry in Europe and America.

In the study done by Ahrendt (1961), more than 200 *Berberis* were introduced as cultivated species, including *B. insignis, B. incrassate, B. hypokerina, B. hookeri, B.* × interposita, *B.* × *bristolensis, B. chrysosphaera, B. osmastonii, B. replicate, B. taliensis, B. candidula* and *B. oblonga.* However, at present time there is no information about their cultivation. This indicates the occurrence of gene erosion and a decrease of genetic diversity in barberry species.

Today, the most prevalent use of barberries throughout the world is in the urban landscape. Many species of barberry are used for ornamental aspects of landscaping and as hedges around the world. The most prominent and widespread of them is the Japanese barberry, *B. thunbergii*, which has several cultivars, and it is resistant to wheat black stem rust. Conversely, there are many concerns about the spread of this species as an invasive species (Ward and Williams 2011; Ward et al. 2009).

Barberries are renowned for their high antioxidant properties, with secondary compounds in their roots and stems showing potential therapeutic benefits. Numerous studies have explored the antioxidant and medicinal properties of barberries, emphasizing their potential in treating various diseases (Kong et al. 2022; Salehi et al. 2019; Singh and Mahajan 2013; Wang et al. 2019; Patel 2021; Nooreen et al. 2024).

Moreover, barberry genotypes exhibit substantial genetic diversity, presenting opportunities for breeding programs to develop cultivars with enhanced medicinal properties. This section provides an overview of recent advancements in barberry botany, active compounds, domestication, cultivation, crossbreeding techniques, genetic mapping, and tissue culture.

2.1.1 Botanical Classification and Distribution

Barberries belong to the Berberidaceae family. In this family there are more than 13–19 genera and it contains three subfamilies including *Berberidoideae* (x = $7^?$), *Nandinoideae* (x = 8 and 10), and *Podophylloideae* (x = 6). Only the subfamily *Berberidoidea* has a sensitive flag that bend to the stigma when contacted (seismonasty). There are two tribes in this subfamily including *Ranzanieae* and *Berberideae. Ranzanieae* are herbaceous plants that have the single genus *Ranzania.* Berberideae are woody shrubs, according to the latest research of scientists have four genera: *Berberis, Mahonia, AllBerberis* and *Moranothamnus.* Genus *Berberis* with dimorphic stems and almost always spiny unifoliate leaves (compound leaves with one leaflet); Genus *AllBerberis* with thorn-less dimorphic stems and compound leaves with 3–9 leaflets proposed by Dr. Peter H. Raven, to emphasize its distinctness (Chen 2017), and has been classified into multi-leafed *Berberis* in a previous study (Kim et al. 2004). The genus *Mahonia* has compound leaves with terminal leaflets and 3–9 leaflets, while the last genus, *Moranothamnus*, has uni-leaves, 7-foliolate, and a thornless stem. *Mahonia* and *Berberis* are the more renowned of the genera. New research has been conducted with the plastomes and nuclear ribosomal DNA markers, suggesting that *AlloBerberis* and *Moranothamnus* could have emerged as a result of the reciprocal hybridization between *Berberis* and *Mahonia* (Hsieh et al. 2021). The

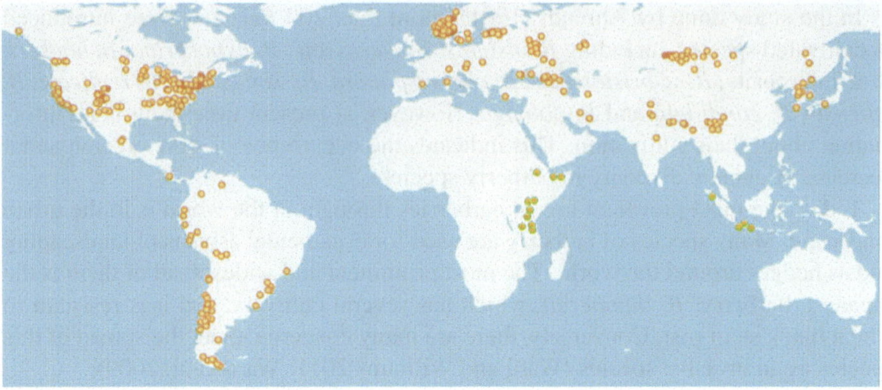

Fig. 2.1 Schematic map of worldwide distribution of *Berberis* spp (*source* www.gbif.org)

genera *Mahonia* and *Berberis* are very close related to each other and in the earliest literature regarding barberries they are classified in the same genus, *Berberis*. For more than 200 years, there has been a debate on whether or not to separate *Mahonia* from *Berberis* (Fedde 1902; Hsieh et al. 2021; Kim et al. 2004; Yu and Chung 2014).

Berberis has both evergreen and deciduous forms and can be found in the temperate regions of the Northern Hemisphere extending from parts of Asia, Europe, and South America to the highlands of East Africa (Fig. 2.1). Many of *Berberis* spp from Eurasia are deciduous species and the vast majority of evergreen *Berberis* spp are American (Ahrendt 1961).

2.1.2 Importance (Economic and Health Benefits)

Barberry has long been used as an herbal remedy in the traditional medicinal practices of Iran, India, China, and Native Americans (Arayne et al. 2007; Rahimi-Madiseh et al. 2017; Salehi et al. 2019). The roots and stems of barberries in the form of extracts or raw powders, and can be used for treatments in defined doses externally or consumed (Ali et al. 2008). Barberries have been used in the traditional medicinal Hindu, called Ayurveda, to treat fever, sores, inflammation, infection, skin problems, eye related disorders, diarrhea, malaria, liver diseases, jaundice, and as a relief of health problems such as heart failure (Ali et al. 2008). In Nepal, *Berberis* are used in ophthalmic medicine and as treatments for jaundice, malarial fever, diarrhea, and peptic ulcers (Manandhar 2002). Fruits of Nepalese barberries are also eaten fresh and are sometimes used locally for making alcoholic drinks (Manandhar 2002). In Europe, the common barberry, *B. vulgaris*, was frequently used as long hedges along fields before barberry eradication programs were implemented to control cereal black-rust during the nineteenth-century. Its berries have been used as a sour and sweet seasoning for meat dishes, especially chicken, and for preparing jams and beverages (Barnes et al. 2020). In Iranian traditional medicine, barberry has been also used as a blood purifier, anti-malarial drug and antiarrhythmic agent and it apply to treat jaundice, liver healthy, skin rashes (Moghaddam et al. 2007).

Today, barberries are still used in the food and pharmaceutical industries (Table 2.1).

An outstanding seedless barberry cultivar '*Zereshk Bidaneh*' is cultivated commercially ~ 20,000 ha in Ghaenat and Birjand of Iran. 22,000 MT dried fruit from cv '*Zereshk Bidaneh*' are annually produced in Iran. It's dark-red dried berries are used as a food additive and flavoring, along with saffron, in local foods such as Zereshk-Polo and Tah-Chin. Additionally, the fresh fruit is utilized in making jams, fruit juices, and syrups. In recent years, '*Zereshk Bidaneh*' has also found its way into some famous Iranian candies like Pulki and Sohan Asaly. Beyond local consumption, extensive research has explored the creation of carbonated beverages, jelly sauces, new candies, pastilles, and edible colored powders from seedless barberry fruits, with some of these innovations reaching industrial production (Alemardan et al. 2013; Rahimi-Madiseh et al. 2017).

Barberries have recently been gaining interest for their use in medicine and in the pharmaceutical industries. The medicinal uses are compelling enough for barberry to be grown in a wide variety of areas and to be harvested from their natural habits. Isoquinoline alkaloids are the main bioactive compounds in barberry plants. Berberine, a yellow chemical with a bitter taste, is the most abundant alkaloid found within *Berberis* species. See (https://pubchem.ncbi.nlm.nih.gov/compound/2353). Berberine has been isolated from the root, stem, bark and rhizome of *B. vulgaris* (barberry), *B. aristata* (tree turmeric), *B. aquifolium* (Oregon grape), *B. lyceum*, and *B. crataegina* (Singh and Mahajan 2013).

The effectiveness of barberry species in maintaining heart health has been demonstrated by their ability to improve hypertension, ischemic heart disease, cardiac arrhythmias and cardiomyopathy (Kong et al. 2022). Imenshahidi and Hosseinzadeh

Table 2.1 A broad spectrum of traditional and modern medicinal usage of *Berberis* spp

Berbreis in traditional medicine	Treatment of liver diseases
	Treatment of astrointestinal tract diseases
	Treatment of urinary system diseases
	Antibacterial, antifungal and antiprotozo applications
	Anti-inflammatory and antiallergenic activity
Berberis in modern medicine	Treatment cardiovascular diseases: control triglycerides, and cholesterol, reducing blood pressure
	Anti-diabetic function and effectiveness in metabolic syndrome
	Antioxidant and anticancer activity
	Anti-viral activity of berberine (HIV-1 and SARS-CoV-2)
	Anti-obesity effect of berberine

(2019) systemically reviewed 77 clinical trials studies on human subjects by using of berberine and barberry and confirmed the wide range of therapeutic applications of them including lipid-lowering, insulin-resistance improving actions, anticancer, and so on, without major side effects.

Additional beneficial compounds such as protoberberines, berbamine, tetrandrine, and chondocurine, which have been known for their anti-inflammatory and immunosuppressive properties, have been detected by phytochemical analysis of the root and stem back extracts of *B. vulgaris* (Kong et al. 2022).

Barberry has an array of other purposes outside of its direct consumption. *Berberis* and *Mahonia* that have multiple colors in their leaves and fruits have a special impact on landscape design. Subsequently, in nearly all parts of the world different species of barberries and *Mahonia*, mainly *B. thunbergii* cultivars, are used as ornamental shrubbery. Barberry is attractive for hedging and separating farms because it grows densely, and has sharp thorns that prevent any animals and people from passing through.

The roots of barberries are utilized to produce a natural yellow dye, and the fully grown branches of the *Berberis* shrub are noted for their lightness and strength, which renders them excellent for creating hand tools.

B. boliviana is a wild species from Peru, commonly grown around fields as a protective live fence due to its sharp thorns. Its fruits are also used as a natural colorant for fibers, imparting a soft purple hue (Del Carpio Jiménez et al. 2011).

Recently, barberry fruit extracts have become more prominent in the food and pharmaceutical industries. Barberry fruits are rich in bioactive compounds such as phenolic compounds, anthocyanins, flavonoids, carotenoids, alkaloids and vitamin C. Dark blue berries of *B. microphylla* have an antioxidant capacity 10 times greater than apple, orange, and pear fruit and 4 times greater than that of blueberry (Rodoni et al. 2014). *B. boliviana* L. fruit anthocyanin content (1.47 g cy-3-glu/100 g of fresh berries) was higher than black chokeberries (560 mg/100 g), elderberry (450 mg/100 g), black raspberries (300–400 mg/100 g) and bilberry (300–320 mg/100 g) (Del Carpio Jiménez et al. 2011).

Chlorogenic and syringic acid have been found to be the most abundant phenolic compounds in *Berberis* fruit extracts. Gholizadeh-Moghadam et al. (2019) reported that gallic acid and p-coumaric acid were the most commonly found phenolic compounds in the extracts of *Berberis* fruits. The levels of phenolic compounds were reported to be gallic acid (334.82 mg/L), caffeic acid (51.78 mg/L), chlorogenic acid (119.53 mg/L), p-coumaric acid (257.09 mg/L), cinnamic acid (0.57 mg/L), rutin (7.61 mg/ L), apigenin (4.44 mg/L), and quercetin (37.20 mg/L). Barberry fruits have been reported to have strong antioxidant activity (Hoshyar et al. 2016). In addition to the fruits: roots, twigs, and leaves of barberries also provided antioxidant activities (Hadaruga et al. 2010). The antioxidant activity was positively correlated with the content of main plant anthocyanins, phenols, and flavonoids in barberry fruits (Hadaruga et al. 2010; Končić et al. 2010). The levels of anthocyanins and phenolic compounds in barberry are high enough that they can be used as a natural source for industrial usage. Twelve anthocyanins were identified in the purified extract of

B. lycium berries. Two anthocyanins delphinidin-3-glucoside (35.3%) and cyanidin-3-glucoside (47.2%) were characterized as major components (Pradhan and Saha 2016). Furthermore, the antioxidant and antibacterial activity of barberry juice extract has suggested its possible use as a natural preservative in food industry (Aliakbarlu et al. 2014). New research has been done on the production of barberry juice powder and tablets with high antioxidant properties and tablets containing 60% barberry fruit powder can be offered as a 'ready to drink' product (Naji-Tabasi et al. 2021). The oil extracted from barberry seeds delayed the oxidation of soybean oil more effectively than commercial antioxidants. *Berberis* seed oil improve oxidative stability of soybean oil (Tavakoli et al. 2017).

The color of barberry berries ranges from red to dark blue and blackish purple. Barberry with blackish purple fruits are a richer source of anthocyanins and total phenol, while barberries with red fruits contain high radical scavenging activity and angiotensin-converting enzyme inhibition properties (Şensu et al. 2021). The barberry fruit has an effective role in remediating high blood pressure and cardiovascular disease. The red and black (dark-blue or purple-black) colors of barberry berries are used as a natural color in many foods and rice cooking in Iran and Arabian countries (Alemardan et al. 2013). The researchers showed that dark-colored berries have much greater color stability (Özgen et al. 2012). Çoban et al. (2021) showed that the fruit color of *B. crataegina* DC. is more stable than beet juice concentrate as the main supplier of natural red dye. As so, it has been proposed that Barberry fruit extract be used in the confectionery production industry to produce items such as color jellies and marshmallows, as barberry fruit extract acts as a natural colorant with good antioxidant properties (Çoban et al. 2021).

2.1.3 Domestication, Selection and Early Improvements

Berberis spp. serves as a valuable source of several drugs known as natural products or secondary metabolites. However, the majority of these compounds are still harvested from the wild or cultivated in small areas. Unfortunately, due to the high demand for barberry products, over-exploitation has led to the endangerment of useful *Berberis* species, loss of biodiversity, adulteration of plant materials and products, and negative impacts on ecosystems. Efforts have been made to meet the demand for barberry raw materials by implementing propagation of them at natural sites (Jabeen et al. 2015; Verma et al. 2021). This problem intensifies when the roots of the plant, which contain the highest concentration of alkaloid compounds, are harvested. This illegal harvesting of barberry is primarily taking place in the Indian subcontinent and only worsens the issues associated with the overharvesting of barberries (Jabeen et al. 2015; Khan et al. 2016).

Given that there are no known or reported medicinal cultivars among the numerous barberry species, the domestication of plants using new technologies must address the challenge of ensuring uniform raw material supply and production. To successfully adapt a plant for cultivation, comprehensive information about its reproductive biology, fruit set, and propagation is essential. Reproductive biology and pollination ecology of *B. lycium* was precisely studied by Verma et al. (2021). They demonstrated the presence of three variants of *B. lycium* detected in the North-Western Himalayas in India. They found that these variants cannot reproduce naturally and are at risk of elimination. Additionally, seedless fruits are produced due to predation by insect larvae, leading to low fruit set in some variants. Verma et al. (2021) recommend implementing a planting strategy for conservation and management to address these reproductive issues.

Evergreen Indian Barberry (*B. lycium*) seeds germinate without requiring cold stratification. When placed on wet filter paper, they begin germinating within 4–8 days. In contrast, seeds sown directly in soil take 9–15 days to germinate (Verma et al. 2021). On the other hand, Rezaei and Balandary (2012) found that low-temperature stratification is crucial for the seed germination of deciduous wild Iranian barberries (*B. integerrima* and *B. crataegina*). These species require 90 days of stratification at 5 °C. Acids scarification and GA_3 treatments were shown to improve seed germination percentage up to 90% (Rezaei and Balandary 2012). It seems that the seed chilling requirements differ greatly between the evergreen and deciduous *Berberis* species. Deb et al. (2017), when studying the seed biology of *B. manipurana*, showed that seed germination occurred in both 4 and 25 ºC stratification, however, seeds treated with cold stratification treatment (4 ˚C) gave a better response in terms of breaking dormancy. Most of the seeds of *Berberis* species stratified two to six weeks at 4 ºC, while some species may require stratification for up to two to three months. It's extremely important to remove all pulp from the seeds as seedlings are very susceptible to damping-off and some even to black stem rust. In fact, 50% of barberry germinated seeds and seedlings could be lost in the early stages of growth due to damping off diseases (Rezaei and Balandary 2015). The culturing time and medium has remarkable impacts on the seedling survival percentage, for example both cocopeat and perlite substrates under the mist system increase the survival rate of barberry seedlings. The best time to plant barberry seeds (after cold-stratification) is early spring, or even in greenhouse conditions (Rezaei and Balandary 2015). *Berberis* seeds are orthodox in nature and they remain viable for at least 4 years when kept at 1–3 ˚C in a sealed container (Deb et al. 2017). Verma et al. (2021) showed that the seed germination percentage varied greatly between wet filter paper and in soil conditions in the *B. lycium*. Seeds failed to germinate and emerge out of the soil when sown in soil and low germination of seeds sown in soil can be attributed to non-availability of light to the germinating seeds. The seed vigor and germination index of Indian barberry decline with prolonged storage and aging.

Some *Berberis* species can be successfully propagated clonally by stem cuttings (Puczel 2008). Cuttings taken from spring to fall can be rooted under mist conditions and the usage of rooting hormones (i.e. IBA) accelerate root initiation, improve rooting uniformity, increase the number of produced roots, and reduce shrink and rooting time. However, rooting of shoot-cuttings in some species has been unsuccessful due to high amounts of secondary compounds. *B. lycium* shoot-cuttings failed to initiate roots irrespective of season and auxin application (Joshi et al. 1992). Failure of stem cuttings in some *Berberis* species could be mostly attributed to the anatomical barriers and the absence of rooting co-factors. The presence of hetero-cellular two and three-row core rays in the structure of the stem characterizes the barberry as a difficult-to-root species (Andersen 1986). '*Zereshk Bidaneh*' is a challenging species to propagate via stem cuttings. In Iran, the primary propagation method for seedless barberry (cv. '*Zereshk Bidaneh*') involves separating the suckers or performing bush division.

Given the shrubby nature and large number of trunks of barberries, grafting operations have garnered less attention. A new study on barberry found that grafting methods, such as cleft and omega, were not very successful, while some success—less than 50%—was observed in T- and chip-budding techniques performed during spring seasons. The stem anatomical obscure plays an important role in preventing grafting success in barberry (Safari and Rezaei 2021).

2.2 Cultivars Characterization and Phylogeny

The majority of known barberry cultivars are ornamental shrubs. The medicinal *Berberis* species, such as *B. lycium* and *B. aristata*, are harvested directly from nature or occasionally cultivated by the locals in their native habitats. Only one cultivar is grown on a large commercial scale for food seasoning, known in Iran as the cv '*Zereshk Bidaneh*'. Interestingly, this cultivar is a direct selection from wild barberry species found in the northeastern mountainous areas of Iran and most likely belongs to *B. integerrima* (Heidary et al. 2009; Rezaei et al. 2011). The *Zereshk Bidaneh* cultivar was widely cultivated by local farmers due to its high yield, advantageous environmental adaptations, and seedless berries. According to historical manuscripts, the cultivation of this cultivar began only 200 years ago (Alemardan et al. 2013). The superior characteristics of this barberry cultivar include seedless berries, adaptability to arid climates, and very high productivity. In fact, when the berries ripen and turn red, the leaves and branches surrounding the fruit are completely covered and not visible (see Fig. 2.2a and b). This crop is exceptionally well-suited to adverse environmental conditions found in arid climates, where water resources are limited and hot, dry summers prevail (see Fig. 2.2c and d). '*Zereshk Bidaneh*' barberry produces an acceptable crop with low water requirements, making it a valuable choice in areas where few other trees can thrive.

2.2.1 'Zereshk Bidaneh' *Cultivation Practices*

'*Zereshk Bidaneh*' is extensively cultivated in the marginal desert areas of South Khorasan province in Iran, covering approximately 20,000 ha of production areas across various settlements, from rural and smallholder lands to large-scale orchards (Fig. 2.2d). These cultivation areas in Iran plateau at an altitude ranging from around 1200 to 1500 m and latitude of 32–34° N. This region experiences long, hot, and dry

Fig. 2.2 '*Zereshk Bidaneh*' Barberry in South Khorasan, Iran. **a** Top right: the dense berry clusters in one year old bearing shoot, **b** Top left: the mature shrubs in harvesting time. **c** and **d** Two down picture: typical commercial barberries orchards in arid climate of Birjand, Iran (the photos were taken by the author)

summers, followed by cold and wet winters. The fruit quality is notably higher in mountainous areas with higher altitudes and a greater temperature difference between day and night, particularly towards the end of the fruit ripening stage in autumn. The chilling requirement for flowering buds in '*Zereshk Bidaneh*' is 1400 h (Montazeran et al. 2018). As a result, the blooming of barberry flower buds occurs in May, much later than pome fruits like apples and pears, and spring frosts seldom damage the barberry. The growing degree days from the flowering stage to fruit ripening are approximately 170 days. '*Zereshk Bidaneh*' cultivar thrives in well-lit areas with deep soil and adequate air circulation to produce a substantial quantity of high-quality berries. Barberry shrubs exhibit good resistance to wind and are commonly used as windbreaks. However, hot winds, particularly during flowering, significantly reduce fruit set and berry quality, while excessive summer heat in lowland areas leads to juiceless berries with white berry disorder.

'*Zereshk Bidaneh*' berries are seedless, primarily due to simulative parthenocarpy and self-incompatibility (Ebadi et al. 2010). This cultivar bears numerous fruit clusters, such that during the bearing season, the small red berries cover the entire branch (Fig. 2.2a and b), a characteristic less common in wild species. Nonetheless, it appears that horticultural practices such as training, pruning, and nutrition significantly influence barberry productivity.

Propagation of '*Zereshk Bidaneh*' shrubs is typically carried out through layering and sucker separation. Suckers taken from mature shrubs, aged two or three years, have shown higher success rates in orchard establishment. Although minor differences exist among Iranian seedless barberries in cultivated areas, they all appear to originate from the same variety, with variations mostly related to climatic conditions influencing berry quality and quantity (Moghaddam et al. 2013). However, the possibility of clonal mutations in seedless barberry orchards cannot be dismissed. The origin and domestication of '*Zereshk Bidaneh*' remain unknown. The entire domestication process is based on the indigenous knowledge of traditional farmers, and in recent decades, new research has been conducted on the application of modern horticultural practices such as nutrient fertilization, pruning, pest management, irrigation systems, and harvesting methods.

The water requirements of seedless barberry are lower than those of similar fruit trees or shrubs, but during the initial growth stages, weekly irrigation with modest water is necessary. Rainfed cultivation of this crop has not been very successful in the desert areas of Iran due to very low rainfall. Although barberry shrubs exhibit excellent tolerance to water deficiency, fruit set is significantly reduced under such conditions. As the shrubs mature, the irrigation cycle increases, and it has been observed that local farmers can produce an acceptable crop within six months with only three or four irrigations during the growing season. '*Zereshk Bidaneh*' thrives in relatively light loamy-sandy soils with good drainage, although it also grows well in calcareous soils and tolerates high electrical conductivity (EC) and pH in poor soil.

One crucial aspect of domesticating and commercially harvesting '*Zereshk Bidaneh*' shrubs is proper training and regular yearly pruning. '*Zereshk Bidaneh*' has a shrubby nature and produces a large number of root and crown suckers. In the training process of these shrubs, the number of main trunks is typically reduced to 3–5, with proper distribution around, and additional suckers are removed. If suckers are needed for propagating new seedlings, a maximum of two suckers per shrub is retained for this purpose. Annual pruning involves removing dry and diseased shoots, water sprouts, and suckers, typically performed between November and February. The primary goal of annual pruning is to reduce the compactness of the middle parts of the shrubs and promote openness. Barberry shrubs left unpruned produce numerous suckers in the middle parts, leading to compact branching. This, coupled with the presence of large thorns, reduces access to the inner parts of the shrub and significantly decreases yield. Highly compact bushes also increase the activity of the barberry stem borer, *Syrista parreyssi*, one of the few pests causing significant damage to barberry orchards.

Harvesting berries is one of the most challenging and costly operations in '*Zereshk Bidaneh*' production due to the thinness of the berries' peel, the abundance of shoot thorns, and the density of branches. Barberry shrubs typically begin bearing fruit 3 to 4 years after planting. The optimal harvesting time is when the berries are completely uniform in color, sweetness, and juiciness, usually in early autumn when the weather cools down (Moghaddam et al. 2013). Harvest timing significantly affects fruit quality, with delayed harvests resulting in sweeter and tastier fruit, particularly if autumn rains allow for a delay. Slight autumn frosts towards the end of the ripening stage can enhance sugar content and juiciness, but there is a risk of damage from frost.

Hand harvesting of fruit in '*Zereshk Bidaneh*' is challenging due to the numerous thorns on the plant. Common harvesting methods include picking clusters by hand, shaking the branches to dislodge the fruit, cutting branches carrying fruit clusters, and a combined method involving both picking and shaking. Harvesting fruiting branches is more prevalent but leads to heavier annual burdens on barberry bushes. The yield of '*Zereshk Bidaneh*' depends on various factors, including management practices, shrub size and age, and harvesting date and method. Proper harvesting at the right time is crucial for producing high-quality berries.

After harvesting, barberry fruits should be processed promptly as they can only be stored fresh for about two weeks before fermenting and rotting due to their high moisture content. Some harvested berries are used to produce fruit juice concentrate, while fresh berries can be frozen at – 18 °C for long-term storage, retaining their antioxidants and vitamins. However, most Iranian barberries are dried before being marketed. Drying prevents fermentation and enables long-term storage in low-temperature, low-humidity conditions. Storage at high temperatures (25 °C and above) for extended periods can cause color changes, browning, and a decrease in quality.

The drying method significantly affects the quality of Iranian barberry production and depends on the harvesting technique. Berries harvested using the picking cluster and shaking methods are sun-dried for 10 to 15 days before being cleaned and packaged. Higher-quality dried berries, known as Puffy Zereshk, are obtained by drying in shade houses. Fruit-bearing typically begins economically in the seventh year, with an average yield of 2 tons per hectare. As the shrub ages, the yield increases, reaching its maximum between fourteen and twenty-five years. Barberry shrubs can continue to produce fruit for up to 85 years. On average, 4 kg of fresh berries yield 1 kg of dried berries.

2.2.2 Current Agricultural Challenges

The cultivation of barberries faces several challenges, typical of many agricultural crops, including water and resource instability, high production costs, climate change, supply–demand imbalances, and limited access to agricultural technologies. These challenges must be addressed within the context of barberry cultivation, offering opportunities for learning and improvement. Specific challenges unique to barberry cultivation will be discussed in the following sections. Notably, no breeding cultivars of barberry have been released solely for medicinal purposes.

2.2.2.1 Concerns on Barberries Growing (Wheat Black-Stem-Rust Problem and Invasive Risk)

Barberry species have faced declining popularity due to two critical issues: the spread of black stem rust and their invasive tendencies. Both issues require careful attention in any barberry breeding program. Barberry plays a significant role in spreading black rust in cereals, acting as a secondary host for this fungus. Efforts to eliminate common barberry shrubs were initiated in the 1920s across Europe and the Americas to control wheat black rust disease (Barnes et al. 2020). However, rust-resistant barberry varieties, such as Japanese barberry, continued to be cultivated due to their attractive leaf color and landscape design appeal. Japanese barberry's faster growth rate, facilitated by seed transfer through local birds, led to its spread into natural areas, making some scientists consider it an invasive species (Brand et al. 2012; D'Appollonio 2006; Ward and Williams 2011). Other *Berberis* species, including *B. darwinii* (Darwin's barberry) in New Zealand (McAlpine and Jesson 2007) and *B. aristata* and *B. julianae* in South Africa (Keet et al. 2016), have also been introduced as invasive species. Additionally, naturally occurring hybrids (e.g., *B.* × *ottawensis*) between Japanese barberry and the black rust-susceptible common barberry further contributed to the spread of this species across the Americas, raising concerns about the resurgence of black stem rust in wheat (Connolly et al. 2013; Hale et al. 2015). In the United Kingdom, there are also concerns about the potential invasion of wheat rust due to the common barberry (Barnes et al. 2020).

Through the efforts of scientists, many potential solutions have been created to address these concerns. The three main strategies that can be put into effect to address these barberry issues are as follows: *first*, the cultivation of barberry in areas with a lower risk of spreading buckwheat rust. *Second*, the breeding black-rust resistant cultivars and the establishment of strict quarantine rules to prevent the reproduction and distribution of susceptible species— as specific DNA molecular markers were developed to identify resistant Japonica barberry cultivars from susceptible species. *Lastly*, the use of seedless or sterile cultivars for medicinal or green space purposes.

Wheat black rust epidemics are influenced by climatic conditions and cultivation systems (Chen 2005, 2017). Climatic factors affect the survival, infection, and reproduction of black rust, as well as various stages of its reproductive cycle on barberry (Wang and Chen 2015; Zhao et al. 2016). Studies on rust infection risks in Southeast Asia, Europe, and Central Asia have identified regions with varying levels of risk for barberry infection by rust pathogens (Sinha and Chen 2021).

Canada and the USA have strict laws regarding the importation and cultivation of susceptible *Berberis* species. However, black rust-resistant cultivars, particularly Japanese barberry, can still be cultivated. DNA markers have been developed to identify true rust-resistant cultivars and hybrids, offering hope for transferring resistance genes to beneficial barberry species (Lambert and Obae 2016; Lubel et al. 2008; Obae et al. 2013). The natural hybrids between common barberry and Japonica barberry, which can either be susceptible or resistant to black rust species, show that a significant percentage of the hybrids contain rust-resistant genotypes, which raises hopes that black rust resistant genes, will transfer to beneficial species of barberry.

To prevent the overspreading of barberry species in the natural environment, seedless species with intentional defects in the reproductive system can be used. In a study, sterile clones of Japanese barberry (*B. thunbergii* var. atropurpurea) developed by the induction of tetraploids in meristem of seeds through exposure to Colchicine and Oryzalin (Lehrer et al. 2008). Since the main cause behind the uncontrolled spread of barberries is the consumption of seeded berries by birds and distribution of viable seeds, if seedless varieties or sterile clones are used, the problem of *Berberis* as an invasive species can be solved.

2.2.2.2 '*Zereshk Bidaneh*' Agricultural Challenges

Iranian barberry cv '*Zereshk Bidaneh*,' despite being a direct selection from local wild species, faces numerous agricultural challenges hindering its widespread cultivation. Some of these challenges include:

Thorny shrub and hand harvesting: The presence of large thorns on barberry shoots poses difficulties for horticultural operations and manual harvesting. The sharp thorns not only cause bodily injuries but also hinder mechanization, contributing to the lack of efficiency in fruit harvesting.

Alternate bearing: '*Zereshk Bidaneh*' suffers from the severity of alternate bearing, a common problem in many fruit crops. The practice of cutting off fruit branches entirely leads to a decline in yield the following year due to the formation of flower buds on one-year branches.

Limited climate adaptation: While '*Zereshk Bidaneh*' shrubs are highly adapted to desert conditions, they require specific climatic conditions for optimal fruit production. Cold autumn nights are crucial for fruit juiciness and sugar levels, while extremely hot summers and water stresses can lead to fruit dehydration and quality reduction. Additionally, high humidity increases the risk of black rust and contamination in nearby cereal fields, limiting cultivation to specific geographical areas.

Low prices and underdeveloped processing industry: Despite significant quantities of dried berries entering the market during harvest season, the lack of advanced processing infrastructure and limited consumption have led to declining prices and low profitability. Pruned and harvested branches, valuable sources for berberine extraction, are often underutilized, with no active processing industry. Although barberry beverages hold promise due to their antioxidant properties, the beverage industry's establishment has been slow.

Pests: The rate of pests and diseases in barberry is comparatively less than other fruit crops. However, due to the expansion of the area under cultivation, some pests may be posing a threat in the future. Sawfly, *Syrista parreyssii.*, is one of the most important pests of barberry in Iran. Damages from these pests can be seen throughout most barberry orchards in Iran. The afflicted branches are weakened, and the fruits formed on them in the following year have weak and pale berries that are lower in quality. Other pests that are present but less harmful in barberry are the barberry aphid (*Lisomophis berberidis*), leaf-eating moth *(Pandemic ribena*), two-spotted spider mite (*Tetranychus urticae*), and an earwig (*Forticula tomis*).

2.2.3 Genetic Improvement Objectives

Barberry cultivation has significantly declined due to efforts to control wheat black rust, although notable progress has been made in breeding ornamental cultivars resistant to wheat black rust. With the vast diversity of *Berberis* species in the Northern Hemisphere, the initial step in domesticating wild species involves selecting black rust-resistant cultivars rich in bioactive compounds. Japanese barberry species, along with others, exhibit wheat black rust resistance, offering valuable genetic resources for breeding and domestication. Roelfs (1985) indicated that 51 species of *Berberis*, 12 species of *Mahonia*, and 3 species of × *MahoBerberis* were resistant to wheat and rye stem rust. A list of about 189 species and cultivars of *Berberis*, 31 species and cultivars of *Mahonia*, and 4 species of *MahoBerberis* that are resistant to black rust and are mostly ornamental has been published by the USDA (USDA 2002). Characterization of a natural population of *B.* × *ottawensis* (an interspecific hybrid of black wheat rust-susceptible *B. vulgaris* and black wheat rust-resistant *B. thunbergii*) reveals that 33% are resistant to stem rust and 16% contain intermediate phenotypes caused by *Puccinia graminis* (Bartaula et al. 2019). This indicates that resistant genes of stem black rust in barberry can be transferred through controlled crossing, However, no attempt has been made to develop black rust-resistant barberry cultivars for medicinal purposes.

Reducing thorniness is crucial for handling barberry plants. The type and number of thorns varies from species to species in barberies and mahonia, and thornless genotypes are already found in some wild species, which can be a very good source for breeding. Ahrendt (1961) recorded 17 species of spineless *Berberis*, with the remaining 13 species being from South America. Some *Berberis* species also have weak thorns, which are often solitary, though sometimes threefold, slender, and only 1–9 mm long. Sometimes, the thorns are absent from some parts of the stem. Ahrendt (1961) described 166 species with weak thorns, which were mostly collected from Asia. Among the ornamental Japanese barberry varieties there is only one that is thornless (*B. thunbergii* "Thornless"), despite the fact that most Japanese barberry have thorns. All genetic resources indicate that thornlessness is a trait that is found in the germplasm of barberries and can intentionally be used to produce superior thornless genotype in a breeding program. Mutations can also be an option to change the genes involved in the production of thorns in barberry plants. Modifying growth habits and reducing sucker formation are important breeding goals to lower production costs and facilitate mechanization.

Medicinal barberries have been practically eradicated from production due to efforts to control wheat rust in Europe and America. Nevertheless, the diversity of barberry species in the Northern Hemisphere remains very high. The first step in cultivating barberry globally is to domesticate wild species that possess high levels of bioactive substances and alkaloids. Extensive research has highlighted the significant diversity of antioxidants present in wild barberry. Barberry species, and even genotypes within a single species, exhibit substantial differences in their antioxidant components and alkaloid content. Therefore, a major breeding objective is to optimize the quantities of these valuable compounds.

Increasing fruit marketability is one of the most important objectives in a barberry breeding program. Key factors that affect fruit marketability include fruit size, color, taste index, and juiciness. By focusing on these attributes, breeders can develop barberry varieties that appeal more to consumers and thus enhance their commercial potential. Safari-Khuzani et al. (2020) selected 11 promising progenies with ideal characteristics related to fruit quality, including berry weight, skin color, taste, seedlessness, and the expression of small seeds from superior F_1 segregating populations in Iran. In this *Berberis* population, some genotypes produce berries weighing 0.4 g and reaching 1.5 cm in length.

Barberry fruits are typically very sour, with high acidity being a significant issue, although some consumers enjoy this sour taste. In most wild species, the acid content is so high that the fruit is not edible. The fruit pH in *B. vulgaris* and *B. crataegina* was reported to be between 2.44 and 3.25, indicating notably high acidity levels. Consequently, another breeding objective for barberries is to reduce fruit acidity and increase the taste index. Additionally, improving yield, reducing annual bearing, and increasing the range of cultivar adaptation are prioritized breeding objectives.

2.3 Germplasm Biodiversity and Conservation

Barberry is widely spread throughout the world. *Berberis* shrubs exhibit extraordinary endurance in poor environmental conditions, high adaptability, robust propagation ability, and an ancient evolutionary history. These factors contribute to the high genetic variation found in barberry genotypes. Moreover, some morphological traits, such as leaf characteristics and shrub growth, can show significant changes under various geographical conditions. In *Berberis*, it is very difficult to identify true species within germplasms due to interspecific hybridization and heteromorphy. The widespread consumption of barberry species by birds has facilitated their dispersal in natural environments, resulting in a relatively enormous variety within these settings.

2.3.1 Germplasm Diversity

A unique diversity is observed among barberry species and genotypes in temperate regions of the world, particularly when considering Mahonia within the barberry germplasm and the potential for interspecies hybrids. The *Berberis* family is one of the evolutionary pioneers among dicot plants, exhibiting considerable diversity and varied growth habits. Barberries vary in shoot growth habits, berry shape, berry color, and leaf structure. Some species are evergreen, while others are deciduous, ranging from small to large bushes, with single leaves or multi-leaved structures, red fruits to dark blue and black, and round to oval-shaped berries. Most barberries thrive in mountainous regions at high altitudes, between 600 and 2500 m. The bushes typically feature thorny branches with yellow flowers, although greenish flowers are rare (Li et al. 2017).

Variability in morphological traits: Several studies were conducted on the morphological diversity in West Asia among *B. asiatica*, *B. lycium*, and *B. aristata* in India and Pakistan (Ahmed et al. 2013; Bukhari et al. 2022; Husain and Rao 1998), *B. integerrima* and *B. caratagenia* in Iran (Jannatizadeh and Khadivi-Khub 2016; Rezaei et al. 2020; Tatari et al. 2019), *B. vulgaris*, *B. integerrima* and *B. caratagenia* in Turkey and Kyrgyzstan (Yahya 2017; Yilmaz et al. 2017). A germplasm collection containing 42 Iranian barberry genotypes with repetitions was evaluated by Rezaei et al. (2020) (Fig. 2.3). The study revealed high variability, especially in the number of berries per cluster, clusters per branch, and fruit weight. The number of berries per bunch varied significantly among accessions, from 4.67 to 32.33. Fresh berries weighed between 0.084 and 0.55 g, with berry lengths ranging from 6.20 to 14.5 mm. Juice percentage varied from 13.33 to 53.33%, berry shapes ranged from round to elongated, and seed number per berry varied from 0 to 3.66. Total soluble solids (TSS) ranged from 9.5 to 26%, and vitamin C levels from 29.33 to 76.27 mg/100 ml.

In the Jammu and Kashmir regions of Pakistan, the berry dry weight of Asian barberry species varied from 0.115 to 0.120 g, with fruit colors ranging from purple to dark blue. The number of clusters per plant in Pakistani barberries varied from 54 to 65 (Ahmed et al. 2013), which is less than that of the Iranian seedless barberry, where the cultivar '*Zereshk Bidaneh*' has approximately 32 clusters per branch (Fig. 2.2a).

Berberis genotypes have relatively high morphological diversity within each species. Ardestani et al. (2013) reported that the average berry weight of wild barberry genotypes was ranged from 0.051 to 0.348 g in Iran and Pakistan, while Akbulut et al. (2009) stated that the average fruit weight in Turkey was 0.070 g. Barberry genotypes grown in different parts of the world differ significantly in terms of fruit weight, which may be attributed to varying growing conditions, genetic backgrounds, climates, geographies, cultural practices, and maturation stages. Bukhari et al. (2022) noted a remarkable difference in *B. lycium* plant samples collected from five districts due to variations in altitude, climatic conditions, and soil texture (Fig. 2.3).

In recent years, many plant scientists have been interested in investigating the physicochemical diversity of various barberry species. Extensive research has been conducted to evaluate the levels of antioxidant compounds and alkaloids in different parts of various barberry species. Fruits like berries are recognized as an important component of a healthy diet due to their high antioxidant activity. Some barberry species have demonstrated even higher antioxidant capacity than other antioxidant-rich fruits such as blueberries (Rodoni et al. 2014). Research has revealed significant variation in phenolic and anthocyanin compounds within barberry species (Aliakbarlu et al. 2014; Hassanpour and Alizadeh 2016; Končić et al. 2010; Pradhan and

Fig. 2.3 Variability of color, shape and size of berries, seeds and thorns in 42 *Berberis* genotypes in a germplasm collection from the northeastern regions of Iran (the photos were taken by the author)

Saha 2016; Salehi et al. 2019; Srivastava et al. 2015). This diversity may be of great interest to breeders as it could result in the development of higher quality fruit.

Variability in bioactive components: The phenolic content in barberry species depends on various factors, including genotype, plant growth stages, seasons, agronomic practices, maturity level at harvesting time, postharvest storage, and environmental conditions (Benvenuti et al. 2016; Yilmaz et al. 2009). Additionally, the final reported results can vary based on extraction methods, analytical techniques, and measuring units, leading to potential errors and confusion.

In Iran, Motalleb et al. (2005) reported that the phenolic content in Iranian wild barberry fruits was 3450 mg GAE/100 g. Ozgen et al. (2012) found that the total phenolic content of barberry fruits in juice varies between 2565 and 3629 mgGAE/l among 6 barberry genotypes grown in the inner part of Turkey. Ardestani et al. (2013) also showed higher phenolic content in *B. integerrima* (47.80 g/100 g) compared to *B. vulgaris* (27.99 g/100 g). In contrast, Hassanpour and Alizadeh (2016) reported higher phenolic content for *B. vulgaris* (10.46–18.43 mg GAE/g DW) than *B. integerrima* (5.98–14.16 mg GAE/g DW). Nilda et al. (2018) showed that phenolic compounds in *B. vulgaris* ranged from 2281 to 3462 mgGAE/l in fruit juice. Şensu et al. (2021) reported *B. integerrima* with purple berry contained the highest amount of phenolic (81.42 mgGAE/gDW) followed by *B. vulgaris* (74.74 mgGAE/gDW) with red berries. Gıdık (2021) reported that the total phenolic content of Turkish

Berberis samples ranged from 10.84 to 28.92 mg GAE/g, with the lowest and highest levels observed in *B. crataegina* and *B. vulgaris* L. (wild).

In addition to the fruit, the leaves and shoots of *Berberis* also contain phenolic components. Şensu et al. (2021) showed that the total phenolic contents in leaf samples of *Berberis* species ranged from 24.81 to 42.62 mgGAE/g DW, while in shoot samples, it varied between 21.36 and 15.02 mgGAE/g DW. The leaves of *B. aristata* and *B. thomsoniana* contain 50.44 and 65.30 mgGAE/g DW, respectively (Lok Ranjan et al. 2018). The main phenolic compound in barberries is chlorogenic acid, ranging from 22.42 to 32.9 mg/g DW, with Gundogdu (2013) reporting chlorogenic acid as the dominant phenolic compound in B. vulgaris (0.752 g/kg FW).

Anthocyanins have also garnered significant interest in barberries in recent years. Özgen et al. (2012) reported that the total monomeric anthocyanin content was between 506 and 803 mg/L in fruit juice as cyanidin 3-glycoside among six barberry genotypes was sampled from the middle part of Turkey. Ardestani et al. (2013) found that anthocyanin levels in *B. vulgaris* ranged from 22.96 to 65.28 mg cyanidin 3-glycoside/100 g DW, while in *B. integerrima*, the levels ranged from 39.15 to 417.07 mg cyanidin 3-glycoside/100 g DW in Iran. Total anthocyanin content in *B. vulgaris* genotypes varied from 360 to 874 mg/L of fruit juice as cyanidin 3-glycoside (Nilda et al. 2018). Şensu et al. (2021) reported total anthocyanin content in *B. integerrima* and *B. vulgaris* as 455.91 mg and 366.52 mg cyanidin 3-glycoside/100 g DW, respectively. Yildiz et al. (2014) observed anthocyanin levels in a *B. vulgaris* genotype from Turkey ranging from 35.1 to 100.4 mg cyanidin 3-glycoside/100 g DW.

The total flavonoid content in barberry fruits ranged from 160.82–170.4 mg Rutin equivalents/g DW for *Berberis* spp in Turkey. The purple-colored barberry fruit had a higher flavonoid content (Şensu et al. 2021). Gıdık (2021) reported that the fruit flavonoid content in Turkish wild *Berberis* had a range of 0.41–2.20 mg quercetin / g. Additionally, the total flavonoid content in the leaves of *B. aristata* and *B. thomsoniana* was 7.82 and 16.46 mg quercetin/g DW, respectively (Lok Ranjan et al. 2018).

The taste of *Berberis* berries is important for fresh fruit consumption. The taste is determined by a combination of aroma and the flavor index, which is the ratio of total soluble solids (TSS) to total acidity (TA). In a collection of 42 *Berberis* accessions, the TSS ranged from 9.5 to 26, with a mean of 19.73% in Iran (Rezaei et al. 2020). Özgen et al. (2012) reported a TSS that ranged from 6.95 to 20.85% in *B. vulgaris* genotypes. Ardestani et al. (2013) found an average TSS value of 17.33% in barberry genotypes, while Ahmed et al. (2013) reported TSS values between 23.60 and 27.63% in barberry genotypes.

Barberry fruits are known for their sour taste, and their acidity levels are typically higher than those of other fruits. The total acidity of wild-type barberry in Iran ranged from 2.68% to 8.44%, with an average of 5.19% (Rezaei et al. 2020). Malic acid is the dominant organic acid in the fruits of all barberry genotypes, followed by citric and tartaric acids. The high organic acid content of barberry fruits contributes significantly to their taste and flavor profile. Among *B. vulgaris* genotypes from Antalya, Turkey, the malic acid content ranged from 3.41 to 8.81 g/l, and the citric acid content ranged from 1.12 to 1.91 g/l (Nilda et al. 2018). Additionally, Akbulut et al. (2009) reported an average pulp ratio of 75.59% in barberry genotypes grown in Turkey.

Alkaloids in Berberis: Several researchers have reported that alkaloids are the main bioactive chemical constituents of *Berberis* species. Major alkaloids in *Berberis* species include berberine, berbamine, palmatine, columbamine, jatrorrhizine, and oxyacanthine (Bhardwaj and Kaushik 2012). Among these, berberine and berbamine have received the most attention in scientific literature due to their significant pharmacological properties. *Berberis* alkaloids are primarily located in the cortical tissues of the roots and stems, with the bark of old roots containing the highest concentrations of these compounds. The berberine content varies significantly among different *Berberis* species and tissues (Table 2.2). The most extensive research on berberine content has been conducted on well-known *Berberis* species such as *B. aristata, B. asiatica, B. lycium* and *B. vulgaris.*

The maximum range of berberine content is found in the root parts (0.69–13.39%) of most *Berberis* species. However, there is inconsistency in the reported concentrations of berberine, likely due to differences in plant species, locations, and analytical techniques used (Singh and Mahajan 2013). For example, Srivastava et al. (2004) reported higher content in *B. aristata* (2.8%) in comparison to *B. asiatica* (2.4%), while Andola et al. (2010) found that *B. asiatica* (4.3%) had higher berberine content than *B. lycium* (4.0%) and *B. aristata* (3.8%). Sehrawat et al. (2009) reported that berberine content in the roots of *B. aristata* from 11 locations ranged from 0.82 to 1.86%, with the highest content observed in winter samples. The berberine alkaloid is also found in other plant species, for instance the total berberine content (6.34 ± 0.14) in the dried roots (% w/w) of Goldenseal (*Hydrastis canadensis*) was significantly higher than the root berberine content in *B. thunbergii* (1.38 ± 0.14) and *B. vulgaris* (0.43 ± 0.02) (Villinski et al. 2003). A progressive increase in the berberine content of the plants has also been observed with an increase in age (Singh and Mahajan 2013). The maximum berberine concentration, 13.39% in Dw, was reported in roots of *B. lycium* (Katare et al. 2022).

Beside genotypes, the extracted bereberine are highly depending on tissue, age of plant, season, environmental and extraction and detection techniques (Patel 2021). It seems that the extraction strategy has a great impact on the amount of berberine yield. Katare et al. (2022) obtained 13.39% berberine in root of *B. lycium* by optimizing the berberine extraction process while the previously reports only showed maximum ~ 4% (Garhwal 2010).

Variability in DNA genome: Since the morphological and biochemical traits are often misleading due to excessive environmental impact, stable and easily observable DNA markers are essential to characterizing a diversity of germplasms. DNA markers such as random amplified of polymorphic DNA (RAPD), amplified fragment length polymorphism (AFLP), inter simple sequence repeats (ISSR), simple sequence repeats (SSR), and Internal transcribed spacer (ITS) were used for assessing the genetic diversity, phylogenetic relationships, and identification of cultivars among *Berberis* genotypes (Table 2.3).

A phylogeny constructed by the ITS sequences from 79 taxa represent a high diversity of *Berberis* (Kim et al. 2004) and show that most of the Schneider (1905) and

Table 2.2 The highest reported berberine content in different tissues of *Berberis* species

Species	Tissue	Detection method	The highest reported berberine content (W/W) %	Source
B. koreana	Root	HPLC/UV	1.32	(Yoo et al. 1986)
	Stem		0.1	
	Leaves		0.005	
B. vulgaris	Root	HPLC/UV	0.43	(Villinski et al. 2003)
B. thunbergii			1.38	
B. chitria	Root	HPTLC	3.16	(Srivastava et al. 2006)
B. tinctoria	Root	HPTLC	3.36	(Rawat and Srivastava 2007)
B. vulgaris	Stem	HPLC /UV	1.42	(Kosalec et al. 2009)
B. caroatica	Root		1.21	
B. aristata	Root	HPLC/Uv	1.86	(Sehrawat et al. 2009)
B. vulgaris	Root	HPLC DAD	12.9	(Končić et al. 2010)
B. caroatica	Root		7.29	
B. aristata	Raw herb	HPLC/UV	2.7	(Pasrija et al. 2010)
B. aristata	Root	HPTLC	3.2	(Andola et al. 2010a)
B. asiatica	Root	HPLC/UV	4.3	(Garhwal 2010)
	Stem bark		3	
B. aristata	Root		3.8	
	Stem bark		2.6	
B. lycium	Root		4	
	Stem Bark		2.8	
B.pseudumbellata	Root	RP-HPLC	3.60	(Andola et al. 2010b)
	Stem bark		2.82	
B. aristata	Fruit	HPTLC	0.033	(Kamal et al. 2011)
B. umbellata	Root	HPTLC	3.05	(Singh et al. 2012)
Wild *Berberis* Species from Patagonia	Fruit	HPLC/UV	0.001	(Ruiz et al. 2014)
B. soulieana	Root	HPLC/UV	1.48	(Li et al. 2014)
B.triacanthophora			1.47	
B. bergmanniae			2.76	
B. gagnepainii			2.29	

(continued)

Table 2.2 (continued)

Species	Tissue	Detection method	The highest reported berberine content (W/W) %	Source
B. henryana			0.69	
B. soulieana	Stem		0.25	
B.triacanthophora			0.22	
B. bergmanniae			0.84	
B. gagnepainii			1.07	
B. henryana			1.01	
B. vulgaris	Root	HPLC/UV	10.29	(Ahmad et al. 2019)
B. lycium	Root	HPLC	13.39	(Katare et al. 2022)

Ahrendt (1961) traditional geographical groups (such as groups *Australes*, *Septentrionales*, *Orientales*) schemes were mostly congruent with the ITS phylogeny at higher levels. However at a lower taxonomic level, the ITS phylogeny did not recognize any sections or subsections that were presented by Schneider (1905) or Ahrendt (1961). Feng et al. (2021) reported the complete chloroplast genome (166,275 bp containing 147 genes) of *B. weiningensis*, demonstrating that *Berberis* species closely cluster with *Mahonia* species in phylogenetic analyses. Molecular DNA markers in studies of wild barberry species from a range of geographical areas indicate high genetic diversity inter and intra species of *Berberis* (Akishev et al. 2012; Rezaei et al. 2011; Safamanesh et al. 2017; Tripathi and Goswami 2013).

These genetic distances can aid scientists in selecting parent plants for breeding programs. Parents with high genetic distance are more likely to produce a highly segregated offspring population, increasing the chance of selecting superior genotypes for cultivation and breeding purposes.

2.3.2 *Cultivars Identification*

Identifying barberry cultivars is crucial, particularly for Japanese barberry, to prevent the spread of black stem rust by avoiding the use of common barberry as a secondary host (Table 2.3). his issue gained importance with the development of natural hybrids between common barberry and Japanese barberry, known as *B.* × *ottawensis*, and other hybrid cultivars sometimes sold as Japanese barberry. In Canada, out of 60 Japanese barberry cultivars, only 11 are licensed for use in the region.

Japanese barberry cultivars are generally distinguishable from others during the growing season when plants are fully grown and leaf color, flower shape, or fruit color are visible. However, seedlings sold in nurseries are typically leafless during

Table 2.3 Molecular markers used for genetic characterization in *Berberis* spp

No	Marker type	Purpose	Genotype and Species	Location	References
1	AFLP	Phylogenetic relationships	13 Patagonian species of *Berberis*	Patagonia, Argentina	(Bottini et al. 2002)
2	ITS	Taxonomic and phylogeny	79 taxa of *Berberis*	South America-Old World (mainly Asia)	(Kim et al. 2004)
3	ITS	Phylogenetic relationships	13 Patagonian species of the genus *Berberis*	Patagonia, Argentina	(Bottini et al. 2007)
4	AFLP	Cultivar identification	*B. thunbergii*	Canada	(Côté and Leduc 2007)
5	AFLP	Cultivar identification	*B. thunbergii*	USA	(Lubel et al. 2008)
6	AFLP	Genetic diversity and phylogeny	8 native populations of *Berberis* species	Iran	(Heidary et al. 2009)
7	ITS	Genetic variability and taxonomic status	*B. aetnensis*, *B. croatica* and *B. vulgaris*	Croatia	(Karlović et al. 2009)
9	ITS	Species distinguish	*B. lycium* and *B. asiatica*	India	(Balasubramani et al. 2011)
10	SSR	Genetic diversity and phylogeny	47 genotypes of *Berberis* species	Iran	(Rezaei et al. 2011)
11	RAPD	Genetic diversity	3 native populations of *B. iliensis*	Kazakhstan	(Akishev et al. 2012)
12	RAPD	Genetic diversity	50 accessions of *B. lycium*	India	(Tripathi and Goswami 2013)
13	SSR	Cultivar identification	*B. thunbergii*	USA	(Lambert and Obae 2016)
14	ISSR and SSR	Genetic Variation	20 ecotypes of *Berberis*	Iran	(Safamanesh et al. 2017)
15	ISSR	Genetic diversity and the relationship	*B. crataegina* *B. sphaerocarpa*	Turkey & Kyrgyzstan	(Pinar et al. 2021)
16	Universal Plant DNA barcoding regions	Genetic diversity and Phylogenetic study	*B. lycium*	Pakistan	(Bukhari et al. 2022)

(continued)

Table 2.3 (continued)

No	Marker type	Purpose	Genotype and Species	Location	References
17	ISSR and RAPD	Genetic diversity	*Berberis* genotypes selected from an F1 population	Iran	(Safari and Rezaei 2023)
18	ITS2	Species identification	Iranian *Bereberis* spp	Iran	(Gharamanlu et al. 2023)

the dormant season and in the early growth stages, leading to the dispersal of some banned cultivars. Several studies have utilized DNA markers to identify legal cultivars of ornamental Japanese barberry (Table 2.3). Initial identification efforts used AFLP markers. With the development of macro-satellite markers, which offer significant advantages, true cultivars have been more accurately identified (Côté and Leduc 2007; Lubel et al. 2008). Although AFLP markers are highly accurate for detection and validation, they require high-quality DNA. A study using these markers revealed that some ornamental Japanese barberry cultivars introduced by importers were incorrectly labeled (Lubel et al. 2008). To address this issue, Obae et al. (2013) developed 25 markers of AFLP to identify and differentiate the 11 approved *B. thunbergii* cultivars within Canada. Additionally, Lambert and Obae (2016) identified ten of the 11 approved *B. thunbergii* cultivars in Canada by seven SSR loci.

These advancements in DNA marker technology ensure accurate identification and help maintain the integrity of licensed cultivars, thereby preventing the spread of black stem rust through mislabeled barberry plants.

2.3.3 Genetic Resources Conservation Approaches

Barberry is a resilient species capable of both sexual and asexual propagation. However, its genetic resources are threatened by urbanization, climate change, disease and pest outbreaks, deforestation, and particularly the eradication efforts to control wheat black rust (Barnes et al. 2020). The uncontrolled harvesting of barberries for alkaloid extraction and medicinal use also poses a significant threat (Jabeen et al. 2015). Severe genetic erosion has been observed in several *Berberis* species, including *B. lycium*, *B. asiatica*, and *B. aristata* in India and Pakistan, *B. canadensis* in North America, and *B. holstii* in Malawi, Africa (Nyirenda 2008). Barberries can be propagated both sexually and clonally, thus both conservation strategies for vegetatively propagated plants and seed crops are applicable. Conservation strategies are divided into in-situ and ex-situ methods, seed banks, in vitro tissue culture, cryopreservation, and EST- and genomic DNA databases.

There are many species and varieties of *Berberis* in famous botanical gardens and nurseries. More than 825 species, varieties, and cultivars of barberry can be found in the Royal Horticulture Society (RHS), UK, with the aid of nurseries and suppliers and most of them are used for ornamental purposes.

A germplasm collection enables more accurate evaluation of accessions. With agricultural care provided for genotypes and no competition for water and nutrients, the domestication process can be studied, allowing for the selection of superior genotypes as parents in breeding programs. Two notable examples of such conservation strategies are found in Iran. The first collection, established in 2001 at the Mashhad Food Research Institute, includes 42 genotypes (Rezaei et al. 2020). The second, at the Birjand Agricultural Research Center, was developed in 2010 and consists of 20 genotypes propagated through suckers (Zeratgar 2011).

In vitro conservation techniques include slow growth and cryopreservation methods for storing and exchanging germplasm. An in vitro collection from 36 accessions of *Berberis* plant for the creation of a cryopreservation was developed in Kazakhstan (Romadanova et al. 2016). The MS medium (Murashige and Skoog 1962) with 30 g/l sucrose included double the concentration of iron chelate, 0.8 mg/l BAP, 0.1 mg/l GA3, 0.02 mg/l IBA, 1 mg/l ascorbic acid, 2 mg/l calcium pantothenate, 1.75 g/l Gelrite, 4 g/l agar in pH 5.7 was proposed for optimal *Berberis* micropropagation (Romadanova et al. 2016).

Cryopreservation involves protecting plant tissues from ice formation using cryoprotectants or dehydration. Common methods include vitrification, encapsulation-dehydration, slow freezing, and droplet freezing, with the PVS2 (Plant Vitrification Solution 2) method being widely used for *Berberis* (Rajasekharan and Sahijram 2015). PVS2 consists of 30% glycerol, 15% ethylene glycol, 15% dimethylsulfoxide (DMSO) (all v/v), and 0.4 M sucrose is the most commonly applied solution. This solution is more effective for dehydration and less toxic to plant material (Bekheet et al. 2020). *Berberis* shoot tips need PVS2 treatment at 0 °C for 80 min for 50–70% regrowth after cryopreservation. Barberry also requires a cold acclimation before tissue culture for cryopreservation. For barberry shoot tips, PVS2 treatment at 0 °C for 80 min is necessary for 50–70% regrowth post-cryopreservation. Cold acclimation prior to tissue culture also enhances regrowth rates.

A seed cryobank was established for 20 forms of *B. integrifolious* and 3 forms of *B. iliensis* in the Kazakhstan barberry germplasm for long-term storage at − 20 °C and − 196 °C (Romadanova et al. 2016).

DNA banking preserves DNA for future analysis, offering a safeguard against crop diversity loss. Easily extracted from various plant tissues, DNA can be stored long-term in DNA banks. These banks routinely store DNA from nuclei, mitochondria, and chloroplasts, as well as derivatives like RNA and cDNA. Preservation methods include liquid nitrogen, ethanol, and silica drying.

Prominent institutions like the Royal Botanic Gardens maintain over 20,000 DNA specimens from all plant families. The Missouri Botanical Garden holds 20,000 plant tissue samples for DNA extraction in conservation research. Similarly, the Kew DNA and Tissue Collection houses around 50,000 plant genomic DNA samples and over 15,000 silica-dried tissue samples. These repositories are vital resources for future genetic and conservation studies.

2.3.4 Cytogenetics

Berberis is predominantly diploid (2n = 28), though some species are tetraploid (2n = 56), including *B. buxifolia, B. heterophylla, B. turcomanica*, and *B. integerrima* (Bottini et al. 1999; Dermen 1931; Heidari et al. 2010). The Iranian barberry cultivar '*Zereshk Bidaneh*' is also tetraploid(Heidari et al. 2010). Bottini et al. (1999) suggest a link between ploidy levels and species distribution, noting that polyploid species are more widely distributed and tolerant of harsh conditions. However, Adhikari et al. (2014) argue that polyploidy may not have significantly driven the speciation and diversification of *Berberis* in the Nepal Himalayas.

Several karyotype analysis were performed on *Berberis* (Liu and Luo 2019; Varsha et al. 2014). The karyotype analysis precisely shows that the basic chromosome count (x) in *Berberis* is fourteen. The chromosomes' relative lengths varied from 4.22 (chromosome 1) to 2.73 (chromosome 28), and the actual lengths for the chromosomes ranged from 2.75 to 1.82 µm. Flow cytometry is also used to determine ploidy levels, offering ease in sample preparation and rapid, accurate genome size determination using various plant tissues. Rounsaville and Ranney (2010) found that genome sizes in diploid *Berberis* range from 2.48 to 3.36 pg, and in diploid Mahonia from 1.71 to 2.66 pg. Their analysis indicated that *Berberis* has a significantly larger genome (~18%) compared to *Mahonia*, suggesting genome expansion during evolution. For tetraploid *Berberis*, flow cytometry showed genome sizes ranging from 5.8 to 6.15 pg (Rounsaville and Ranney 2010).

2.4 Traditional Breeding

Traditional breeding methods remain promising for the improvement of medicinal and aromatic plants, including barberries, especially since many of these plants have only recently been introduced to cultivation. These methods benefit from a high degree of polymorphism, allowing for direct selection based on phenotypic diversity and the correlation of morphological and economically valuable traits. This approach facilitates the selection of superior genotypes with high productivity based on easily observable traits. The primary goal of breeding medicinal plants is to enhance both the yield of raw plant materials and the concentration of bioactive components. Currently, the number of cultivars developed through the breeding of medicinal plants, including barberries, is minimal. Breeding efforts focus on increasing the quantity and quality of bioactive components that are valuable in the pharmaceutical industry. Traditional breeding methods for medicinal plants include selection, hybridization, polyploidy induction, and mutations. Modern methods include micro-propagation and tissue culture, molecular breeding, and genetic engineering.

In barberries, the main bioactive compounds of interest are alkaloids, especially berberine, which has proven pharmaceutical benefits. Additionally, barberries contain phenolic compounds, anthocyanins, and flavonoids, which possess antioxidant properties and are used in the health industry to strengthen the human immune system.

2.4.1 Improvement of Strategies

Diversity and selection are the two main pillars of breeding, and the genus *Berberis* responds well to breeding methods due to the significant naturally-occurring diversity in its populations. The first step in breeding various species and genotypes of barberry is to evaluate the genetic diversity across different regions and establish a germplasm collection from superior wild genotypes. These genotypes are then assessed for growth, fruit quality, productivity, bioactive components, alkaloids, and antioxidant properties in a controlled agricultural environment. Following this, the domestication ability of selected genotypes is evaluated. Repeated trials in different locations can confirm the superiority of selected genotypes. Promising genotypes are subsequently planted on a commercial scale to evaluate their economic value before being introduced to local farmers.

In addition to domestication, interspecific and intraspecific hybridization strategies can be employed to improve barberry. This process involves collecting and evaluating germplasm, selecting suitable parents, hybridizing selected parents, collecting seeds, raising seedlings in a nursery, and transplanting them to progeny row-trial plots for initial screening and selection. Hybrids with desirable traits identified through progeny row trials are further evaluated in primary yield trials for important agronomic, fruit, and biochemical qualities, including alkaloids. A small percentage (5–10%) of hybrids are selected based on these qualities.

Next, in a final yield trial using 3–5 replications, plants are thoroughly assessed for productivity, environmental adaptation, susceptibility to pests and diseases, rooting ability, and response to agronomic practices. The best hybrid is selected and vegetatively mass multiplied for further testing in different regions. Trials are conducted at multiple locations, and hybrids that perform consistently well across all seasons, locations, and years are ultimately selected.

2.4.2 Traditional Breeding Methodologies and Limitations

Barberries have traditionally been collected directly from nature. Breeding programs, except for one case, primarily focus on ornamental cultivars (Cadic 1992; Brand and Durocher 2022). In ornamental species, the main breeding objectives include developing various leaf colors and bicolors, leaf persistence, and cultivars with sterile seeds, often achieved through interspecific hybrids. A crucial aspect of crossing species involves pollen grain collection and the emasculation of barberry flowers.

Barberry flowers produce very small amounts of pollen grains, and their aggregation properties, similar to other tree species, make collection challenging. Pollen is collected from mature inflorescences using the n-pentane method (Cadic 1992). Flowers are soaked in an n-pentane solution for about one minute. Each flower is then held by the pedicle with forceps, swirled in the solution to release pollen grains, and finally removed and discarded. The n-pentane solution is allowed to volatilize, leaving only the pollen grains in a petri dish (Brand and Durocher 2022; Safari-Khuzani et al. 2020). For genotypes with a lot of nectar in the flower, pollen grains may mix with remaining sugar syrup. In such cases, using filter paper before the n-pentane volatilizes is recommended (personal experience).

Barberry is a self-pollinating plant. During flower anthesis, with minimal external stimulation, the filaments bend towards the stigma suddenly, depositing pollen on the stigma surface. Therefore, emasculation must be performed before pollen grains fall, which coincides with the balloon stage when the yellow petals are visible but the flowers are not yet open. Barberry flowers at this stage are small with sensitive tissue that can break and fall off easily. Hence, emasculation is time-consuming and requires great care, which can be done with fine pliers. At this stage, the petals and the flag are removed (Fig. 2.4). Pollination can be performed the next day when the stigma surface is slightly wet and can be repeated 24 h later to ensure success. Pollen grains can survive in a dry environment at 4 °C for up to a week. Pollination is conducted on two or three flowers in each cluster, with the rest removed to enhance fruit set. The flowers should be bagged after pollination, and the bags removed after initial fruit set. Once the fruits ripen, the seeds are collected and require scarification and cold stratification for germination.

Important traits selected by barberry breeders include fruit size, seedless fruits, thornless stems, black rust resistance, high antioxidant properties of fruits, adaptability, pest and disease resistance, and improved vegetative propagation ability.

2.4.3 Mutation Breeding

Mutation induction involves treating plant propagules with chemical or physical mutagens, such as radiation. This technique is used in plant breeding to develop new varieties with specific traits. Mutations can occur in nuclear DNA, chromosomes, or extra-nuclear (cytoplasmic) genetic systems. In barberry, mutation breeding has primarily focused on inducing sterile forms in invasive species.

A breeding program by the USDA aimed at developing non-invasive nursery crops has developed procedures for inducing sterility through mutation treatments in *B. thunbergii* (Japanese barberry) and *Miscanthus* spp. They established radiation sensitivities for gamma and fast neutron treatments of seeds and unrooted cuttings for *B. thunbergii* DC, identifying individual plants with novel traits, including seedlessness (Smith and Noyszewski 2018).

Fig. 2.4 Stages of emasculation and pollination in barberry **a** Flowers of barberry s in the balloon stage (ready to emasculation). **b** and **c** Emasculated flowers of barberry. **d** Bagging of pollinated flowers (the photos were taken by the author)

2.4.4 Hybridization

2.4.4.1 Conventional Hybridization

Due to the long juvenile period and high heterozygosity of barberries, traditional breeding primarily relies on F1 hybrid populations, followed by screening and selection. The process begins with selecting parent plants based on specific breeding objectives and crossing them during their regular flowering seasons. Seeds are collected after the fruit ripens, planted in nurseries, and the resulting seedlings are transplanted to the field for further screening and selection. Hybrids with the desired traits are vegetatively propagated to maintain the unique combination of traits.

One notable breeding program focused on crossing barberry for fresh fruit production was conducted by Safari-Khuzani et al. (2020). In this program, 42 indigenous barberry genotypes from the Northeast region of Iran were selected and grown in a germplasm collection with three replications of each genotype. After the plants began fruiting, each genotype was evaluated based on growth habits and fruit quality. Genetic distances were determined using morphological traits and SSR markers

Fig. 2.5 The pictures of selected superior genotype and berries of intraspecific F1 progenies of barberry (the photos were taken by the author)

(Rezaei et al. 2011, 2020). Concurrent research examined the mechanisms of seedlessness and the modes of pollination and fruit set in the seedless barberry cultivar '*Zereshk Bidaneh*' (Ebadi et al. 2010). Three selected genotypes (R4N1, R2N1, and R5N1) and '*Zereshk Bidaneh*' were crossed in a diallel cross design. The R2N1 and R4N1 genotypes (*B. integerrima*) have large fruits, while the R5N1 genotype (*B. crataegina* × *B. integerrima*) is notable for high vitamin C and anthocyanin content (Rezaei and Balandary 2015).

The breeding program resulted in 90 mature shrubs of F1 interspecific hybrids available for selection. Despite the limited number of hybrids, significant variation was observed, ranging from completely seedless genotypes to those with various colors and shapes of large-sized fruits (Fig. 2.5) (Safari-Khuzani et al. 2020). The primary aim of selecting barberry fruits is to develop currant barberry cultivars with large, black fruits, which are more desirable due to their higher antioxidant properties. Interestingly, some interspecific hybrids produced fruits with a round shape, a trait not exhibited by either parent, suggesting that round fruit shape may be controlled by recessive genes.

2.4.4.2 Hybrid Cultivars

Berberis hybrids have been produced with the usage of cultivation practices since the early 19th Century, including broad-ranging crosses between simple and compound-leaved species (Ahrendt, 1961). Even the intergeneric hybrids between *Mahonia* and *Berberis* (*MahoBerberis* C.K. Schneid.) were also common (Laar 1975; Wyman 1958). Naturally occurring hybridization in wild populations of *Berberis* species appears to be relatively common, as evidenced by specimens that appear to be morphologically intermediate between species (e.g., Landrum 1999). However, almost all available hybrid cultivars have usages within landscapes. Several ornamental *Berberis* hybrids are known to stay very small, even without much pruning, which is known as a 'dwarf barberry'. Hybrid dwarf barberries are used as low-lying hedges and make wonderful shrub beds. The two dwarf barberry hybrids are 'Corallina Compacta' and 'William Penn'. 'Corallina Compacta' is raised by the Daisy Hill nursery (Forrest 2010), and is a *B. x stenophylla*, a garden hybrid between *B. darwinii* × *B. empetrifolia*,. It has narrow leaves, like those of rosemary, grows to around 45 cm and has wonderful orange-red flowers. The 'William Penn' barberry belongs to the *B.* × *gladwynensis* barberry hybrids, which are derived from crossing *B. verruculosa* and *B. gagnepainii*, and is a broadleaf evergreen cultivar (Sagers 2005). The other interspecific barberry hybrids that have dual use for both medicinal and ornamental purposes includes *B.* × *carminea* (uncertain parentage), *B. gagnepainii* 'Georgeii' (uncertain parentage), *B.* × *lologensis* (*B. darwinii.* × *B. linearifolia*) 'Apricot Queen' an ornamental cultivar, *B.* × *lologensis* (*B. darwinii* × *B. linearifolia*), *B. rubrostilla* (*B. aggregata* × *B. wilsoniae*), and *MahoBerberis aquisargentii* (*Mahonia aquifolia* × *B. sargentii.*) (Bown 1995). *MahoBerberis*, rarely have flowers or fruit and tend to exhibit numerous leaf morphs among single plants (Dirr 2009). In addition, all *MahoBerberis* hybrids have been comprised of only one species of *Mahonia* (*M. aquifolium*), and the crossing appears uni-directional as *Mahonia* only functions as the maternal parent (Dirr 2009).

2.4.5 Role of Biotechnology

Barberry breeding programs stand to benefit greatly from the integration of biotechnological tools, which offer efficient and effective means of enhancing breeding efforts. Extensive research has been conducted to assess the antioxidant potential and effective alkaloids in various barberry genotypes. For instance, the biosynthesis of berberine, a benzylisoquinoline alkaloid, has been thoroughly elucidated in *Berberis* species, with the enzymes and genes involved fully characterized (Facchini 2006; Karp et al. 2019). Berberine biosynthesis, derived from tyrosine via specific enzymes present in vesicles within *Berberis* and *Coptis* genera, has been extensively studied (Facchini 2006). Roy et al. (2021) recently reported the full-length cDNA sequencing of the berberine biosynthesis pathway in different parts of *B. koreana*, offering insights into potential targets for genetic engineering to enhance berberine production.

Some of the important contributions of biotechnology in *Berberis* include using tissue culture for clonal propagation (Molinar et al. 1996; Sazmand and Safarnezhad 2016), berberine biosynthesis by cell suspension culture (Alvarez et al. 2009; Breuling et al. 1985), and hairy root induction by *Agrobacterium rhizogenes* (Brijwal et al. 2015b). In vitro techniques have been utilized to develop tetraploids sterile *Berbers* with colchicine and oryzalin treatments (Lehrer et al. 2008).

2.5 Molecular Breeding

2.5.1 Molecular Marker-Assisted Breeding

The utilization of molecular markers in DNA to select and screen crops within breeding programs offers numerous advantages, enhancing the efficiency and accuracy of classical plant breeding through Marker-Assisted Selection (MAS). The identification of Quantitative Trait Loci (QTL) for various plant species has facilitated the establishment of molecular marker-gene relationships. However, in the case of barberry, the focus on classical breeding methods for medicinal objectives has been limited, resulting in the relatively nascent stage of MAS development for its important traits.

Due to the lack of extensive conventional breeding programs for medicinal usage, the MAS for the barberry is still in its beginning stages. Attempts have been made to use molecular markers to identify true cultivars or species, most of them for the identification of ornamental cultivars, especially Japanese barberries cultivars. Balasubramani et al. (2011) used ITS markers for the development of barcoding DNA for the detection of three Indian *Berberis* species. It could be said that the leading molecular breeding research that has been done on the barberry is the gene mapping of the black stem rust resistant genes. In this study, Bartaula et al. (2019) developed the first genetic maps for two *Berberis* species (*B. thunbergii* and *B. vulgaris*). Genetic linkage maps were developed for non-host resistance to the stem rust pathogen in both *B. thunbergii* and *B. vulgaris* from a single interspecific F1 mapping population. The gene maps constructed by using 184 F1 progeny, *B.* × *ottawensis*, (obtained from the crossing of a susceptible species of barberry, *B. vulgaris,* with a resistant species, *B. thunbergii,*) and 14043 GBS-SNP (Genotyping-by-sequencing from single-nucleotide polymorphism). F1 interspecific hybrids exhibit a range of reactions to stem rust, from fully resistant to fully susceptible, with various intermediate forms. 14 distinct linkage groups, in agreement with the reported chromosomal number, was constructed a 4-point stem rust reaction type described in the linkage maps. A single significant QTL related to non-host resistance to the stem rust pathogen genes (QPgr-3S) was identified in 25 cM from the telomere of a chromosome between its two flanking markers, M411 and M969 (Bartaula et al. 2019). The position of QTL region QPgr-3S is indicated alongside the chromosome 3 linkage map for the *B. thunbergii* parent. This 13 cM QTL region, QPgr-3S, was found to span 13 physical contigs and contain a total of 576 high-confidence genes. From them, seven have relatively high priority candidate genes, including three exhibiting homologies to genes in public databases, including dormancy-associated auxin repressor proteins (TR20791), zinc ion binding proteins (TR27614), and glutamine synthetase proteins (TR12856). Candidate genes associated with non-host resistance to the stem rust pathogen were TR20791, TR27614, and TR12856 (Bartaula et al. 2019). Dissection of a QTL related to non-host resistance to the stem rust pathogen genes (QPgr-3S) on chromosome 3 was helpful in marker-assisted selection (MAS) breeding. Furthermore, it must be validated that for this trait, no MAS has yet to start in barberries.

2.5.2 Functional Genomics

Functional genomics is a field of molecular biology that attempts to describe gene (and protein) functions and interactions. Functional genomics in barberries is still in its infancy stages. Little work has been done in this area except for gene expression for berberine biosynthesis (Roy et al. 2021) and flavonoid gene analysis in barberries (Thole et al. 2019). The berberine biosynthesis pathway in *B. koreana* starts with two L-tyrosine molecules, which are converted to dopamine and 4 hydrophenylacetaldehyde. These two molecules are combined by (S)-norcoclaurine synthase (NCS) to become (S)-norcoclaurine, an immediate precursor to all benzylisoquinoline alkaloids. (S)-Norcoclaurine is converted to (S)-reticuline via a series of enzymatic reactions. (S)-Reticuline is then transformed into various benzylisoquinoline alkaloids, including berberine, noscapine, sanguinarine, morphine, etc. (Fig. 2.6). The isolated berberine biosynthesis enzymes were Tyrosine decarboxylase(TYDC), L-tyrosine aminotransferase (TyrAT), tyrosine/tyramine 3-hydroxylase (3OHase), (S)-norcoclaurine synthase (NCS), (RS)-norcoclaurine 6-O-methyltransferase (6OMT), (S)-coclaurine N-methyltransferase (CNMT), (S)-N-methylcoclaurine 3′-hydroxylase/N-methylcoclaurine 3′-monooxygenase (NMCH), 3′-hydroxy-N-methyl-(S)-coclaurine 4′-O-methyltransferase (4ÓMT), berberine bridge enzyme (BBE), and (S)-tetrahydroprotoberberine oxidase (STOX), (S)-scoulerine 9-O-methyltransferase (SOMT), and canadine synthase (CAS/CYP719A21). Roy et al. (2021) showed that the genes of TYDC, NCS, SOMT, and NMCH were highly expressed in mature leaves and mature fruit, whereas those of 6OMT and CNMT showed higher expression in young leaves and mature fruit. Genes encoding CAS showed higher expression in young leaves and young fruit and 4′OMT showed no particular expression pattern. *B. koreana* produces berberine by blocking the pathways leading to other benzylisoquinoline alkaloids, like papaverine, morphine, sanguinarine, and noscapine, effectively only allowing the pathway to lead to berberine synthesis. In fact, genes for any of the other benzylisoquinoline alkaloid were not found in *B. koreana*. Perhaps the main gaps in the research of Roy et al. (2021) are from the lack of study of root and stem bark tissues, where most of the berberine is produced.

Roy et al. (2021) sequenced full-length cDNA berberine biosynthesis pathway in *B. Koreana* to determine the number of isoforms of berberine expressed genes. 23,246 full-length unigenes were identified and among them 8479 had more than one isoform ranged from two to thirty-one. Thirteen of the sixteen genes encoding enzymes for berberine synthesis were present in more than one copy. This demonstrates that gene duplication and translation into isoforms may contribute to the functional specificity of the duplicated genes and isoforms in plant alkaloid synthesis.

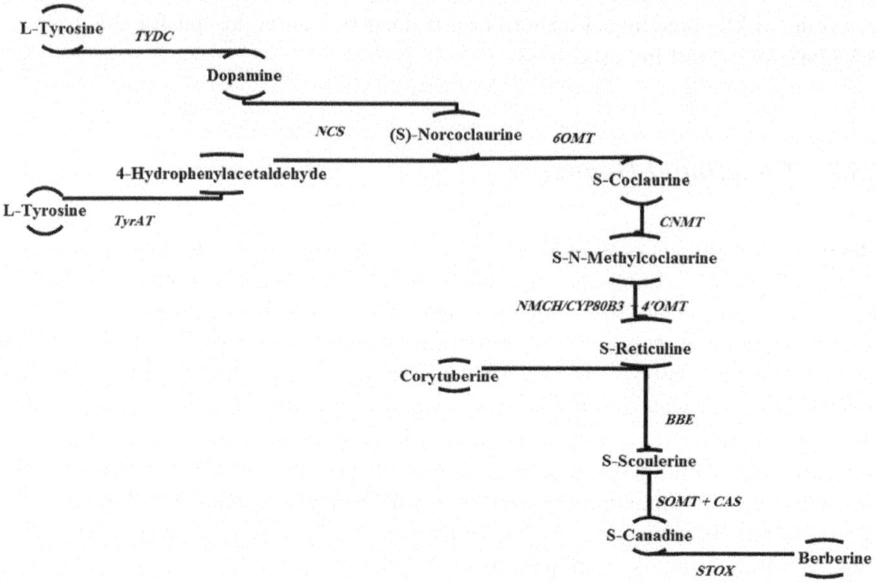

Fig. 2.6 A brief schematic scheme of Berberine biosynthesis pathway in *B. koreana* (modified from Roy et al. 2021). Enzymes (abbreviation): tyrosine aminotransferase (TyrAT); tyrosine decarboxylase (TYDC); (S)-norcoclaurine synthase (NCS); (RS)-norcoclaurine 6-O-methyltransferase(6OMT); (S)-coclaurine N-methyltransferase (CNMT); (S)-N-methylcoclaurine3′-hydroxylase/N-ethylcoclaurine 3′-monooxygenase (NMCH/CYP80B3); 3′-hydroxy-N-methyl-(S)-coclaurine4′-O-ethyltransferase (4′OMT); Berberine bridge enzyme (reticuline oxidase) (BBE); (S)-scoulerine 9-O-methyltransferase (SOMT); Canadine synthase enzyme (CAS); and (S)-tetrahydroprotoberberine oxidase (STOX)

The chloroplast genome of *B. weiningensis* was also sequenced, which obviously supports that *Mahonia* and *Berberis* are not monophyletic (Feng et al. 2021). The length of the complete chloroplast genome of *B. weiningensis* was 166,275 bp. It contains 147 genes, of which 101 were protein-coding genes, 8 rRNA genes, and 38 tRNA genes (Feng et al. 2021).

2.5.3 *Bioinformatics*

The DNA database contains genetic information typically stored in a computerized database format. It includes complete genome sequencing, expressed sequence tags (ESTs), and taxonomic DNA codes. These databases serve as valuable resources for research and act as a form of genetic conservation for the future, stored as computer data on database websites. A reference genome, approximately 129 Gb in size, for *B. thunbergii* cv. 'Kobold' (GenBank: GCA_003290165.1) has been constructed (Bartaula et al. 2019). NCBI is one of the most renowned and largest of these databases, offering relatively widespread public access (http://www.ncbi.nlm.nih.gov/). A search for the genus *Berberis* on this website reveals the complete chloroplast genome of *B. fortune* with 165,291 bp and the complete chloroplast genome of *B. weiningensis* (MW018363) with a length of 166,275 bp. Over 14,022 *Berberis*-related nucleotide sequences, 4260 proteins, 1156 genes, and 325 sequences related to *Berberis* taxonomy are recorded on the NCBI website. A publicly available resource of genomic and metabolic data obtained on berries can be found at the BacHBerryGEN database (http://jicbio.nbi.ac.uk/berries). Fruit Transcriptome Sequences of *B. buxifolia* (Calafate) have also been reported within this database.

2.6 Tissue Culture Applications

2.6.1 *Micropropagation Approaches*

The tissue culture and micropropagation of various barberry species represent effective strategies for increasing the reproduction of endangered species. Several tissue culture protocols have been optimized for different barberry species, including *B. microphylla* (McLeod et al. 2017), *B. aristata* (Brijwal et al. 2015a), *B. chitria* (Aseesh et al. 2013), *B. lycium* (Dhar et al. 2012), *B. vulgaris* var asperma ('Bidaneh Zereshk') (Sazmand and Safarnezhad 2016) and *B. buxifolia* (Alvarez et al. 2009; Arena et al. 2005; Arena and Vater 2000). An efficient micro-propagation protocol for *B. asiatica* was optimized by Bisht et al. (2021). Calls formation induced in-vitro grown leaf explants in Murashige and Skoog (MS) medium with 1.0μM BA and 0.1μM 2,4-dichlorophenoxyacetic acid (2,4-D). Shoot proliferation was obtained from callus in MS medium containing 2.0μM BA plus 0.5μM IAA or 6.0μM BA plus 0.5μM NAA. The maximum rooting percentage (70%) and root number per shoot (2.36) were recorded in ½ MS containing 0.05μM IAA. Rooted shoots when transferred to potting medium containing vermiculite, soil, and sand (1:1:1) resulted in 65% survival. Shoot proliferation for *B. trifoliata* (Agarita) was successfully optimized in WPM medium supplemented with 5.5 μM BA. Micro-shoots rooting in 1.0 μM NAA treatment after 6 month (Molinar et al. 1996).

2.6.2 *Cell Suspension*

The development of plant tissue cultures, including organs and cells, to produce secondary metabolites has been ongoing for more than three decades. Plant cell culture is now recognized as a promising, renewable, sustainable, and environmentally friendly alternative to obtaining plant-derived secondary metabolites from wild plants. While there are numerous studies utilizing the production of secondary metabolites from callus culture and differentiated cells/tissues, cell suspension culture from dedifferentiated cells is still predominantly preferred (Motolinía-Alcántara et al. 2021).

In the case of barberry, efforts have been made to culture tissue/cell suspension and produce the berberine alkaloid. Suspension cultures of *B. wilsonae* produce 4 berberine-type alkaloids: berberine, palmatine, columbamine, and jatrorrhizine (Breuling et al. 1985). Notably, the formation of the phenolic alkaloids columbamine and jatrorrhizine, and of berberine, is dependent on the concentration of dissolved oxygen. With higher aeration rates, yields of berberine and jatrorrhizine can be significantly increased. It has been discovered that *Berberis* species have alkaloid yields of more than 3 g/l with 50% dissolved oxygen tension in the medium.

Alvarez et al. (2009) used thidiazuron and picloram to establish cell suspension cultures of *B. buxifolia*. As oxidation processes often pose challenges in callus culture, a two-stage culture system for producing berberine was ultimately developed. In the first stage, thidiazuron or picloram are used for biomass production, followed by transferring the cultures to a berberine production medium. Suspension cell cultures are transferred to an MS liquid medium supplemented with benzylaminopurine (4.4 μM), which induces the immediate production of berberine. The highest berberine yield in cell suspension was 102 mg/g DW, while the in vitro shoot cultures yielded 200 mg/g DW. Both of these were significantly lower than the yields from whole plants in the field (416 mg/g DW). However, it should be noted that the advantage of berberine production in cell suspension cultures lies in their continuous production capability in bioreactors.

2.7 Genetic Engineering

Genetic engineering, particularly transformation technology, creates advancements and opens new pathways for the production of large volumes of pharmaceutical substances. Induction of hairy root cultures by *Agrobacterium rhizogenes* is a transformation technology that enhances the production of secondary metabolites (Hu and Du 2006). The primary advantage of hairy root culture is its ability to grow rapidly without the need for exogenous growth regulators. Hairy roots can produce the same compounds found in parent plants, making them a valuable transgenic tool for producing new proteins and secondary metabolites.

In barberry, where most alkaloids are produced in root tissue, inducing hairy roots with *A. rhizogenes* is a key objective for berberine production. *Agrobacterium*-induced hairy roots are favored due to their genetic stability, rapid growth rate, and ability to produce secondary metabolites consistently over successive generations without genetic degradation or loss of biosynthetic capacity. Their high branching and fast growth rate make them suitable for scaling up to commercial levels, even in bioreactors (Hu and Du 2006).

An efficient protocol for hairy root induction by *A. rhizogenes* strains was established in *B. aristata* (Brijwal et al. 2015b). The highest transformation frequency

(42.59 ± 0.92%) in leaves was observed after 27.66 ± 0.33 days with a 48-h co-cultivation period and a 3-h infection period. Strain 532 was found to be more effective than strain 2364 in inducing hairy roots (Brijwal et al. 2015b). The transgenic nature of the hairy roots was confirmed by the amplification of rol A and rol B genes of *A. rhizogenes* during root formation and in the transformed callus. Brijwal et al. (2015b) also demonstrated that the transformation frequency of callus increased (from 61.11 ± 1.60% to 72.22 ± 1.60%) by using acetosyringone (100 μM) during the 48-h co-cultivation period on semisolid Murashige and Skoog medium.

2.8 Recommendations for Future Research

Success in the widespread medicinal or food use of barberry depends, firstly, on the extensive expansion of the cultivars. Therefore, superior cultivars with the ability to adapt to the climatic conditions of each region should be established by developing breeding programs with government or private support. Secondly, the dependent pharmaceutical industry must be developed alongside its extensive cultivation, and methods of industrial extraction of effective substances must be optimized for both the pharmaceutical and food industries.

The genetic basis for controlling many economically important traits of *Berberis* is not yet known. Molecular markers related to traits (MAS) must be established to increase the efficiency of conventional breeding methods and to save time.

Due to the overharvesting of barberries from nature, the establishment of germplasm collections is required for conservation and primary screening. Selecting superior genotypes can be very important in improving barberries and developing domesticated varieties.

The optimization of bioreactors and the use of transgenic plants with *Agrobacterium rhizogenesis* and suspension cell culture can be used to produce alkaloids. Gene mapping and QTLs for quantitative traits should also be identified to improve traits. Additionally, the current impetus for the development of genomic resources and genetic engineering should continue to facilitate the development of new barberry cultivars capable of producing superior quality products.

2.9 Conclusions and Prospects

Barberries have gained significant relevance today due to their antioxidant properties and valuable alkaloids. These alkaloids in barberry exhibit important therapeutic properties, such as anticancer and antimicrobial activities, and are used in the treatment of significant ailments like hypertension, diabetes, cardiovascular disease, and hepatitis. However, there is currently no breeding program developed to generate commercial cultivars, and most usage is related to species collected from their natural habitats. It appears that barberry is still in the domestication phase, particularly regarding medicinal improvement objectives.

There is still no established program for releasing cultivars specifically for medicinal purposes, such as alkaloid production. Many limitations exist in breeding programs aimed at achieving medicinal objectives in *Berberis* species. Currently, there is very little information on the genetic control and heritability of many important traits such as fruit quality and quantity, bioactive components (phenols, anthocyanins, and flavonoids), main alkaloids (berberine), disease resistance, and environmental compatibility. The handling of thorny shrubs, pollen collection, flower emasculation, long juvenile periods, and pruning operations all pose significant challenges to *Berberis* breeding programs. Issues related to the commercial use of bioactive components and the extraction of alkaloids from this valuable medicinal plant are still largely unexplored and remain enigmatic.

Cv '*Zereshk Bidaneh*' is one of the most promising *Berberis* cultivars, cultivated in poor arid regions of Iran, showing potential for more widespread use in other parts of the world. Considering the increasing attention given to the anthocyanin and antioxidant properties of barberry fruits, there is great hope for the more effective utilization of this crop in medicine and the food industry. Conservation of existing species and selection of wild species are important goals in the protection and breeding of barberry species. Additionally, concerns related to barberry cultivation, such as buck stem rust and invasive species, must be addressed to confidently allow the widespread use of this valuable shrub worldwide.

Global warming and climate change, driven by human activities, are among the most pressing issues threatening human life in the future. These changes have caused some desert areas to become drier and present the threat of submerging some coastal cities. Additionally, temperature and seasonal rainfall changes are increasingly evident. The impact of climate change on agricultural crops has been more pronounced in perennial tree and shrub species compared to annual crops. For instance, the percentage of spring frost damage, especially for fruit trees, has significantly increased in recent years. Therefore, cultivating crops with higher climate adaptability will be a key priority moving forward.

Arena et al. (2015) demonstrated that the high phenotypic plasticity of *B. microphylla* leaves allows them to adjust their morphology to new cultural conditions, such as higher temperatures and lower irradiance. This plasticity is an advantageous strategy for survival under changing environmental conditions. Barberry, with its numerous species, appears to exhibit high plasticity to climate changes and should be given more attention in future agricultural planning.

Cv '*Zereshk Bidaneh*' has thrived in the central desert conditions of Iran. However, it is important to note that barberry flower buds have high chilling requirements. As a late-flowering plant, barberry blooms later than apples, thereby reducing the risk of spring frosts. This characteristic makes barberry a promising crop for regions experiencing climate variability, providing a potential solution to some of the challenges posed by global climate change.

References

Adhikari B, Pendry C et al (2014) New chromosome counts of *Berberis* L. (Berberidaceae) suggest that polyploidy does not play a significant role in the diversification of the genus in the nepal himalaya. Edinb J Bot 71(3):297–308

Ahmed M, Anjum MA et al (2013) Characterization of indigenous barberry germplasm in Pakistan: variability in morphological characteristics and nutritional composition. Fruit 68(5):409–422

Ahmad S, Hussain A et al (2019) Quantification of berberine in *Berberis vulgaris* L. Root extract and its curative and prophylactic role in cisplatin-induced in vivo toxicity and in vitro cytotoxicity. Antioxidant 8(6):185. https://doi.org/10.3390/antiox8060185

Ahrendt LWA (1961) *Berberis* and Mahonia. Bot J Linn 57(369):1–410

Akbulut M, Çalisr S et al (2009) Some physicomechanical and nutritional properties of barberry (*Berberis vulgaris* l.) fruits. J Food Process Eng 32(4):497–511

Akishev ZD, Smekenov IT et al (2012) Rapd analysis of genetic diversity in *Berberis iliensis* a species endemic to kazakhstan. Int J Biol Chem 3:48–53

Alemardan A, Asadi W et al (2013) Cultivation of Iranian seedless barberry (*Berberis integerrima* 'Bidaneh'): a medicinal shrub. Ind Crops Prod 50:276–287

Ali M, Malik A et al (2008) Vegetative propagation of *Berberis aristata* DC. An endangered himalayan shrub. J Med Plant Res 2(12):374–377

Aliakbarlu J, Mohammadi S et al (2014) A study on antioxidant potency and antibacterial activity of water extracts of some spices widely consumed in Iranian diet. J Food Biochem 38(2):159–166

Alvarez MA, Eraso NF et al (2009) Two-stage culture for producing berberine by cell suspension and shoot cultures of *Berberis buxifolia* lam. Biotechnol Lett 31(3):457–463

Andersen AS (1986) Environmental influences on adventitious rooting in cuttings of non-woody species. New root formation in plants and cuttings. Springer, pp 223–253

Andola HC, Gaira KS et al. (2010a) Habitat dependent variations in berberine content of *Berberis asiatica* Roxb. Ex. DC. in Kumaon, Western Himalaya. Chem Biodivers 7(2):415–420

Andola HC, Rawal R et al (2010) Variations of berberine contents in *Berberis pseudumbellata* a high value medicinal shrub of West Himalaya, India. Med Plants—Int J Phytomed Relat Ind 2(2):111–115

Arayne MS, Sultana N et al (2007) The *Berberis* story: *Berberis vulgaris* in therapeutics. Pak J Pharm Sci 20(1):83–92

Ardestani SB, Sahari MA et al (2013) Some physicochemical properties of Iranian native barberry fruits (Abi and Poloei): *Berberis integerrima* and *Berberis vulgaris*. J Food Pharm Sci 1(3):1–10

Arena M, Radice S (2015) Environmental effect on the leaf morphology and anatomy of *Berberis microphylla* G. Forst. Int J Plant Biol 6:5677

Arena ME, Pastur GM et al (2005) Polyamines and inhibitors used in successive culture media for in vitro rooting in *Berberis buxifolia*. N Z J Bot 43(2):373–380

Arena ME, Vater G (2000) In vitro propagation of *Berberis buxifolia* lam. Biocell 24(1):73–80

Aseesh P, Brijwal L et al (2013) In vitro propagation and phytochemical assessment of *Berberis chitria*: an important medicinal shrub of Kumaun Himalaya India. J Med Plant Res 7(15):930–937

Balasubramani SP, Goraya GS et al (2011) Development of its sequence-based markers to distinguish *Berberis aristata* DC. from *B. lycium* Royle and *B. asiatica* Roxb. Biotechnology 1(1):11–19

Barnes G, Saunders DG et al (2020) Banishing barberry: The history of *Berberis vulgaris* prevalence and wheat stem rust incidence across Britain. Plant Pathol 69(7):1193–1202

Bartaula R, Melo ATO et al (2019) Mapping non-host resistance to the stem rust pathogen in an interspecific barberry hybrid. BMC Plant Biol 19(1):319

Bekheet SA, Sota V et al (2020) Cryopreservation of shoot apices and callus cultures of Globe artichoke using vitrification method. J Gen Eng Biotech 18(1):2

Benvenuti S, Bortolotti E, Maggini R (2016) Antioxidant power, anthocyanin content and organoleptic performance of edible flowers. Sci Hortic 199:170–177

Bhardwaj D, Kaushik N (2012) Phytochemical and pharmacological studies in genus *Berberis*. Phytochem Rev 11(4):523–542
Bisht A, Giri L et al. (2021) In vitro propagation and antioxidant potential of *Berberis* asiatica from western Himalaya. Plant Biosystems 1–7
Bottini M, Greizerstein E et al (1999) Ploidy levels and their relationships with the rainfall in several populations of Patagonian species of *Berberis*. Caryologia 52(1–2):75–80
Bottini MCJ, De Bustos A et al (2002) AFLP characterization of natural populations of *Berberis* (Berberidaceae) in Patagonia, Argentina. Plant Syst Evol 231(1):133–142
Bottini MCJ, De Bustos A et al (2007) Relationships in patagonian species of *Berberis* (Berberidaceae) based on the characterization of rDNA Internal Transcribed spacer sequences. Bot J Linn 153(3):321–328
Bown D (1995) The Royal Horticultural Society encyclopedia of herbs & their uses. Dorling Kindersley Limited
Brand MH, Durocher SN (2022) Four sterile or near-sterile cultivars of Japanese barberry in three foliage colors. HortScience 57(4):581–587. https://doi.org/10.21273/HORTSCI16488-22
Brand MH, Lehrer JM et al (2012) Fecundity of Japanese barberry (*Berberis thunbergii*) cultivars and their ability to invade a deciduous woodland. Invasive Plant Sci Manag 5(4):464–476
Breuling M, Alfermann AW et al (1985) Cultivation of cell cultures of *Berberis wilsonae* in 20-l airlift bioreactors. Plant Cell Rep 4(4):220–223
Brijwal L, Pandey A et al. (2015a) In vitro propagation of the endangered species *Berberis aristata* DC. Via leaf-derived callus. In Vitro Cell Dev Biol Plant 51(6):637–647
Brijwal L, Tamta S (2015) *Agrobacterium rhizogenes* mediated hairy root induction in endangered *Berberis aristata* DC. Springerplus 4(1):443
Bukhari SMF, Ali G et al (2022) Morphology and phylogeny based on matK, rbcL, ITS and trnH-psbA of important medicinal plant *Berberis lycium* Royle from Azad Jammu and Kashmir. Pak J Bot 54(2):507–517
Cadic A (1992) Breeding for ever-red barberries (*Berberis* spp.). In: International symposium on selection and breeding of woody ornamentals 320, pp 85–90
Chen X (2017) Stripe rust epidemiology. Stripe rust. Springer, pp 283–352
Chen XM (2005) Epidemiology and control of stripe rust [*Puccinia striiformis* f. Sp. Tritici] on wheat. Can J Plant Pathol 27(3):314–337
Choudhary S, Kaurav H et al (2021) Daruharidra (*Berberis aristata*): review based upon its Ayurvedic properties. Int J Appl Sci Biotechnol 8(2):98–106
Çoban B, Bilgin B et al (2021) Utilization of the barberry extract in the confectionery products. LWT 145:111362
Connolly BA, Anderson GJ et al (2013) Occurrence and fertility of feral hybrid barberry *Berberis* × *ottawensis* (Berberidaceae) in Connecticut and Massachusetts. Rhodora 115(962):121–132
Côté M-J, Leduc L (2007) Molecular identification of Japanese barberry (*Berberis thunbergii*) cultivars using Amplified Fragment Length Polymorphism. HortScience 42(3):478–482
D'Appollonio J (2006) Regeneration strategies of Japanese barberry (*Berberis thunbergii* DC.) in coastal forests of Maine
Deb CR, Sangtam TL et al (2017) Seed biology of *Berberis manipurana* Ahrendt: a threatened natural dye yielding plant. Am J Plant Sci 8(06):1285
Del Carpio JC, Serrano Flores C et al (2011) Characterisation and preliminary bioactivity determination of *Berberis boliviana* Lechler fruit anthocyanins. Food Chem 128(3):717–724
Dermen H (1931) A study of chromosome number in two genera of Berberidaceae: *Mahonia* and *Berberis*. J Arnold Arbor 12(4):281–287
Dhar S, Sharma Y et al (2012) In vitro plant regeneration system for *Berberis lycium* using cotyledonary node explant. J Trop Med Plants 13(1):51–55
Dirr M (2009) Manual of woody landscape plants: their identification, ornamental characteristics, culture, propagation and uses, 4th edn. Stipes Publishing Co.
Ebadi A, Rezaei M et al (2010) Mechanism of seedlessness in Iranian seedless barberry (*Berberis vulgaris* L var. *asperma*). Sci Hortic 125(3):486–493

Facchini PJ (2006) Regulation of alkaloid biosynthesis in plants. Alkaloids Chem Biol 63:1–44
Fedde F (1902) Versuch einer monographie der gattung mahonia. Bot Jahrb Syst 31:30–133
Feng T, Xiao Q-Y et al (2021) The complete chloroplast genome of *Berberis weiningensis* (Berberidaceae), an endemic and traditional Chinese medicinal herb. Mitochondrial DNA Part B 6(3):1175–1177
Forrest M (2010) Nurseries and nurserymen in Ireland from the early eighteenth to the early twenty-first century. Stud Hist Gard des Landsc 30(4):323–336
Garhwal S (2010) Analysis of berberine content using HPTLC fingerprinting of root and bark of three Himalayan *Berberis* species. Asian J Biotechnol 2(4):239–245
Ghahramanlu A, Rezaei M, Heidari P, Balandari A (2023) Species survey of Iranian barberry genotypes using ITS2 sequences and basic local alignment search tools. Erwerbs-Obstbau 65(6):2491–2499
Gholizadeh Moghadam N, Hosseini B et al (2019) Classification of barberry genotypes by multivariate analysis of biochemical constituents and HPLC profiles. Phytochem Anal 30(4):385–394
Gıdık B (2021) Antioxidant, antimicrobial activities and fatty acid compositions of wild *Berberis* spp by different techniques combined with chemometrics (PCA and HCA). Molecules 26(24):7448
Hadaruga DI, Hadaruga NG et al (2010) *Berberis vulgaris* extract/β cyclodextrin nanoparticles synthesis and characterization. Rev Chim (Bucharest) 61:669–675
Hale IL, Connolly BA et al (2015) The occurrence of hybrid barberry, *Berberis* × *Ottawensis* (Berberidaceae), in New Hampshire and Rhode Island. Rhodora 117(971):384–387
Hassanpour H, Alizadeh S (2016) Evaluation of phenolic compound, antioxidant activities and antioxidant enzymes of barberry genotypes in Iran. Sci Hortic 200:125–130
Heidari S, Marashi H et al (2010) Cytogenetic and pollination studies in seedless barberry (Berberidaceae) and its relative species in eastern Iran. In: 28th international horticultural congress
Heidary S, Marashi H et al (2009) Assessment of genetic structure and variation of native *Berberis* populations of Khorasan provinces [Iran] using AFLP markers versus morphological markers. Iranian J Biotechnol 7(2):101–107
Hoshyar R, Mahboob Z et al (2016) The antioxidant and chemical properties of *Berberis vulgaris* and its cytotoxic effect on human breast carcinoma cells. Cytotechnology 68(4):1207–1213
Hsieh C-L, Yu C-C et al (2021) *Mahonia* vs. *Berberis* unloaded: generic delimitation and infrafamilial classification of berberidaceae based on plastid phylogenomics. Front Plant Sci 12:720171
Hu Z-B, Du M (2006) Hairy root and its application in plant genetic engineering. J Integr Plant Biol 48(2):121–127
Husain T, Rao R (1998) *Berberis lycium* Royle complex in India. Feddes Rep 109(5–6):385–392
Imenshahidi M, Hosseinzadeh H (2019) Berberine and barberry (*Berberis vulgaris*): a clinical review. Phytother Res 33(3):504–523
Jabeen N, Saleem A et al (2015) *Berberis* lycium Royle (Royle 1837): a threatened medicinal plant and its biological activities. EC Agric 1(2):100–108
Jannatizadeh A, Khadivi-Khub A (2016) Morphological variability of *Berberis integerrima* from Iran. Erwerbs-Obstbau 58(4):247–252
Joshi N, Sharma S et al (1992) Studies on the effect of auxin and season on rooting stem cuttings of some important shrubs in nursery beds. Indian for 118(12):893–900
Karlović K, Kremer D et al (2009) Intra- and interpopulation variability and taxonomic status of *Berberis* croatica horvat. Plant Biosyst 143(1):40–46
Karp PD, Billington R et al (2019) The BioCyc collection of microbial genomes and metabolic pathways. Brief Bioinformatics 20(4):1085–1093
Kamal Y, Singh M et al (2011) Quantitative analysis of berberine in *Berberis aristata* fruits and in a traditional anti-inflammatory unani formulation by use of a validated HPLC method. Acta Chromatogr 23(1):157–168

Katare AK, Singh B et al (2022) Rapid determination and optimisation of berberine from Himalayan *Berberis lycium* by Soxhlet apparatus using CCD-RSM and its quality control as a potential candidate for covid-19. Nat Prod Res 36(3):868–873

Keet JH, Cindi DD et al (2016) Assessing the invasiveness of *Berberis aristata* and *B. julianae* (Berberidaceae) in South Africa: management options and legal recommendations. S Afr J Bot 105:288–298

Khan T, Khan IA et al (2016) Conservation status evaluation of *Berberis* species across the Karakoram mountain ranges, Pakistan using IUCN red list categories and criteria. J for Res 27(6):1385–1390

Kim YD, Kim SH et al (2004) Taxonomic and phytogeographic implications from ITS phylogeny in *Berberis* (Berberidaceae). J Plant Res 117(3):175–182

Kosalec I, Gregurek B et al (2009) Croatian barberry (*Berberis croatica* Horvat): a new source of berberine-analysis and antimicrobial activity. World J Microbiol Biotechnol 25(1):145–150

Končić MZ, Kremer D et al (2010) Evaluation of antioxidant activities and phenolic content of *Berberis vulgaris* L. and *Berberis croatica* Horvat. Food Chem Toxicol 48(8–9):2176–2180

Kong Y, Li L et al (2022) A patent review of berberine and its derivatives with various pharmacological activities (2016–2020). Expert Opin Ther Pat 32(2):211–223

Laar H (1975) Mahonia en Maho Berberis. Dendroflora 11(12):19–33

Lambert KA, Obae SG (2016) Molecular characterisation of *Berberis thunbergii* cultivars using microsatellite markers. J Hortic Sci Biotechnol 91(2):156–160

Landrum LR (1999) Revision of *Berberis* (Berberidaceae) in Chile and adjacent southern Argentina. Ann Mo Bot Gard 793–834

Lehrer JM, Brand MH et al (2008) Induction of tetraploidy in meristematically active seeds of Japanese barberry (*Berberis thunbergii* var. *atropurpurea*) through exposure to colchicine and oryzalin. Sci Hortic 119(1):67–71

Li L, Long W et al (2014) Studies on quantitative determination of total alkaloids and berberine in five origins of crude medicine "Sankezhen." J Chromatogr Sci 53(2):307–311

Li X, Zhang L et al (2017) New taxa of *Berberis* (Berberidaceae) with greenish flowers from a biodiversity hotspot in Sichuan province, China. Plant Divers 39(2):94–103

Liu J, Luo X (2019) First report of bicolour FISH of *Berberis diaphana* and *B. soulieana* reveals interspecific differences and co-localization of (AGGGTTT) 3 and rDNA 5s in *B. diaphana*. Hereditas 156(1):1–8

Lok Ranjan B, Bina W et al (2018) Antioxidant activity, total phenolic and flavonoid content of *Berberis aristata* DC. and *Berberis thomsoniana* C.K. Schneid. from Sagarmatha National Park, Nepal. Pharmacogn J 10(6):S167

Lubel JD, Brand MH et al (2008) AFLP identification of *Berberis thunbergii* cultivars, inter-specific hybrids, and their parental species. J Hortic Sci Biotechnol 83(1):55–63

Manandhar NP (2002) Plants and people of Nepal. Timber Press

McLeod C, Ojeda A et al (2017) Seed germination and in vitro propagation of different *Berberis microphylla* G. Forst accessions from Magellan's region in Chile. Acta Hortic 1172:225–230

McAlpine KG, Jesson LK (2007) Biomass allocation, shade tolerance and seedling survival of the invasive species *Berberis darwinii* (Darwin's barberry). N Z J Ecol 1–12

Moghaddam PR, Fallahi J et al (2013) Effects of harvest date, harvest time, and post-harvest management on quantitative and qualitative traits in seedless barberry (*Berberis vulgaris* L.). Ind Crops Prod 42:30–36

Moghaddam PR, Huda A et al (2007) Indigenous knowledge in agriculture with particular reference to medicinal crop production in Khorasan, Iran. WASD 105–115

Molinar F, Mackay WA et al (1996) Micropropagation of Agarita (*Berberis trifoliata* Moric.). HortScience 31(6):1030–1032

Montazeran A, Khadivi A et al (2018) The first report: chilling and heat requirements of seedless barberry (*Berberis vulgaris* L. var. asperma). Sci Hortic 231:188–193

Moran R (1982) *Berberis claireae*, a new species from Baja California; and why not Mahonia. Phytologia 52:221–226

Motalleb G, Hanachi P, Kua SH, Fauziah O, Asmah R (2005) Evaluation of phenolic content and total antioxidant activity in Berberis vulgaris fruit extract. 648–653

Motolinía-Alcántara EA, Castillo-Araiza CO et al (2021) Engineering considerations to produce bioactive compounds from plant cell suspension culture in bioreactors. Plants 10(12):2762

Murashige T, Skoog F (1962) A revised medium for rapid growth and bio assays with tobacco tissue cultures. Physiol Plant 15(3)

Naji-Tabasi S, Emadzadeh B et al (2021) Physico-chemical and antioxidant properties of barberry juice powder and its effervescent tablets. Chem Biol Technol Agric 8(1):1–11

Nilda E, Muhammed K et al (2018) Physicochemical diversity among barberry (*Berberis vulgaris* L.) fruits from eastern Anatolia. Not Bot Horti Agrobot Cluj Napoca 46(2):336–342

Nooreen Z, Rai AK et al (2024) A review on folklore uses, phytoconstituents and pharmacological activities of the genus Berberis. Curr Org Chem 28(6):405–432

Nyirenda CP (2008) The conservation biology of *Berberis* holstii Engl. In Nyika National Park, Malawi. http://www.pearl.plymouth.ac.uk

Obae SG, Brand MH et al (2013) Using amplified fragment length polymorphism markers to confirm identity and correct labeling of Japanese barberry (*Berberis thunbergii*) cultivars in the market. Hort Sci 48(2):150–157

Özgen M, Saraçoğlu O et al (2012) Antioxidant capacity and chemical properties of selected barberry (*Berberis vulgaris* L.) fruits. Hortic Environ Biotechnol 53(6):447–451

Pasrija A, Singh R et al (2010) Validated HPLC-UV method for the determination of berberine in raw herb Daruharidra (*Berberis aristata* DC), its extract, and in commercially marketed Ayurvedic dosage forms. Int J Ayurveda Res 1(4):243–243

Patel P (2021) A bird's eye view on a therapeutically 'wonder molecule': Berberine. Phytomedicine plus 1(3):100070

Pradhan PC, Saha S (2016) Anthocyanin profiling of *Berberis lycium* Royle berry and its bioactivity evaluation for its nutraceutical potential. J Food Sci Technol 53(2):1205–1213

Pinar H, Yahya HN et al (2021) Molecular characterization of barberry genotypes from Turkey and Kyrgyzstan. Erwerbs-Obstbau 63(4):403–407

Puczel U (2008) Effect of cutting type on the rooting process in selected species of evergreen barberries. Lietuvos Agrarinių Ir Miškų Centro Filialo Sodininkystės Ir Daržininkystės Instituto Ir Lietuvos Žemės Ūkio Universiteto Mokslo Darbai 27(4):27–34

Rahimi-Madiseh M, Lorigoini Z et al (2017) *Berberis vulgaris*: specifications and traditional uses. Iran J Basic Med Sci 20(5):569

Rajasekharan PE, Sahijram L (2015) In vitro conservation of plant germplasm. In: Bahadur B, Venkat Rajam M et al (eds) Plant biology and biotechnology: volume II: Plant genomics and biotechnology. Springer, New Delhi, pp 417–443

Rawat AKS, Srivastava SK (2007) Pharmacognostic evaluation of the roots of *Berberis tinctoria* Lesch. Nat Prod Sci 13(1):27–32

Rezaei M, Balandary A (2012) Effects of gibberellic acid (GA3), stratification and acid scarification on the seed germination of three Iranian barberry genotypes. Seed Res J (in Persian). 2(3):11–18

Rezaei M, Balandary A (2015) Study of seed set in crossing among seedless barberry and wild type barberry genotypes. Iranian J Hortic Sci 46(2):323–331

Rezaei M, Ebadi A et al (2011) Molecular analysis of Iranian seedless barberries via SSR. Sci Hortic 129(4):702–709

Rezaei M, Sarkhosh A et al (2020) Characterization of valuable indigenous barberry (*Berberis* sp.) germplasm by using multivariate analysis. Int J Fruit Sci 20(1):1–19

Rodoni LM, Feuring V et al (2014) Ethylene responses and quality of antioxidant-rich stored barberry fruit (*Berberis* microphylla). Sci Hortic 179:233–238

Roelfs AP (1985) 1—Wheat and rye stem rust. In: Roelfs AP, Bushnell WR (eds) Diseases, distribution, epidemiology, and control. Academic Press, p 3–37

Romadanova N, Mishustina S et al (2016) In vitro collection of cultivars and wild forms of *Berberis* spp for cryobank creation. J Biotechnol 231:S39

Rounsaville TJ, Ranney TG (2010) Ploidy levels and genome sizes of *Berberis* L. and Mahonia Nutt. species, hybrids, and cultivars. HortSci 45(7):1029–1033

Roy NS, Choi I-Y et al (2021) Gene expression and isoform identification of PacBio full-length cDNA sequences for berberine biosynthesis in *Berberis koreana*. Plants (Basel, Switzerland) 10(7):1314

Ruiz A, Zapata M et al (2014) Flavonols, alkaloids, and antioxidant capacity of edible wild *Berberis* species from Patagonia. J Agric Food Chem 62(51):12407–12417

Safamanesh B, Esmaeilzadeh Bahabadi S et al (2017) Investigation of genetic variation in *Berberis* vulgaris using ISSR and SSR molecular markers. JCMR 9(1):23–34

Safari-Khuzani A, Rezaei M et al (2020) Screening of the superior F1 segregating populations of barberry (*Berberis* spp.) for cultivation based on phenotypic characterizations. Ind Crops Prod 158:113054

Safari A, Rezaei M (2023) Assessing genetic diversity of Berberis genotypes selected from an F1 population using molecular markers. J Berry Res 13(4):343–354

Safari M, Rezaei M (2021) Grafting success of *Berberis integerrima* cv Bidaneh on wild type barberries. Int J Fruit Sci 21(1):1030–1039

Sagers LA (2005) Waterwise shrubs under six feet. www.digitalcommons.usu.edu

Salehi B, Selamoglu Z et al (2019) *Berberis* plants-drifting from farm to food applications, phytotherapy, and phytopharmacology. Food 8(10):522

Sazmand M, Safarnezhad A (2016) Callus induction, regeneration and proliferation of *Berberis vulgaris* var. asperma using in-vitro technique. IJRFPBGR 24(1):92–101

Schneider C (1905) Die Gattung *Berberis* (Eu*Berberis*): Vorarbeiten für eine monographie. Romet

Sehrawat R, Rajasekaran A et al (2009) Quantitative estimation of berberine in roots of different provenances of *Berberis aristata* DC by HPLC and study of their antifungal properties. Pharmacogn Mag 5(20):355–358

Şensu E, Kasapoğlu KN et al (2021) Orange, red and purple barberries: effect of in-vitro digestion on antioxidants and ACE inhibitors. LWT 140:110820

Singh R, Tiwari SS et al (2012) Botanical and phytochemical studies on roots of *Berberis umbellata* Wall. Ex g. Don. Indian J Nat Prod Resour 3(1):55–60

Singh IP, Mahajan S (2013) Berberine and its derivatives: a patent review (2009–2012). Expert Opin Ther Pat 23(2):215–231

Sinha P, Chen X (2021) Potential infection risks of the wheat stripe rust and stem rust pathogens on barberry in Asia and southeastern Europe. Plants 10(5):957

Smith AG, and Noyszewski AK (2018) Mutagenesis breeding for seedless varieties of popular landscape plants. Acta Hortic 1191: 43–52. https://doi.org/10.17660/ActaHortic.2018.1191.7

Srivastava SK, Rawat AKS et al (2006) Pharmacognostic evaluation of the roots of *Berberis chitria* lindl. Nat Prod Sci 12(1):19–23

Srivastava S, Srivastava M et al (2015) A review on biological and chemical diversity in *Berberis* (berberidaceae). EXCLI J 14:247–267

Srivastava SK, Singh Rawat AK et al (2004) Pharmacognostic evaluation of the root of *Berberis* asiatica. Pharm Biol 42(6):467–473

Tatari M, Ghasemi A et al (2019) Assessment of genetic diversity of barberry germplasm (*Berberis* spp.) in central regions of Iran by morphological markers. J Hortic Res 27(1)

Tavakoli A, Sahari MA et al (2017) Antioxidant activity of *Berberis integerrima* seed oil as a natural antioxidant on the oxidative stability of soybean oil. Int J Food Prop 20(Suppl 3):S2914–S2925

Thole V, Bassard J-E et al (2019) RNA-seq, de novo transcriptome assembly and flavonoid gene analysis in 13 wild and cultivated berry fruit species with high content of phenolics. BMC Genom 20(1):995

Tripathi V, Goswami S (2013) Assessment of genetic diversity in *Berberis lycium* Royle complex using RAPD markers. J Cell Biol Genet 3(1):1–13

USDA (2002) Black stem rust; identification requirements and addition of rust resistant varieties. Fed Regist 67(32)

Varsha S, Kumkum M et al (2014) Karyotypic analysis in western Himalayan species of *Berberis* L. Res J Pharm Biol Chem Sci 5(2):1244–1251

Verma S, Wani IA et al (2021) Reproductive biology and pollination ecology of *Berberis* lycium Royle: a highly valued shrub of immense medicinal significance. Plants 10(9):1907

Villinski J, Dumas E et al (2003) Antibacterial activity and alkaloid content of *Berberis thunbergii*, *Berberis vulgaris* and *Hydrastis canadensis*. Pharm Biol 41(8):551–557

Wang J, Jiang Y et al (2019) A review on analytical methods for natural berberine alkaloids. J Sep Sci 42(9):1794–1815

Wang MN, Chen XM (2015) Barberry does not function as an alternate host for *Puccinia striiformis* F. Sp. *tritici* in the U. S. Pacific Northwest due to teliospore degradation and barberry phenology. Plant Dis 99(11):1500–1506

Ward J, Williams S (2011) Controlling an invasive shrub, Japanese barberry (*Berberis thunbergii* DC), using directed heating with propane torches. Nat Areas J 31(2):156–162

Ward JS, Worthley TE et al (2009) Controlling Japanese barberry (*Berberis* thunbergii DC) in southern new England, USA. For Ecol Manag 257(2):561–566

Wyman D (1958) Two new maho *Berberis* hybrids. Arnoldia 18(2):9–12

Yahya H (2017) The morphological and molecular characterization of barberry genotypes from Turkey and Kyrgyzstan. agris.fao.org. 56

Yildiz H, Ercisli S et al (2014) Some physicochemical characteristics, bioactive content and antioxidant characteristics of non-sprayed barberry (*Berberis vulgaris* L.) fruits from Turkey. Erwerbs-Obstbau 56(4):123–129

Yilmaz KU, Ercisli S, Zengin Y, Sengul M, Kafkas EY (2009) Preliminary characterisation of cornelian cherry (Cornus mas L.) genotypes for their physico-chemical properties. Food Chem 114(2):408–412

Yilmaz KU, Abdullaev A et al (2017) Morphological characteristics of different wild *Berberis* sphaerocarpa Kar. Et Kir. genotypes grown in Jeti-Oguz district (Issyk-Kul Province) of north eastern Kyrgyzstan. Acta Sci Pol Hortorum Cultus 16(2):123–132

Yoo S-J, Lee K-B et al (1986) Studies on the seasonal variation of berberine contents in *Berberis koreana*. Korean J Pharmacogn 17(2):123–128

Yu C-C, Chung K-F (2014) Systematics of *Berberis* sect. Wallichianae (Berberidaceae) of Taiwan and Luzon with description of three new species, *B. schaaliae*, *B. ravenii*, and *B. pengii*. Phytotaxa 184(2):61–99

Yu C-C, Chung K-F (2017) Why Mahonia? Molecular recircumscription of *Berberis* SL, with the description of two new genera, Allo *Berberis* and Moranothamnus. Taxon 66(6):1371–1392

Zeratgar H (2011) Investigation of genetic diversity of barberry germplasm (seeded and seedless) in South Khorasan and establishment of a collection orchard. In: Esfehan 7th Iranian Cong Hortic Sci 174671

Zhao J, Wang M et al (2016) Role of alternate hosts in epidemiology and pathogen variation of cereal rusts. Annu Rev Phytopathol 54(1):207–228

Chapter 3
Piper betle (Linn.): An Insight into Breeding and Biotechnological Strategies for Achieving Higher Productivity

Sony Kumari, Madhusmita Borthakur, Rony Bhowal, Puja Roy, and Arun Jerang

Abstract Betel vine being dioecious, is native to East Asian countries, including India, with a chromosomal number (x = 13). Being perennial and a vital cash crop in both national and internal markets, it is widely cultivated by various communities in a modified artificial structure, "Boroj", which provides support and shades for growth and development. Ten different varieties of betel vine are known to be consumed by the ethnic communities of India and are named based on the landraces. In spite of high cultivation and high medicinal prospects, the betel vine is not much scientifically explored in terms of its genetic studies and breeding. However, molecular markers like Random Amplified Polymorphic DNA, Amplified fragment length polymorphism (AFLP), Inter Simple Sequence Repeat (ISSR), and Simple Sequence Repeat (SSR) have been used to determine the germplasm diversity to enhance the breeding and cultivation of the betel vine. Sequence Characterized Amplified Regions (SCAR) is also used for sex determination in betel vine, a medicinal plant with phytochemicals, and other secondary metabolites. A resistant variety of the betel vine, IIHR PBH 09–16 hybrid was developed, which is resistant to both biotic and abiotic stress. Lack of awareness among the poor cultivars, the use of the same conventional breeding method, and existing of diverse synonyms of the same varieties constitute a significant voids in the recent breeding techniques. The current review has attempted to focus on and highlights the comprehensive account of genetic breeding and biotechnological intervention, with an aim at complementing the conventional approach of the breeding techniques. The study will lead to the improvement of the betel vine along with the socio-economic development of the poor betel vine cultivars among different landraces in the country.

S. Kumari (✉) · M. Borthakur · R. Bhowal · P. Roy · A. Jerang
Department of Applied Biology, School of Biological Sciences, University of Science and Technology, Meghalaya, Baridua, India
e-mail: sonykumari_15@yahoo.com

S. Kumari · M. Borthakur · P. Roy · A. Jerang
Applied Biology, University of Science and Technology, Meghalaya, Ri Bhoi, India

J. M. Al-Khayri et al. (eds.), *Biodiversity and Genetic Improvement of Medicinal and Aromatic Plants I*, Advances in Plant Breeding Strategies 9,
https://doi.org/10.1007/978-3-031-81846-2_3

Keywords Perennial climber · Boroj · Molecular techniques · Micropropagation · Pleiotropic effect

3.1 Introduction

Betel vine (*Piper belle* Linn.) of *Piperaceae* family is an evergreen, perennial, dioecious creeper comprises of 10 genera and 2000 species (Chakraborty and Shah 2011). It is native to tropical South-East Asia, including the Philippines, Malaysia, Indonesia, and mainly cultivated in India including Assam, West Bengal, Bihar, Andhra Pradesh, Karnataka, Kerala, Madhya Pradesh, Uttar Pradesh, Maharashtra, Orissa and Tamil Nadu (Guha 2006). The Betel vine cultivating countries in the world include, mainly India, Sri Lanka, Thailand and Bangladesh (Fig. 3.1) and the farming areas in India is highlighted in Fig. 3.2. The betel vines are mainly consumed by people of various regions within Indian Territory and are named based on the landraces. The dark green heart-shaped leaves are consumed primarily on their raw texture aided with lime, areca nut, etc. The betel vine is famous as paan (Hindi) in India, which is known as "Parrut" in Sanskrit. The cultivation of the betel vine can be seen throughout India, with the highest production and consumption in North East corridor of the country. High cultivation and productivity in this region depend on the higher consumption rate by the ethnic tribes of the area. The ethnic communities of Assam and Meghalaya have an inherent custom of consuming betel nut and betel vine and also considered a mark of respect. The betel vines are named Nagaballi, Nagurvel, Saptaseera, Sompatra, Tamalapaku, Tambul, Tambuli, Vaksha Patra, Vettilai, Voojangalata, etc., in different parts of the country (Guha and Jain 1997). The consumption of the betel leaf is inherited among the tribes since the last fifteenth century during the arrival of Europeans to India.

The betel vines, along with areca nuts, are considered to be elite for the society. They are often seen in use to welcome the royal guests along with traditional cultural rituals (Ahuja and Ahuja 2011). The betel leaf is often seen consumed along with betel nuts enhances the taste. It is considered as an excellent mouth freshener, and mild vitalizer, ethnically consumed and served on different rituals during social, cultural, and religious occasions (marriage, festivals and other rituals) (Ahuja and Ahuja 2011).

3.2 Origin and Distribution

Betel vine is extensively cultivated in hotter and damper conditions, generally found in India, and Sri Lanka, and extending to Peninsular Malaysia. Because of its ritual importance for India, it has been under cultivation in centuries. Chakraborty and Shah, 2011 reported 10 genera of betel vines, over 2000 species approximately, with its aromatic nature being consumed since ancient times. Itis cultivated over, for

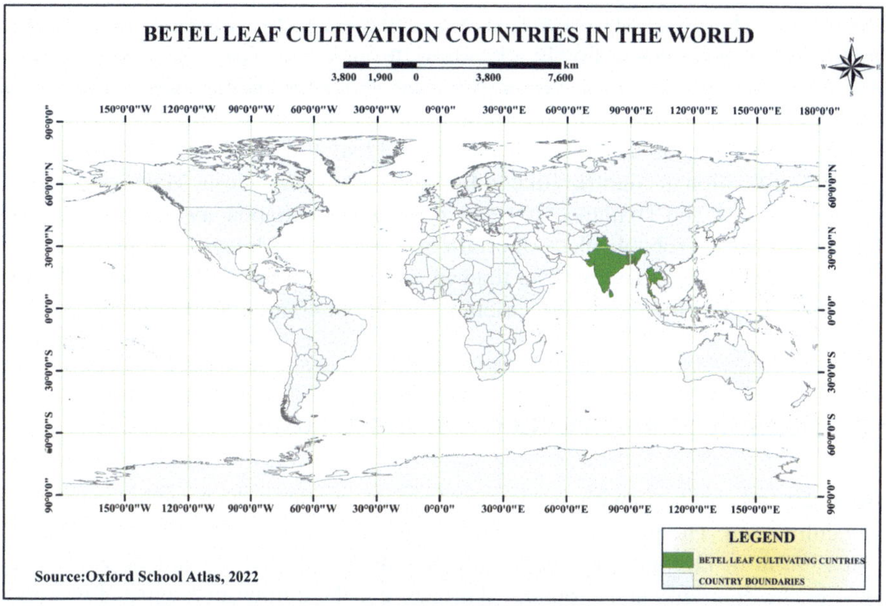

Fig. 3.1 Betel vine cultivation

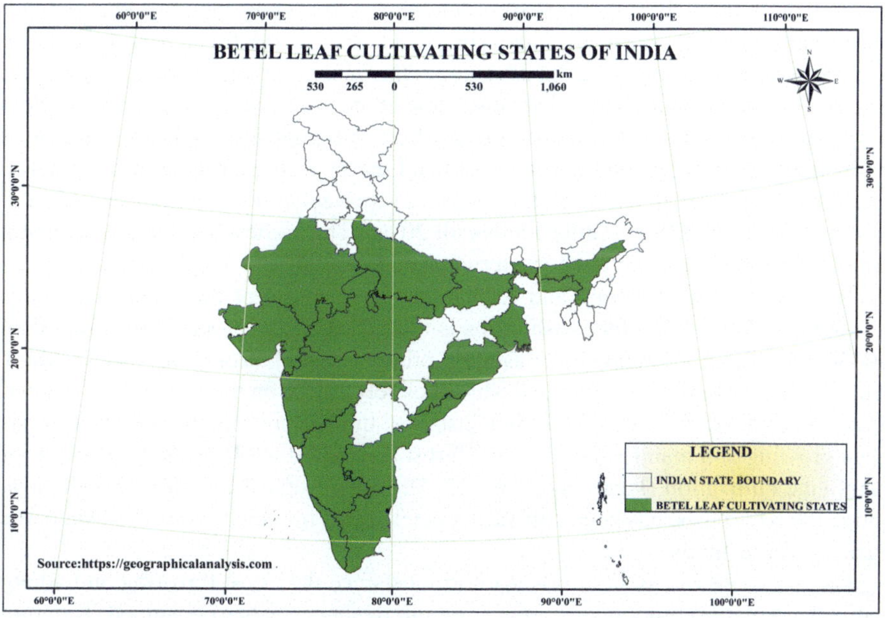

Fig. 3.2 Betel vine cultivation in different States of India

2500 years and is believed to be native to central and eastern Malaysia. *The* vine first emerged in tropical Asia, slowly spreading through East Africa and Madagascar. The plant's leaf is eaten raw and has different vernacular names depending on the geographical region and cultivation condition (Biswas et al. 2022). Guha (2006) mentioned the geographical distribution of betel leaf almost all over the world, with majority cultivation in countries like India, Sri Lanka, Bangladesh, Malaysia, Burma, Nepal, Myanmar, and Thailand. However, a study by Chattapdayay and Maity (1967) suggested that the probable place of origin of betel vine is Malaysia. At present time, the plant is more popular in India than in any other country of the world. It was reported that there are about 100 varieties, and about 40 are found in India.

3.3 Domestication, Selection and Importance

Black soil and clay loam with maximum of organic matter is the best choice for betel vine cultivation. The growth of the vine occurs from sea level to 1000 m and more than 179 cm of rainfall is necessary for its cultivation. Shady tropical forest with considerable humidity and enough soil moisture is the minimum required for proper growth of the vine. Both open and closed systems are in practiced for its cultivation. Betel vine is a commercial crop, which is wildly consumed in the form of betel quid due to its mouth freshener and its stimulant nature. It is consumed in India together with a variety of spices, sauces, flavourings, and other mouth fresheners. In addition to being edible, it has many health advantages that increase its nutritional advantages. Bangladesh, India, and Thailand are just a few of the tropical nations where betel leaf is grown. Betel leaf is often referred to as Paan in Bangladesh and India and eaten with areca nut. 200,000 metric tonnes of betel leaves were produced in Bangladesh in 2019 (Mahfuza et al. 2020). Betel leaves are reportedly grown in India each year at a cost of about $9 billion (Guha and Nandi 2019). The highest betel leaf production areas in India include the north-eastern states of Assam and Meghalaya as well as the southern states of Kerala and Karnataka. One feature of the plant that makes it more profitable than other cash crops is that, if cared for properly for more than 10–30 years, a single plant can constantly produce leaves throughout its life. Many ethnic tribes of Indian and Bangladeshi tribes rely solely on the production of betel leaves as a means of subsistence. For instance, the Khasia tribe in Bangladesh has been growing and maintaining massive forest area near the Moulvibazar district for betel vine cultivation (Islam and Nath 2014). In the tropical forests of Meghalaya, India, the Khasi tribe has been cultivating betel leaves for generations by sustainable agricultural methods.

The relevance of essential oils is enormous since they give flavoring and, along with other volatile components, serve as mouth fresheners when consumed (Kaintura et al. 2020). Additionally, betel leaves are a significant source of phenolic chemicals, including hydroxy-chavicol, which contributes to their fragrant quality (Kaintura et al. 2020). Betel leaf is renowned for its rich reservoir of phytochemicals, each

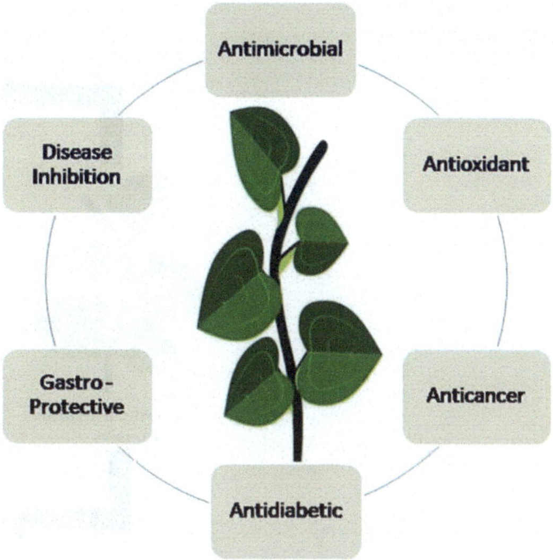

Fig. 3.3 Betel leaf importance in various sectors

boasting distinct health benefits such as antimicrobial, antioxidant, gastroprotective, anticancer, and anti-diabetic properties (Roy and Guha 2021). These properties collectively contribute to its esteemed status in traditional and modern medicinal practices as shown in Fig. 3.3. The phytochemicals, and bioactive compounds fall in the groups of phenols and flavanoids, etc. These compounds allow the betel leaf to be considered as one of the best examples of traditional phytomedicine. The chemical constituents of betel leaf are mainly Chavibetol, Chavibetol acetate, Allypyrocatechol diacetate, Campene, Chavibetol methyl ester, Eugenol and A-pinene (Shah et al. 2016), which makes betel leaves a health beneficial plant source.

3.4 Current cultivation and Traditional breeding Practices with Challenges

3.4.1 Cultivation and Traditional Breeding

Betel vine thrives in warm, humid climates, preferring well-drained soil with a pH of slightly acidic to neutral. It needs partial shade and adequate water supply. Propagation methods include stem cuttings, layering, and seeds. Soil preparation is crucial, and vines can be grown vertically or horizontally. Regular pruning manages growth and encourages branching, while adequate spacing prevents overcrowding and disease spread. They possess green leaves of heart shape and is a perennial climber plant grown under shade conditions. The length and width of the leaves are approximately 5–15 cm and 3–10 cm, respectively. Lateral veins are observed

Fig. 3.4 Betel vine cultivation practices

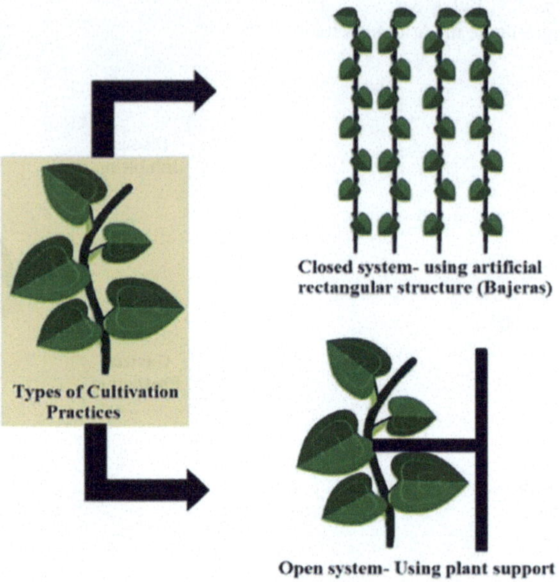

from the bottom to the apex of the leaf, which are five to seven in number. The conventional growth of crops needs a proper arrangement of environmental situation like soil moisture, humidity, and temperature. If the optimum growth condition is not found in the growing state, it will be detrimental to the crop. To avoid such problems, betel leaf cultivation is done in rectangular-shaped artificially constructed structures; the hut-like structure maintains the plant growth conditions (Guha 2006). Figure 3.4 represents the method of betel vine cultivation by open system and closed system respectively.

Support system is not required in open system; however, closed system involves the support. The structure made for providing such environmental conditions in India is known as Boroj, made up of straw (paddy), sticks (jute), stems (bamboo), leaves, and petioles (banana). Boroj (2–3 m high, 10–20 m long, and 5–15 m wide) supports the proper setup required for cultivation by providing shade, and humid conditions. The main explants; used for the cultivation is stem cuttings (3–5 years) which grow through the vegetative propagation mode of reproduction. The growth and yield of the plant depend on appropriate maintenance of environmental conditions. The primary sequence of tasks followed for cultivation is demonstrated in Fig. 3.5.

As mentioned, betel vine can be grown in two different manners; one in an open system and the other type of growth is under a closed system. Both methods are the traditional mode of betel vine cultivation. The first type, i.e., the open system cultivation method, is practiced in highly humid and moderate sunshine natural environments that persist throughout the year (Ullah et al. 2020). The second type is the cropping of vines in partially controlled conditions, commonly practiced high humid and low sunshine conditions that do not prevail throughout the year. Hence, the crops

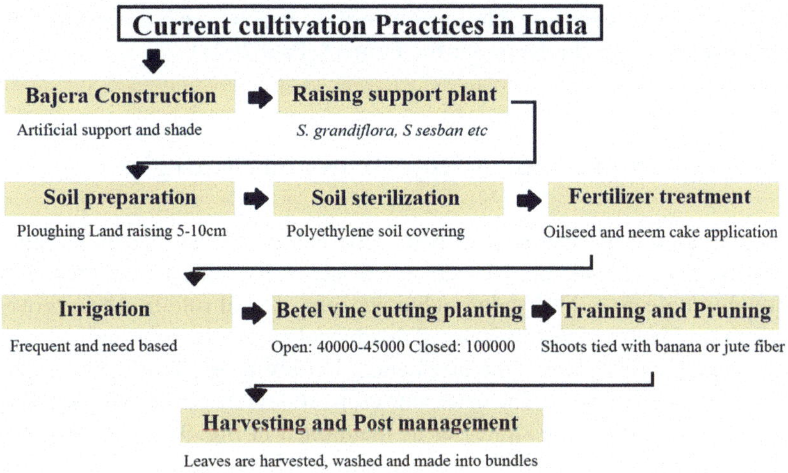

Fig. 3.5 Current cultivation practices in India

are to be protected and shaded from extreme sunny and dry air conditions. When vine grows up, they are trained on some support, mainly; plants that can provide shade to the grown vines, and increase the humidity necessary for better crop cultivation. This type of cultivation done in closed environmental conditions helps in moisture retention and microclimate conditions adding to the proper development of the betel leaf. The growth of a healthy vine depends on the proper irrigation and fertilizer used during cultivation. After the vine reaches the proper height, they are plucked and made into bundles for selling.

3.5 Current Agricultural Problems and Challenges

Due to increased demand for betel leaf by consumers in the national and international markets, the healthy plant should be grown on a large scale. It should have resistance to various environmental stress conditions. The challenge of producing a high yielding crop with proper current practices makes use of fertilizers. However, fertilizers have adverse effects on the soil, as well as the crop (Tirkey et al. 2019). There are different challenges including agronomic and, biotic and abiotic stresses. As described, the main problem is vine disease; lack of improved variety and proper management practices leads to enormous economic loss to the farmers. The main category of challenges, which contributes the crop damage and loss in betel vine, is the capital problem, natural calamity, irrigation problem, pest, and microbial infection, lack of knowledge in the use of fertilizer and pesticide, etc. Betel vine cultivation at a large scale is not an easy task. All the problems mentioned have to be overcome for proper cultivation. Some of the challenges are overcome by a few government bodies by providing loans to the farmers involved in such crop cultivation; however, returning loan is also a challenge for the farmers (Paul 2021).

3.6 Genetic Improvement Objectives and Strategies

Plant breeding and improvement in betel vine is the science of manipulating and changing its genetic makeup. The change is generally done to get crop with for better quality objective. Molecular techniques play a vital role in the improvement and selection of better plants. It is noted that recent molecular tools are yet not been employed in the case of betel vine production. However, they have been used in sex determination and selection. The betel vine crops should be grown with broad objectives comprising three main strategies as mentioned in Fig. 3.6. The improvement should include the objectives, of betel leaf better yield and quality.

Traditional cultivation practice of betel vine has been used for many years at a small scale by the farmers. The use of advanced methods to increase the yield due to lack of proper research is lacking. To meet the market demand, major strategies have to be made through hybridization technology. Measures should be taken by various government and non-government agencies to provide proper arrangements and facilities for betel vine large-scale production. Some of the parameters that should be considered for better production include, cold storage setup, proper irrigation systems, research canter development for variety improvement, soil improvement, improved cultivation practices. Crop varietal improvement is an essential requirement at the current time. Biotechnology may contribute to quality improvement by employing tools and techniques of genetic engineering. Micropropagation can contribute to the mass production of improved crops (Ranade et al. 2002). Espinosa-Leal et al. (2018) reported that in vitro propagation methods in crops are less expensive and are the most common method employed for mass production in plants. However, the technology is limited in the case of betel vine. It is essential to employ recent techniques for in vitro and genetic manipulation and make the betel leaf quality better than the traditionally cultivated types. To fulfil the consumer's demand, betel

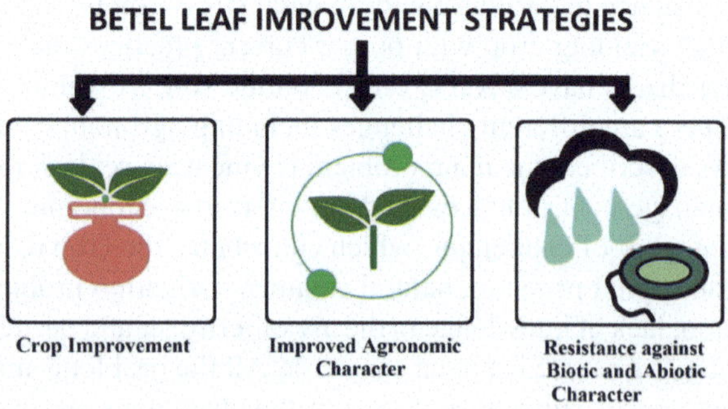

Fig. 3.6 Betel leaf improvement Strategies

leaf should be tasty with a healthy and broad product. The shelf life of the leaves needs to be enhanced by adopting the ICM package (Das et al. 2016). To enhance betel leaf production, an integrated crop management system can be employed and have to be tested in farmer's field.

3.7 Role of Biotechnology

The traditionally used method of betel leaf cultivation under natural and controlled conditions needs to be technically improved, for better growth and yield on a large scale. The Boroj shape of preparation huts artificially enhances the cultivation rate of betel vine (Sahoo and Sahoo 2017). The cultivation, which is done under controlled conditions, is also known as greenhouse cultivation (Ullah ct al. 2020). However, the modern tools and techniques of biotechnology and genetic engineering can offer better scope for developing improved varieties of betel.

The biotechnological tools have been used to improve the qualitative aspects, improvement of crop species, and bioactive compounds such as essential oils. There is scanty information on the biotechnological tools used for the propagation and cultivation of various herbs crops such as betel leaf. Biotechnology has significantly contributed to crops by enhancing their yield, making them more resistant to pests and other disease-causing agents. The technique of biotechnology has contributed for assessing micropropagated plants and their genetic variability of the crop for screening better genotype, using molecular marker technique i.e. using Random amplified polymorphic DNA (Ranade et al. 2002). The in vitro propagation of the plant is less expensive (Espinosa-Leal et al. 2018). Unfortunately, not much work has been done on betel vines employing biotechnology tools. However, considering its role in other agricultural products, it is suggested that biotechnology can play a crucial role in betel vine improvement by enhancing agronomic traits, pest resistance, and medicinal properties. Techniques like marker-assisted selection and genetic engineering can be used to introduce desirable traits like disease resistance, increased leaf yield, and enhanced flavor or aroma. Biotechnology can also enhance the production of bioactive compounds responsible for medicinal properties, such as antimicrobial, anti-inflammatory, and antioxidant effects (Kavi Kishor et al. 2024).

3.8 Germplasm Biodiversity and Conservation

3.8.1 *Germplasm Diversity, Cultivars Characterization, and Phylogeny*

Betel vine has significance in the indigenous ethnic community of Indian territory as a symbol of respect. It is also represented as the neglected green gold of India (Guha 2006). Due to the widespread geographical distribution of the betel vine, the studies on genetic variations are not well. In the coastal areas of Orissa, and West Bengal, PCR techniques like Random amplified Polymorphic DNA (RAPD), Amplified fragment length polymorphism (AFLP), Inter Simple Sequence Repeat (ISSR), and Simple Sequence Repeat (SSR) are used to study diversity in germplasm to enhance the breeding and cultivation (Arumuganathan and Earle 1991; Huang et al. 2009; Werren et al. 2010). The PCR based technique also identifies duplicate germplasm to determine the genetic variations (Virk et al. 1995). Betel vines are dioecious with their characteristics to produce both male and female gametes directly by biotic or abiotic factors.

To determine the genetic variation between male and female betel vines, along with the relatedness of various cultivated betel vine leaves of different landraces, the RAPD has a significant role. As reported in the analysis of RAPD in two closely related betel vines, Bangla and Kapoori, Kapoori were found to be more heterogeneous than Bangla (Verma et al., 2004).

Since little research is done on the genetic variations of betel vine leaf due to its diversified variation distributed throughout, the use of cheaper PCR-based RAPD techniques can reveal the variations also among the identical groups more efficiency. The genome size and chromosome numbers also vary when analyzed through RAPD. As reported by Ranade et al. (2002), the Bangla and Kapoori varieties were closer and clustered together than any other landraces betel vine on RAPD, proving RAPD to be an efficient technique for genetic variations even to a more identical landrace of betel vine varieties. A better understanding of genetic variations, both intra, and inter-species, along with genetic diversity, might help the germplasm breeders to enhance their conservation strategies along with efficiently utilizing the resources.

3.8.2 *Genetic Resources Conservation Approaches*

The conservation, preservation, and sustainable use of genetic resources of the betel leaf are essential due to the growing threats of climate change, urbanization, the emergence of diseases including the multidrug resistance antibiotics, bioaccumulation for toxic xenobiotic compounds in the ecosystem, along with rapid industrial growth, and global warming. A conservation strategy of betel vine also includes in situ and in-vitro preservation, cryopreservation and DNA gene bank accession deposits. For in situ conservation, ecological conditions like a shaded tropical forest with an average rainfall of up 2250–4750 mm, high humidity up to 80%, and a temperature range from 15 to 40 °C are of utmost necessary (Mridha et al. 2005). As reported by CSIR, 1969, a well-drained fertile sandy to loamy soil is favorable for its growth with pH ranges from 5.6 to 8.2. For vegetative propagation, "Boroj," a small hut-like structure with 2 m height and 0.02 ha in area, partially shaded in a humid environment, is often preferred (Guha 2006). The 'Boroj' are made from the commercially available bamboo stem in the local market along with straw, jute petioles and leaves of plants, etc. The betel vine climbers are mostly grown on a horizontal bed mimicking the similar ecological parameters suitable for the growth of the plant. Due to excessive earthing, the height of the elevated bed frequently increases. Moreover, the explants from 3 to 6 years old betel vines are replanted in 8–10 cm narrow, deep furrows of sterilized soil with spacing ranges from 50 to 60 cm × 10–20 cm mainly in the rainy season. The in situ conservation of the betel vine is labor intensive and requires a rapacious supplement of feeder including 200–300 kg of P_2O_5/ha, 400–600 kg of N/ha, and 250 kg K_2O/ha that are supplemented externally either by oil cakes or through the use of synthetic fertilizer (Guha and Jain 1997; Maity 1989). The in situ production of the betel vine is more susceptible to pest and viral diseases and requires a broader range of supplements. Using fungicides like 0.05% of Endosulphan, 0.05% of Malathion, 0.1% of Bavistin, and 0.5–1.0% Bordeaux is utmost necessary for the growth of betel vine leaf in situ in later phases which act as insect resistive fertilizer (Maity 1989). However, the use of the supplements varies from region to region, seasonally dependent, variety to variety, and several other parameters, but the scientific data are scarce due to the diversification of betel vine leaf. The in situ cultivation of the betel vine is cultivated by poor farmers using traditional methods, which also varies from region to region. But scientific documentation on the grown betel vine leaf is scarce and requires in-depth knowledge through research.

In research sponsored by ICAR-All India coordinated Research Project on Medicinal and Aromatic Plants & Betel vine (AICRP-MAP&B) by ICAR-Indian Institute of Horticulture Research (IIHR), Bengaluru over the use of betel vine hybrid at Jalgaon district of Maharashtra (India), IIHR PBH 09-16 hybrid was developed with better growth traits of Kapoori type betel vine. The objective of the hybrid development is to develop a resistant variety of the betel vine leaf to both biotic as well as abiotic conditions along with high productivity. In a study by IIHR PBH, 09-16 hybrid varieties have better productivity and quality as compared to the local available Kapoori variety with light green color leaf and mild pungency, attracting higher productivity and market price than any other available local variety (https://icar.org.in/content/farmer's-feedback-betelvine-hybrid-received). The use of the hybrid variety leads to increased productivity range from 25–30% and 40% more income generation. This in-situ conservation has turned out to be a boon for the poor betel vine cultivars, thereby raising the socio-economic prospects of the region.

For the in-vitro cultivation of betel vine, young shoot and leaf tissues are used, which are initially surface sterilized with mercuric chloride to remove any ectomicrobiota and are inoculated in plant-based basal media like Murashige and Skoog's media (MS) supplemented with Indole acetic acid, N6-benzyl amino purine (BAP), α-napthaleneacetic acid (NAA) for better vegetative propagation (Elamathi, 2016).

However, the in vitro cultivation of the betel vine also depends on several factors like culture media sterilization, regenerating media with the supplement of the hormone, etc. Micropropagation is influenced by the high growth of bacterial contamination. Overuse of the surface sterilizers, like the high use percentage of mercuric chloride along with the use of phenolic compounds, also limits the regeneration ability of the propagated explants. The explants parameters are also responsible for the proper growth, which includes the use of upper nodes, internode, and leaf as an optimal segment for the regeneration of the betel vine (Elaih 2015). The micropropagation of the betel vine is attributed to the use of McCown's Woody plant medium (WPM) supplemented with various concentrations of chitosan, naphthaleneacetic acid (NAA), benzyl adenine (BA), indoleacetic acid (IAA), kinetic (KIN) along with a wide range of micronutrients like ammonium nitrate, ammonium sulfate, potassium nitrate, potassium sulfate, calcium nitrate, magnesium sulfate along with micronutrients like sodium EDTA, zinc sulfate, copper sulfate, boric acid at a varied concentration which enhance the growth of the explants to callus (Babu et al. 1992).

For prolonged sustainable use, cryopreservation is an effective method. The use of − 196 °C liquid nitrogen for cryogenic preservation is cheaper and safe for the easy maintenance of cryptic bioresources like betel vine, etc. The betel vine germplasm in liquid nitrogen ceases the cellular division to long term storage and risk of losing the accessions either due to contamination or any human errors during the micropropagation techniques. Cryopreservation techniques have been developed for a few species like betel vine, ginger, cardamom, and capsicum, whereas the National Bureau of Plant Genetics (NBPGR), a unit of Indian Council of Agricultural Research, Department of Agricultural Research and Education (DARE) and the Ministry of Agriculture, Government of the India, maintain the germplasm in cryo gene banks preserves the germplasm with accession numbers along with the cultivar's identity and geographical source. A large number of germplasm are also stored by the All India Co-ordinated Research Project on Betel vine (AICRPB) to support and sustain the betel vine farmers over some time. A few germplasms with the accession number are summarized in Table 3.2.

3.9 Cytogenetics

Betel vine leaf is being dioecious with high variation in chromosomal number ranges from $2n = 26$, 32, 52, 58, 62, 78, and 195 (Pawar et al. 2021). The most common and frequent number is $2n = 78$ for most landraces distributed diversely (Jose and Sharma 1988). In the study of Phurailatpam et al. (2018), the primary chromosome in the dioecious betel vine i.e., $x = 13$, originates from the lower chromosome with numbers 6 and 7. The ploidy level has a vital role in the evolution of the betel vine. Before the emergence of the dioecious nature of the betel vine, it was believed that only the male varieties were cultivated in India, which was later reported to be sexually dioecious (Bajpai et al. 2012). The morphological structure of the male plant differs from that of the female being narrowly ovate leaves in males to that

Table 3.2 Different types of cultivated varieties of betel vine isolated from distinctive locations (Patra et al. 2011)

Accession number of germplasm	Cultivated varieties of betel vine	The primary source of betel vine collection
Apb-2	Awani	Jorhat (Assam)
Apb-5	Gandhi pana	Jorhat (Assam)
Opb-1	Godibangala	Kakatapur (Orissa)
Opb-2	Nuabangala	Chandanpur (Orissa)
Opb-3	Balipana	Puri (Orissa)
Opb-8	Kali mahata	Puri (Orissa)
Opb-7	Birkoli	Balangir (Orissa)
Kpb-10	Halisaharsanchi	24 Pragana (West Bengal)
Opb-9	SGM-1	Sirugamani (Tamil Nadu)
Cpb-10	Bangalamandasorechinthalpudi	Chinthalpudi (Andhra Pradesh)
Cpb-12	Karapakubangala	Baputal (Andhra Pradesh)
Cpb-13	Andaman local	Andaman Nicobar
Rpb-3	Maghai	Jutalpur (Madhya Pradesh)
Rpb-1	Ramtekbangala	Ramtek (Madhya Pradesh)
Kpb-4	Kali Bangala	Howrah (West Bengal)

of cordate leaves in females (Maiti and Biswas 1991). Due to high variations in the chromosomal number, and smaller size, makes the cytogenetic study more difficult in betel vine (Biswas et al. 2022). Alternate use of flow cytometry to determine the cytogenetics and the ploidy level of the dioecious betel vine has been studied due to the presence of the nuclear genome, which also can choose the ploidy nature of the plant (Dolezel 1997). Flow cytometry also determines and cluster the male dioecious to that of the female dioecious. In a flow cytometric study by Phurailatpam et al. (2018), the control male has 39 chromosomes ($2n = 3x = 39$) to that female cultivar with 52 number of chromosome ($2n = 4x = 52$). Hence, the ploidy level in the dioecious climber betel vine can also be determined also with flow cytometry, which needs an extensive research in the upcoming future due to high diversity and distribution throughout the territory.

3.10 Molecular Breeding

Breeding of Betel vine enhances the quality and quantity aspects of the product. The selection of the improved variety is made by an appropriate marker system. The phenotypic selection method relies on DNA-based genetic markers as evidenced for authentication, for example, higher leaf yield selection obtained from improved cultivars (Verma et al. 2004). The genetic marker is the most advanced technology which can be used for any change that occurs at the genetic level in betel vine. Improvement may be for a rate of gene gain, leaf yield, propagation efficiency, stability, survival, etc., which can be assessed by a molecular marker, which is a trait linked to type of marker system. Molecular markers such as RAPD, SSR, and ISSR have been used in betel, which is in hybrid selections. These markers are very popular and can distinguish between hybrid and non-hybrid types. Such linkage maps-based markers are parental dependent as they are sparsely distributed in the genome. Therefore, these markers in betel vine can be used only when specific crosses are validated by supporting genetic backgrounds to saturate them. This unique molecular marker can help in finding markers that are linked to parents by the determination of sex in both male and female plants. Betel vine character can be checked depending on the variation and polymorphism of individual plants. Samantaray et al. (2012) found that studies on polymorphism can have a significant role in studying the characteristics in betel nut. In the study, four primers showed polymorphisms in the different sex types that were used to re-test the DNA amplification with individual plants of each sex type. The amplification pattern was consistent and reproducible with 11 male and 8 individual female plants. The size of the unique bands was found to be bp 1400 and 850 bp in male individuals; this size pattern was absent in the females. However, in females, another unique band of size was found to be 980 bp which was not seen in males. The study proves that RAPD is significant marker that are confirmed as a putative sex-linked marker for male and female plants of betel leaf (Samantaray et al. 2012).

In the in silico study, Sekar et al. (2022) has shown a positive response and relation of hydroxychavicol, a phenolic compound in betel leaf with COVID-19-associated maiming mycotic infections. The work has reported that ADME screening of the compound proved to have explicit reciprocity, and it has been found to have a significant binding affinity with the target protein of the infection. The free binding energy of betel leaf against SARS-CoV-2 Mpro (6LU7) is represented in Table3 (Sekar et al. 2022).

3.11 Tissue Culture Application

Tissue culture is an extensively used method to promote plant growth in a nutrient-containing in vitro method, where the entire plant can be grown. The process of tissue culture was first attempted by German botanist, to cultivate plant cells using in vitro technique on an artificial medium (Ali and Ray 2018). The tissue culture method is beneficial for propagating plants to increase yield and genetic modification. India has a broader variety of cultivars compared to other countries. The betel vine micropropagation is yet to be explored to a large extent. The plant has the full potential to be grown under in-vitro conditions for micropropagation like other related crops.

Table 3.3 Betel leaf phytoconstituents binding free energy against SARS-CoV-2 Mpro (6LU7)

Ligands	Binding affinity (kcal/mol)
$C_{28}H_{43}NO_6$	– 11.5
$C_{28}H_{25}N$	– 8.5
$C_{35}H_{48}O_3$	– 8.1
$C_{33}H_{47}ClO_7$	– 8
$C_{20}H_{28}O_6$	– 7.9
$C_{24}H_{34}O_6$	– 7.9
$C_{30}H_{40}O_6$	– 7.8
$C_{22}H_{20}O_S$	– 7.8
$C_{15}H_{28}O$	– 7.8
$C_{24}H_{24}$	– 7.6
$C_{19}H_{13}BrClNO_3$	– 7.4
$C_{27}H_{36}O_8$	– 7.4
$C_{16}H_{11}ClN_{2O}$	– 7.3
$C_{19}H_{24}0_6$	– 7.3
$C_{22}H_{22}N2O_2S$	– 7.3
$C_{23}H_{32}O_5$	– 7.2
$C_{15}H_{24}O$	– 7.1
$C_{23}H_{32}O$	– 7.1
$C_{25}H_{36}O_5$ $C_{25}H_{34}O_7$	– 7.1
$C_{15}H_{26}O$	– 7
$C_{18}H_{16}ClNO_3$	– 7
$C_{26}H_{44}O_5$	– 7

3.12 Micropropagation Approaches

The approach of micropropagation is indicated in Fig. 3.7 which includes multiplication by axillary buds/apical shoots (by multiplying them in an induced in vitro technique to develop a plant from the meristem, buds, and shoot tip culture), adventitious shoots, organogenesis, somatic embryogenesis.

Embryo rescue, an in vitro culture technique, can be used to protect weak, immature, and hybrid or viable embryos to prevent degradation. This method is used mainly to improve the excising weak immature embryos by culturing them on an artificial medium (Akmal 2021). The decrease in betel vine yield is contributed by various diseases, and biotic and abiotic stresses, and hence to overcome the crop loss, inter-specific and, inter-genic hybridization can be applied. Once the intergenic hybridization technique, the hybrid embryos are carefully removed and cultivated on the nutrient medium plant growth regulators (PGPs) along with specific temperature

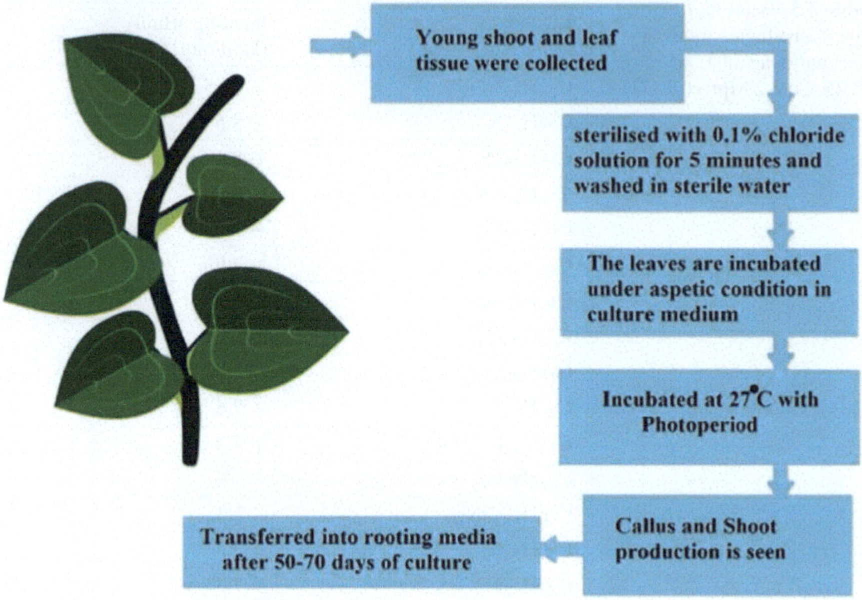

Fig. 3.7 Micropropagation of betel vine for callus and shoot growth

and light. In some cases, embryo degeneration is minimized by nurse endosperm for embryo transplant (Kumari and Thaneshwari 2018).

Babu et al. (1992) reported in the case of CV Lakkuvalli that regeneration can be obtained through the intermediary callus phase or directly from the tissue explants from the shoot, leaf, and root. In addition to the type of explant used, the hormonal role is also essential. Plant growth hormones like IBA, kinetin, NAA, etc., play a vital role as a supplement for somatic embryogenesis and regeneration. Callus formation was seen in shoot tip explants grown in MS medium supplemented with NAA 3 mg 1-1, BAP 0.05 mg I-I. In another study by Patil et al. (2013), kinetin and IAA were used in the amount of 0.9 and 0.5 mg with sucrose 20 g for the growth of embryonic tissue (cotyledonary). The complete plantlet was developed in 9–10 months. The regeneration was also reported in shoot buds by using internodes and nodes. However, when the same condition was used for root and leaf explants by betel vine, organogenesis was not observed. Another report by Kulkarni et al. (2007) suggests that the protocol for the regeneration of betel leaf from callus tissue is already developed, and the plantlets derived by such a method of tissue culture method were potted for hardening successfully.

Synthetic seeds commonly known as artificial seeds were produced by encapsulating of plant tissue (Qahtan et al. 2019). The method is helpful for seedless plants to propagate and multiply genotypes of commercially essential and medicinal plants. The plant propagules can be cultivated in an in vitro culture containing a semi-solid medium or planting regulators/substrates. Synthetic seeds can be produced using somatic embryos or other plant materials. In a Betel vine, a vegetative part of plants is used for the development of artificial seeds. Primarily the plant propagate is desiccated using polyoxyethylene (Polyox) for disinfecting microorganisms, and by using nutrient media, somatic embryo or plant shoots is grown, for encapsulation using Ca-alginate. These synseeds may be directly grown in the soil/ stored for long term & short term for plant formation. The production of synthetic seeds of betel vine is still to be explored, but this technique can have potential applications for propagation.

3.13 Genetic Engineering

To combat rapidly changing climatic conditions and increase production to meet the demands of an ever-increasing population, there is a need for advanced techniques for developing highly resilient varieties. When it comes to genetic engineering and gene editing, biotechnology has a wide range of applications as it enables limitless gene exchange between various species, leading to the development of noble and superior varieties of plants. Despite its widespread application, genetic engineering has social, ethical and biosafety concerns.

Even though it has medicinal qualities like anti-inflammatory, antioxidant, anti-cancer, and ant-microbial properties, betel leaf is primarily used for chewing in India. Betel vine is a commercially viable and economically important plant with many applications. As a result, betel leaf oil extract is an effective Ayurveda medicine. Biotechnological tools can be used to understand genetic diversity, and enhance the production of betel vine. Genetically engineered crops have particular agronomic traits to address the needs of growers and consumers. There are numerous studies conducted on betel vine phytochemical composition to study its various attributes that have therapeutic and economic values. However, research in transgenic betel vine production is limited and there is a great scope for quality improvement. Growing and analyzing large populations of crops over several generations is a common practice in conventional breeding, which is a time-consuming and labor-intensive (Dong and Ronald 2019). Compared to crops that rely on sexual reproduction, vegetatively propagated crops that use genetic engineering may be quickly developed into better varieties. For fruit trees, the vegetative propagation method makes it possible for engineered cultivars to cover a more significant area. Transgenic cultivars include tomato and pepper, which are resistant to cucumber mosaic virus, squash which is resistant to potyviruses and papaya, which is resistant to the papaya ringspot virus. Most transgenic crops were developed to increase agronomic productivity by conferring pest or disease resistance or delayed ripening (Nayaka et al. 2021).

3.14 Mutation Breeding

The mutation is the sudden change in genetic material, defined as the alteration mainly induced at the gene level, i.e., DNA or chromosome. Both the chemical and physical mutagens can be used as they can change the genotype and phenotype of the plant. Physical mutagens, for example, ultraviolet light, X-rays, and neutrons, are the most common in plants (Koornneef et al. 2002). On the other hand, nitrous acid, ethyl methane sulfonate and diethyl sulfate are examples of chemical mutagens commonly used for inducing variations. The report of Till et al., 2004 suggests that the majority of the changes (70–99%) in EMS-mutated populations are the transition of GC to AT (Till et al. 2004). In the case of plants, the primary purpose of mutation breeding is to increase quality and productivity to develop modified and improved varieties (Penna and Jain 2017). The plants are induced first and are selected for better quality. The current data and status of mutation breeding for crops imply that the betel vine is a type of underutilized crop that is consumed locally and traditionally, but, not exploited for mutation breeding. Mutation breeding will be a good method of choice for betel nut varieties plants with improved traits like disease resistance. The traits that can be targeted for improvement in betel vine are their high yield productivity, abiotic, and biotic stress resistance, nutrition and functionality, maturity, agronomy traits and traits involved in post-harvest of betel vine. In vitro mutagenesis can be used and applied for vegetative crops. After mutagenesis, the selection is done to screen the mutant population. In other crops, the technique has proved to be a powerful, and beneficial method for producing improved variety. Mutation breeding has played an essential role in crop improvement. Suprasanna et al. (2012) and Penna et al. (2023) highlighted its importance in the commercial cultivation of different crops.

3.15 Molecular Analysis

Molecular studies in betel vine are limited as their geographical distribution is vast. Most of the studies focus on biochemical and health studies. The molecular-based markers, and the biochemical studies in betel vine, can make the results more authentic in terms of betel vine diversity and other related studies. In the case of betel vine, PCR based marker, namely RAPD, has proved to be significant tool to differentiate Bangla, and Kapoori varieties at genetic level (Ranade et al. 2002). Heterozygosis diversity study has proved that Kapoori cultivars are more heterogeneous than Bangla cultivars. The number of bands produced in the gel was more than 70% in both types. In addition to RAPD, ISSR markers were useful in comparing and establishing the genetic identity and diversity among 15 betel vine cultivars. In Balipana and Birkoli landraces of betel vine maximum similarity was observed; however, the minimum similarity was, between Banglamandesore Chitalpudi and Halisahar Sanchi landraces (Patra et al. 2011).

Sex determination in betel vine is another type of study been done through molecular tools. Sequence Characterized Amplified Regions (SCAR) are an example of advanced sex-specific markers. Its role has been reported in a study to determine the sex type of betel vine and designing breeding programs (Khadke et al. 2012). A sex-related genetic survey has not been conducted before, only the work of Khadke et al. (2012) and it is possible to differentiate the sex in betel vine. The sex type is correlated to the amounts and level of phytochemicals and secondary metabolites found in a betel vine type. For instance, more phenols and antioxidants are observed in females than in males, as reported (Bajpai et al. 2012).

Improved varieties of betel vine have been developed in India. SGM-1 and DPB-6 types are examples that grown in southern States and Maharashtra state. Bidhan pan is another high-quality variety grown in West Bengal and Orissa. Some of the varieties which are cultivated by include, Bangla, Meetha, Sanchi, Karpoori, Kashi, Tellaku, Mahai, Kariyale, Deshawari, etc., by India; Ratadalu, Gelathoda, Kahaneru, etc., by Sri Lanka; Sireh China and Sireh by Malaysia and Sireh Hitam, Sireh Buah, etc., by Indonesia.

3.16 Hybridization Techniques

Betel vine can also be propagated vegetatively, but the variation among the newly produced clones is minimal. The characteristics of following-generation plants are similar to their first-generation parents, with little or no variation. Betel vine is propagated through stem cutting and conventionally can be hybridized to get the hybridized progenies. Samantaray et al., 2012 reported the flowering process in male and female vines along with fruit setting in Bangalore. It is suggested that heterosis is fixed and exploited easily in the case of betel vine. Hybrids of betel vine were formed, and under the experimental condition of the "All India Network Research Project" on betel vine, four hybrids, namely GN, GB, PI, and P2, were found to be developed. Out of all, GN is under a multi-location testing level, as per the report of Samantaray et al. (2012). In India, betel vine is cultivated as a cash crop at different locations. Indian varieties are in great demand in foreign countries; the supply has to be supplemented with hybridization and cultivation. Hence, the cultivation of betel leaves is exceeding and has taken its place globally. The report suggests that a value of Rs. 26.8 crores was earned for the year 2017–18 by betel leaf export to other countries. The maximum contribution was from Odisha, with a value of Rs. 10–12 crores, including both domestic and international export.

3.17 Conventional Breeding and Cultivation

The conventional method of betel vine breeding is essential for its large-scale production. The crop is grown in soil with support as betel leaf is a creeper plant. The raw leaf is consumed by people; hence there should not be any chemical or any other type of contamination in the product. The leaf should be free of chemicals, and environment-loving substances like mustard cake, or biofertilizers can be used for proper growth, and yield of betel vine for a contamination- free crop. Hybridization can provide the best quality plant, for which the Indian Institute of Horticultural Research (IIHR), Bangalore, has initiated a program on hybridization for the effective development of betel vine plants. Hybridization becomes advanced by using modern biotechnological tools helping in enhancing betel vine quality. All the advanced techniques of hybridization are genetic hybridization method dependent. The technology is applied to betel vine, and centers for betel improvement have been opened, and work has been initiated for quality improvement. Stem cutting is preferable to any other explant for betel vine cultivation. The best climate for proper growth is high rainfall (Ideal rainfall: 200–450 cm), soil type is clay loam, the best season is November to December, and January to February and place under the shade with a temperature requirement of (10–40 °C). The cultivation method is basically of two types, i.e. open and closed systems. In the open system, some support is used for the proper creeping of the plant, and in the closed system, artificial rectangular structures known as "Boroj" are made. After proper "Boroj" construction, which provides adequate shade and external environmental conditions, betel vines are planted. Proper irrigation is needed immediately after the planting irrigation has to be done every week. Manuring is preferable for high yield and healthy growth of vines. The report of Dutta et al. (2016) suggests that the application of manure (about 150 kg N/ha/year and 100 kg P_2O_5/ha/year) provides healthy growth. Neem cake and urea are generally used for nutrition supply. Under optimum conditions, the vine grows at a height of 3 m/year periods, afterward, lowering is done, and during each lowering, irrigation is needed for further growth. Depending upon the growth pattern and customer requirement, the leaves are harvested.

3.18 Hybrid Cultivars

Betel hybridization is complex and possesses some limitations. It has been assumed that the process of hybridization is betel vine (homozygous plant) is lacking due to its unknown meiotic segregation nature. The improvement of the betel vine is possible through genetic hybridization. This leads to the production and growth of the huge betel vine hybrid population and its selection. Studies like karyomorphology, which includes cell study of parents at the meiotic level, might help in exploring the variation among the hybrid progenies of betel leaf. For improving the betel vine cultivar, attempts have been made to cross pure landraces under the control conditions. The work of Maiti and Satya (2014) has shown that some attempts to cross different landraces under the controlled conditions of hybridization, some resulting in viable seed sets.

3.19 Conclusions and Prospects

Betel vine, the most loved and traditional crop of many people worldwide. Betel leaf (paan) chewing has become so prevalent over centuries, that it has become the symbol of the offering on various occasions, mainly in India. The ethnobotanical property of betel leaf has increased its production status all over the world. Betel leaf possesses pharmacological properties, antioxidant, anti-diabetic, anticancer, etc., activities, because of the presence of high phenolic content. The crop has so much potential in day-to-day life. However, the plant has not been explored for breeding and large scale production suggesting that there is a need for extensive investigations for such unexplored health-beneficial plants. New varieties with improved properties must be developed through biotechnological techniques. The role of molecular markers can be used for diversity analysis and sex differentiation among plants. Betel leaf has tremendous scope for research and development. Some of the important areas in which detail study is needed and betel leaf has a significant future prospects are presented in Fig. 3.8.

Modern techniques of HPLC and gas chromatography can be used for exploring compositional analysis among different landraces. The characterization and authentication can be of the landraces can be done by using various genetic markers. The

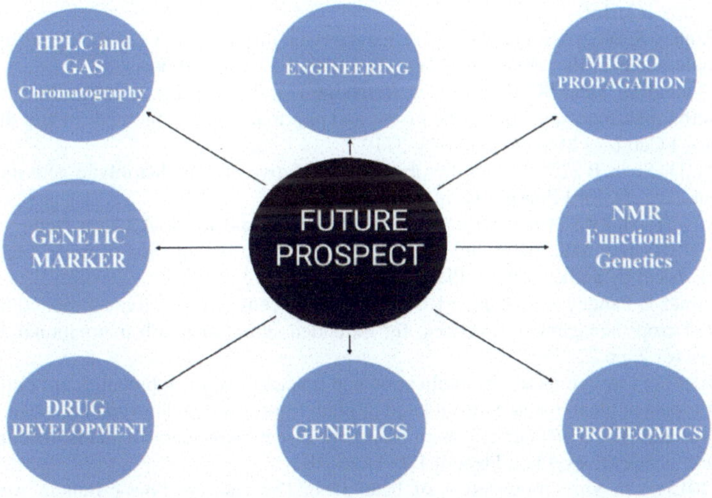

Fig. 3.8 Future prospects and utilization of betel vine

therapeutic and pharmacological properties can be used in drug development after proper characterization and trials. The scope of betel is not limited to characterization and drug development but, can be improved in various aspects for proper growth and cultivation, and increased shelf life and better marketing. Other advanced biotechnological techniques can be useful to discover novel compounds with active bioactivities. The increased demand for betel leaf consumption can be fulfilled by micropropagation and higher biomass varieties. It is required to standardize the quality and quantity of betel vine under various environmental conditions. The products need to be developed on a large scale for the growth and development across different regions, considering all the aspects of betel leaf cultivation.

References

Aboul-Maaty NAF, Oraby HS (2019) Extraction of high-quality genomic DNA from different plant orders applying a modified CTAB-based method. Bull Natl Res Cent 43:25

Adamu A, Aliyu H (2007) Morphogical effects of sodium azide on tomato (*Lycopersicon esculentum* Mill). Sci World J 2(4):9–12

Ahuja SC, Ahuja U (2011) Betel leaf and betel nut in India: history and uses. Asian Agrihist 15(1):13–35

Akmal M (2021) Embryo culture and embryo rescue in Brassica. In: Islam AKMA, Hossain MA, Islam AKMM (ed) Brassica breeding and biotechnology. IntechOpen, London

Ali MN, Ray SS (2018) Plant tissue culture as potential option in developing climate resilient spices. In: Sharangi AB (ed) Indian spices: the legacy, production and processing of India's treasured export. Springer, Switzerland

Arumuganathan K, Earle ED (1991) Nuclear DNA content of some important plant species. Plant Mol Biol Rep 9:208–218

Babu KN, Rema J, Pillai SG et al (1992) Micropropagation of betel vine (*Piper betle* L.). J Spices Arom Crops 1(2):160–162

Bajpai V, Pandey R, Negi MPS et al (2012) Characteristic differences in metabolite profile in male and female plants of dioecious *Piper betle* L. J Biosci 37:1061–1066

Biswas P, Anand U, Saha SC et al (2022) Betelvine (*Piper betle* L.): a comprehensive insight into its ethnopharmacology, phytochemistry, and pharmacological, biomedical and therapeutic attributes. J Cell Mol Med 26(11):3083–3119

Chakraborty D, Shah B (2011) Antimicrobial, antioxidative and antihemolytic activity of Piper betel leaf extracts. Int J Pharm Pharm 3(3):192–199

Chattapdayay SP, Maity S (1967) Diseases of Betelvine and species. ICAR New Delhi

Chaturvedi R, Razdan MK, Bhojwani SS (2003) Production of haploids of neem (*Azadirachta indica A. Juss.*) by anther culture. Plant Cell Rep 21:531–537

Das SN, Kumar P, Pandey AK, Ranjan RD (2016) Enhancement of betelvine production through integrated crop management practices. In: Extended summaries: 4th international agronomy congress, vol 2, pp 22–26

Dolezel J (1997) Flow cytometry, its application and potential for plant breeding. In: Lelley T (ed) Current topics in plant cytogenetics related to plant improvement. Universitätsverlag, Vienna

Dong OX, Ronald PC (2019) Genetic engineering for disease resistance in plants: recent progress and future perspectives. Plant Physiol 180(1):26–38

Elaih M (2015)) In vitro propagation of betel vine. Dissertation, Sher-e-Bangla Agricultural University, Bangladesh

Elamathi M (2016) *In vitro* micropropagation of *Piper betel* L. Int J Curr App Sci 3:44–49

Espinosa-Leal CA, Puente-Garza CA, García-Lara S (2018) In vitro plant tissue culture: means for production of biological active compounds. Planta 248:1–18

Farmer's Feedback on Betelvine Hybrid received. https://icar.org.in/content/farmer's-feedback-betelvine-hybrid-received. Accessed 20 Jul 2022

Guha P, Jain RK (1997) Status report on production, processing and marketing of betel leaf (*Piper betle* L.). Agricultural and Food Engineering Department, IIT Kharagpur, India, pp 15–22

Guha P (2006) Betel leaf: the neglected green gold of India. Hom Ecol 19(2):87–93

Guha P (2007) Extraction of essential oil: An appropriate RuralTech-nology for minimizing wastage of surplus betel leaves. AMA-AGR MECH ASIA AF, 38(4)

Guha P, Nandi S (2019) Essential oil of betel leaf (*Piper betle* L.): a novel addition to the world food sector. Essential oil research: trends in biosynthesis, analytics, industrial applications and biotechnological production, pp 149–196

Huang S, Li R, Zhang Z et al (2009) The genome of the cucumber, *Cucumis sativus* L. Natgenet 41(12):1275–1281

Islam MJ, Nath TK (2014) Forest-based betel leaf and betel nut farming of the Khasia indigenous people in Bangladesh: approach to biodiversity conservation in Lawachara National Park (LNP). J for Res 25(2):419–427

Jose J, Sharma AK (1988) Karyomorphological studies on *Piper betle* L. Acta Bot Indica 16:186–194

Kaintura P, Bhandari M, Kumar R (2020) Medicinal values of betel leaves and its application in food products: a review. J Pharm Innov 9(6):344–348

Kavi Kishor PB, Pullaiah T, Ranga Rao A, Anabela R, Rathnagiri P, Suprasanna P (2024) Plants to pharmacy: recapitulation of natural compounds transmuting human health. In: Kavi Kishor PB, Pullaiah T, Suprasanna P, Rao AR, Romano A (eds) Biotechnological production of bioactive phytochemicals of medicinal value. Elsevier, pp 1–16

Khadke GN, Hima Bindu K, Ravishankar KV (2012) Development of SCAR marker for sex determination in dioecious betelvine (*Piper betle* L.). Curr Sci 103(6):712–716

Koornneef M, Bentsink L, Hilhorst H (2002) Seed dormancy and germination. Curr Opin Plant Biol 5(1):33–36

Kulkarni AA, Kelkar SM, Watve MG, Krishnamurthy KV (2007) Characterization and control of endophytic bacterial contaminants in in vitro cultures of *Piper* spp., *Taxus baccata* subsp. wallichiana, and *Withania somnifera*. Can J Microbiol 53(1):63–74

Kumari P, Thaneshwari R (2018) Embryo rescue in horticulture crops. Int J Curr Microbiol Appl Sci 7(6):3350–3358

Mahfuza EJ, Sayem SM, Hassan MF (2020) Trend of betel leaf production in Bangladesh: prospects and challenges. Res Agric Livest Fish 7(2):209–223

Maity S (1989) Extension bulletin: the betelvine. All India Coordinated Research Project on betel vine. Indian Institute of Horticultural Research, Hessarghatta, Bangalore, India. Metabolism, pp 885–906

Maiti S, Biswas SR (1991) Sexual dimorphism in betelvine. Ind J Agric Sci 61:656–658

Maiti RK, Satya P (2014) Research advances in major cereal crops for adaptation to abiotic stresses. GM Crops Food 5(4):259–279

Mridha MRH, Jahan H, Murshed MM (2005) Relative profitability of different aged betel leaf production at farm level in some selected areas of Bangladesh. J Bangladesh Agric Univ 3:81–188

Nayaka NMDM, Sasadara MMV, Sanjaya DA et al (2021) *Piper betle* (L): recent review of antibacterial and antifungal properties, safety profiles, and commercial applications. Molecules 26(8):2321

Patra AP, Mukherjee AK, Acharya L (2011) Comparative study of RAPD and ISSR markers to assess the genetic diversity of betelvine (*Piper betle* L.) in Orissa. Am J Biochem Mol Biol 1(2):200–211

Patil JG, Ahire ML, Nitnaware KM et al (2013) In vitro propagation and production of cardiotonic glycosides in shoot cultures of *Digitalis purpurea* L. by elicitation and precursor feeding. Appl Microbiol Biotechnol 97:2379–2393

Paul H (2021) Problems and prospects of betel leaf farming: a case study of Alipurduar District, West Bengal. Int J Creat Res Thoughts 9(3):5300–5305

Pawar S, Bhane P, Jha U (2021) Review on betel pepper leaf for anti inflammatory activity. World J Pharm Res 11(2):1196–1208

Penna S, Jain SM (2017) Mutant resources and mutagenomics in crop plants. Emirates J Food Agric 29:651–657

Penna S, Shirani Bidabadi S, Jain SM (2023) Mutation breeding to promote sustainable agriculture and food security in the era of climate change. In: Penna S, Jain SM (eds) Mutation breeding for sustainable food production and climate resilience. Springer, Singapore, pp 1–23. https://doi.org/10.1007/978-981-16-9720-3_1

Phurailatpam AK, Geetha KA, Maiti S (2018) Ploidy distinction in male and female plants of betelvine (*Piper betle* L.): a study by flow cytometry. Genet Resour Crop Evol 65:1565–1570

Qahtan AA, Abdel-Salam EM, Alatar AA et al (2019) An introduction to synthetic seeds: production, techniques, and applications. Synthetic seeds: germplasm regeneration, preservation and prospects, pp 1–20

Ranade SA, Verma A, Gupta M et al (2002) RAPD profile analysis of betel vine cultivars. Biol Plant 45:523–527

Roy A, Guha P (2021) Traditional and functional uses of betel leaf (*Piper betle* L.) pertaining to food sector: a review. J Postharv Technol 9(1):72–85

Sahoo M, Sahoo DR (2017) Betel leaf cultivation in Odisha: problems and prospects. Asian J Soc Sci 7(1):10–15

Samantaray S, Phurailatpam A, Bishoyi AK (2012) Identification of sex-specific DNA markers in betel vine (*Piper betle* L.). Genet Resour Crop Evol 59:645–653

Sekar V, Ramasamy G, Ravikumar C (2022) *In silico* molecular docking for assessing anti-fungal competency of hydroxychavicol, a phenolic compound of betel leaf (*Piper betle* L.) against COVID-19 associated maiming mycotic infections. Drug Dev Ind Pharm 48(5):169–188

Shah SK, Garg G, Jhade D, Patel N (2016) *Piper betle*: phytochemical, pharmacological and nutritional value in health management. Int J Pharm Sci Rev Res 38(2):181–189

Suprasanna P, Jain SM, Ochatt SJ et al (2012) Applications of in vitro techniques in mutation breeding of vegetatively propagated crops. In: Shu QY, Forster BP, Nakagawa H (eds) Plant mutation breeding and biotechnology. CABI Publishing, Wallingford, England

Till BJ, Reynolds SH, Weil C et al (2004) Discovery of induced point mutations in maize genes by tilling. BMC Plant Biol 4:12

Tirkey A, Ramtake V, Porte SS et al (2019) Stability, correlation and path coefficient analysis for yield and quality traits in betelvine (*Piper betle* L.) genotypes under three different sets of conditions. Indian J Genet Plant Breed 79(02):474–484

Ullah A, Munir S, Badshah SL, Khan N, Ghani L, Poulson BG, Emwas A-H, Jaremko M (2020) Important flavonoids and their role as a therapeutic agent. Molecules 25(22):5243

Verma A, Kumar N, Ranade SA (2004) Genetic diversity amongst landraces of a dioecious vegetatively propagated plant, betelvine (*Piper betle* L.). J Biosci 29:319–328

Virk PS, Ford-Lloyd BV, Jackson MT et al (1995) Use of RAPD for the study of diversity within plant germplasm collections. Heredity 74:170–179

Werren JH, Richards S, Desjardins CA et al (2010) Functional and evolutionary insights from the genomes of three parasitoid Nasonia species. Science 327(5963):343–348

Chapter 4
Breeding Strategies for Improved Production of *Blond plantain (Plantago ovata)*

Shilpa Susan Scaria, Jissa Theresa Kurian, Vijayalaxmi S. Dandin, and Kadanthottu Sebastian Joseph

Abstract *Plantago ovata* is an economically significant medicinal plant that is widely grown in many regions of India, Pakistan, Iran, and some areas of Europe. It has a long antiquity of conventional applications and has therapeutic qualities. Seed husk has a variety of purposes in both health and industry, as well as commercial products. The seed husk, also known as psyllium or isabgol, has nutritional benefits, and is used to treat and prevent gastrointestinal and bowel disorders. *P. ovata* is equipped with a range of pharmacological actions. Both allopathic and ayurvedic therapies benefit from the use of *P. ovata*. Due to its medicinal qualities, such as mucilage, gelling agent, super disintegrant, and suspending agent, as well as its pharmacological activities, such as anti-constipation, anti-diarrheal, wound healer, hypocholestrolemic, and hypoglycaemia, it is frequently employed in numerous medicines. As a result, *P. ovata* can be used in the production of a variety of pharmaceutical goods and as a reliable and effective ethnobotanical treatment for a range of medical conditions. Effective conservation strategies are crucial to safeguard this valuable genetic resource. Ex-situ approaches like seed and field gene banks are well-suited for Isabgol, while cryopreservation and DNA banks offer long-term storage and information archives for future research and restoration efforts. Furthermore, marker-assisted breeding with tools like newly developed SSR markers can propel breeding programs. These tools can improve Isabgol's growth rate, seed yield, and stress tolerance. Through the integration of sophisticated breeding techniques, efficient conservation methods, and an understanding of genetic variety, we

S. S. Scaria · J. T. Kurian · K. S. Joseph (✉)
Department of Life Sciences, Christ University, Bengaluru, Karnataka, India
e-mail: joseph.ks@christuniversity.in

S. S. Scaria
e-mail: shilpasusan.scaria@res.christuniversity.in

J. T. Kurian
e-mail: jissa.kurian@res.christuniversity.in

V. S. Dandin
JSS RSH PU College, Vidyagiri, Dharwad, Karnataka, India

J. M. Al-Khayri et al. (eds.), *Biodiversity and Genetic Improvement of Medicinal and Aromatic Plants I*, Advances in Plant Breeding Strategies 9,
https://doi.org/10.1007/978-3-031-81846-2_4

can fully realise the promise of Isabgol and secure its continued use as a priceless medicinal resource.

Keywords Breeding · *Plantago* · Blond plantain · Ispaghula · Dietary fibre

4.1 Introduction

Plantago ovata, also known as Isabgol (referencing the seed's "horse ear" shape), is a prominent medicinal herb native to Asia, the Mediterranean, and North Africa. It is widely cultivated, with India leading production (Talukder et al. 2016). Ispaghula, a water-soluble dietary fibre derived from *P. ovata* seed husks, is a popular over-the-counter medication for bowel control and intestinal cleansing (Gilani et al. 1998). While some sources differentiate Ispaghula from Psyllium, commercially available supplements typically come from *P. ovata*, making the distinction less relevant (Fernandez et al. 2012).

Blond plantain is a plant with a high concentration of water-soluble fibre that has a lot of health benefits, has medicinal qualities, and has advantageous nutritional effects. Ispaghula is a seed that makes up 10 to 30% of the total seeds and is primarily made up of polysaccharides with strong gel-forming abilities. In addition, it contains tannins, 4-O-methylglucuronic acid, oleic, aucubin, palmitic, campesterol, linoleic, and asparagine acids as well as mucilage, rhamnose, sterol, and β-sitosterol. According to the structural and chemical elucidation studies, the neutral arabinoxylan that makes up the gel-forming polysaccharides contains 22.6% arabinose, 74.6% xylose, and a few trace sugars (Fischer et al. 2004). Because of its accessible availability, hydrophilicity, biocompatibility, biodegradability, bio renewability, and low toxicity, ispaghula polysaccharides have been used in numerous studies to show their promise for hydrogel applications (Thakur and Thakur 2014). Research conducted have shown scientific proof that ispaghula is effective in treating several human disorders, including hypercholesterolemia, inflammatory bowel diseases, constipation, and diarrhoea (Hussain et al. 2016; Shah et al. 2020).

In the traditional Indian medical system (Ayurveda), psyllium is frequently used to prevent skin irritations, haemorrhoids, constipation, and diarrhoea. In recent years, its usage in Europe has expanded, and it is now widely consumed in the USA as well (Fradinho et al. 2015). India dominates the global commercial market for psyllium production and export. The two states that produce the most psyllium in India, Gujarat, and Rajasthan, are where most of the crop is grown. Gujarat is the main state in terms of further processing and manufacture of husk, accounting for 35% of global psyllium production.

By mechanically milling, or grinding the outer coating of the seed, mucilage is produced from the psyllium (*P. ovata*) seed coat. The term "mucilage" refers to a class of transparent, colourless gelling agents that are generated from plants. Blond plantain mucilage has a long tradition as a dietary supplement since it is regarded as a rich source of soluble and insoluble fibres and is described as a naturally occurring polysaccharide that forms a gel. It is a mucilage that gels and is well-known for having a laxative effect because of its large water-holding capacity, which is roughly 80 times its weight (Yadav et al. 2016).

Numerous studies on psyllium's possible health advantages and uses in the food and pharmaceutical industries have been conducted. Constipation, symptoms of irritable bowel syndrome, abdominal discomfort, hypercholesterolemia, cancer prevention, diarrhoea, diabetes, inflammatory bowel disease-ulcerative colitis, and obesity are all conditions that are well-known to be helped by psyllium. Additionally, it has prebiotic, hypocholesterolemic, and satiety-promoting properties (Yadav et al. 2016). Due to its potent hydrophilic and gelling qualities, stabilising, suspending, and emulsifying abilities, psyllium seed husk, an established source for the synthesis of blond plantain hydrocolloid, has additional uses in food in addition to its advantageous health effects. In addition to its beneficial health effects, psyllium seed husk, a well-known source for the manufacture of psyllium hydrocolloid, has additional uses in food due to its potent hydrophilic and gelling properties, stabilising, suspending, and emulsifying capacities. Since gums have a large molecular weight, they should be effective viscosity enhancers. Supplements and food product development have highlighted the benefits of psyllium fibre in developing low-calorie, low-fat foods as well as serving as a gluten alternative in everyday foods like yoghurt and biscuits (Ladjevardi et al. 2015). Psyllium is also used extensively in the textile, paper, printing ink, wax, and army sectors, as well as in cosmetics (for the provision of hair and skin care products).

Isabgol is emerging as a critical crop due to its rising global demand and economic significance. The United States, a major importer consuming 8000 metric tons annually, exemplifies the growing market for Isabgol seeds and husk. This surge in interest is attributed to the increasing popularity of natural dietary fibres. Notably, India, the leading producer, exports roughly 90% of its Isabgol crop to developed nations like the USA and Western Europe, reflecting the international trade value of this crop. The objective of this book chapter is to assess the available scientific information about the primary biological properties, health advantages, and discusses how this fibre is used in the food business and as a medication.

4.1.1 Origin and Distribution

The genus *Plantago* is one of the three largest genera in the Plantaginaceae family, with over 250 species worldwide. Only a few *Plantago* species are found in the tropics, where they are mostly found in temperate zones (between 26 and 36°N, latitude). *Plantago* species that form rosettes are classified as Plantago, while stem-forming *Plantago* species are classified as Psyllium. *Plantago ovata* is an annual herb cultivated in India in the states of Gujarat, Punjab, Maharashtra, Rajasthan; also in West Pakistan, Southern Europe, and North Africa. India, Iraq, Iran, Spain, and the Canary Islands are the places where *P. ovata* originated before spreading across

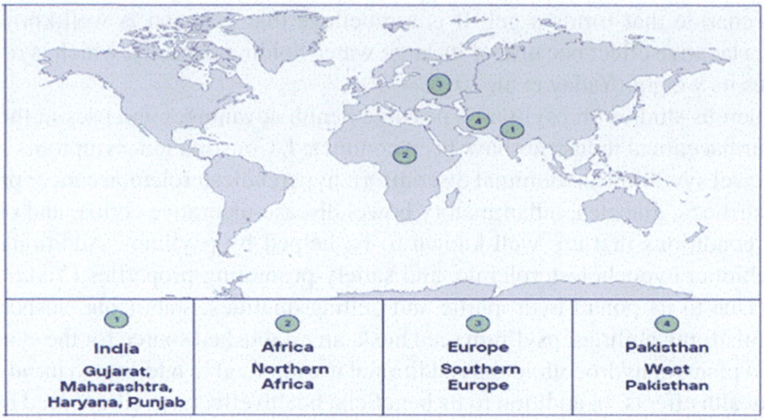

Fig. 4.1 Major regions for the cultivation of *Plantago ovata*

the Southern Mediterranean and Eastern Asia. Commercial production takes place in Pakistan, India, and a number of European nations (Fig. 4.1).

In terms of blond plantain exports and production, India controls the global market. *P. ovata* is mostly grown in North Gujarat, India, as a "Rabi" crop, or post-rainy season crop. A dry, open area is one of this crop's key environmental requirements. Most of the psyllium studies and field tests in the United States have taken place in Arizona and Washington. The mature crop's tendency to shatter is a significant cultural issue that lower Midwest farmers are trying to overcome in order to increase psyllium yield. Crossbreeding high-yielding Indian varieties with more shatter-tolerant hybrids has had some degree of success. The *Plantago* species was brought by Muslims as a medicinal herb, yet it is native to Pakistan and India. As a result, Pakistani cities were the first to introduce it to cultivation. As the need for seed grew, cultivation began; it was initially limited to Lahore and Multan (now Pakistan). India currently produces the most *Plantago* seeds and husks and is the primary provider to the global market; around 90% of the production is distributed (Dagar et al. 2006). At first, it was utilised by residents of Asian nations as a treatment for chronic dysentery and other gastrointestinal issues. Previous research revealed that isabgol has a long history of medicinal applications and was domesticated ages ago by the people for the treatment of illnesses.

India exports about 93% of its total ispaghula husk exports, or around 90% of its total ispaghula gross production. Psyllium husks and industrial powders are shipped to nations like the United States, the United Kingdom, Italy, Australia, Denmark, Pakistan, Indonesia, France, Spain, Germany, Canada, Mexico, Sweden, Korea, Norway, and several Gulf nations. About 75% of all husk exports from India go to the USA, which is also the country's biggest customer of ispaghula.

4.1.2 Botanical Description

The seed from *P. ovata* is referred to as white or blond psyllium, Indian Plantago, or Isabgol in trade circles. The common name for *P. ovata* in India is derived from the Persian terms "isap" and "ghol," which indicate "horse ear" and are descriptive of the seed's form (boat-shaped seeds). The name Ovata refers to the shape of the leaves, whilst *Plantago* is a Latin word that meaning "sole of the foot." Psyllium is a Greek term that alludes to the colour, shape, and size of the seeds. It also means "flea." During the winter, *P. ovata* grows in semi-arid or desert areas when irrigation is accessible. Because the crop requires little soil fertility, it can thrive in sandy-loam to loamy soil with low fertility.

P. ovata is an annual herb that reaches a height of 30–45 cm and has short stems or no stems at all. The leaves are narrow (5–12.5 mm), linear, linear lanceolate or filiform, three-nerved, attenuated at the base, finely acuminate whole or distantly serrated, and covered in soft, fine hairs. From the plant's base, flowering stems emerge, and about 60 days after planting, white flowers with bracts 4 mm wide and measuring 12.5–37.5 mm in length begin to bloom (Fig. 4.2a). Additionally, the flower spikes of ripe psyllium turn a reddish-brown colour. Lower leaves are drying up at this point while top leaves are turning yellow. In spikes, there are 45–69 blooms. Flowers are protogynous, anemophilous, and tetramerous. Numerous, tiny, white flowers in ovoid or cylindrical spikes that are 1.3–3.8 cm long with broad, 4 mm long bracts are present. A free ovary with one or two cells and one or more ovules is attached to the corolla by four protruding stamens. The style is capillary and has a single subulate stigma at the end. The roots display a strong taproot and little fibrous secondary roots.

At maturity, the fruit, 8 mm long, tiny ellipsoidal capsule, opens. The fruit is protected by a persistent corolla. Seeds weigh around 1.5 g. It has up to 8 m (long) × 1 mm (wide), firm, boat-shaped, rosy white or rosy brown seeds (Fig. 4.2b). The seeds have a shiny, glossy exterior. The concave surface has a deep furrow with a hilum, which seems to be a red spot in the centre, while the convex surface has an oval mark in the centre. During milling, the husk, which makes up around 30% of the

Fig. 4.2 **a** *Plantago ovata* **b** seed

weight of the seed and is a tasteless, odourless, translucent, rosy-white mucilaginous membrane, is separated from the seed.

4.2 Cultivation and Traditional Breeding

4.2.1 Cultivation

Isabgol cultivation requires careful management to ensure a successful harvest. The process begins with thorough land preparation. The soil is meticulously harrowed and ploughed to achieve a fine tilth, free of weeds and clods. Beds are arranged for efficient irrigation, facilitating water delivery throughout the growing season. To enrich the soil, farmyard manure is incorporated during the final ploughing, typically at a rate of 10–15 tonnes per acre. Sowing typically occurs during the first week of November. Early sowing allows for better vegetative growth, while late sowing carries the risk of seed loss due to pre-monsoon rainfall coinciding with crop maturity. Healthy, disease-free seeds from the previous year's harvest are preferred for planting. The recommended seeding rate is 3–4 kg/ha, with rows spaced 15 cm apart. A higher seeding rate can increase susceptibility to downy mildew disease. Harvesting takes place approximately 110–130 days after sowing, typically in February or March. Signs of maturity include yellowing of the plant and browning of the spikes. Seeds are ready for harvest when they easily fall out upon gentle squeezing of the spikes. To minimize seed cracking, harvesting is conducted during dry conditions, preferably after 10 am when morning dew has evaporated. The crop is harvested manually, and the seeds are then processed through grinding mills to remove the valuable Isabgol husk. Yields can vary between 700 and 1000 kg per hectare, depending on factors such as cultivar selection, climatic conditions, soil quality, and the effectiveness of crop management practices.

4.2.2 Pest and Disease Management

The primary insect of isabgol crops is the aphid. After seeding, it often appears 50–60 days later. Neem-based formulations are used as a preventative strategy. The three main diseases in Isabgol are rhizoctonia wilt, powdery mildew, and downy mildew. The crop is particularly vulnerable to these diseases when irrigation, seeding rates, and nitrogen application exceed recommendations. In Isabgol manual weeding interventions are advised to be carried out two to three times during the early vegetative stage in order to reduce weed competition and maximise resource utilisation. After 3 weeks of seeding in the field, the first weeding is done.

4.2.3 Breeding

Due to the plant's economic significance, breeding for enhancement in a number of quantitative and qualitative traits must be done. Increased seed size and production, compact, non-shattering spikes, synchronous maturity, dwarf stature, seeds with higher swelling factors, and cultivars tolerant to drought and winter should all be among the breeding goals. Below are some of the improvements to psyllium that have been made.

Genotypic variation was noted in downy mildew index, husk percentage, and seed production. Compared to cultivars from Mehsana, Visnagar, and Sidhpur, those from Palanpur (Banaskantha) are superior. On the other hand, some researchers assert that Patan (Mehsana) strains are of a higher calibre than those grown elsewhere. Gujarat Agricultural University, Anand's Pilwai Research Station has created an enhanced cultivar called Gujarat Isabgol-1. It produces between 10 and 12% more seed than the native cultivars. In India, many exotic psyllium strains were introduced. From the Italian strain EC42706, three novel genotypes were recovered; EC42706-B, EC42706-1, and EC42706-2. The new genotypes were discovered to have more compact spikes, synchronous maturation, and dwarf plant size than previous local genotypes (Dhar et al. 2005).

4.3 Commercial Varieties and Seasonal Availability

The main Isabgol cultivars planted in India include JI-4, RI-87, HI-5, RI-89, GI-1, GI-2, GI-3, MIB-4, HI-34, HI-2, HI-1, AMB-2, and Niharika. A hectare of the Gujarat Isabgol-1 variety produces between 800 and 900 kg of seeds. One thousand kg of seeds per hectare might be produced by the new variety "Gujarat Isabgol-2". Table 4.1 comprises the high yielding varieties of *Plantago.*

Isabgol is a crop that is quite sensitive to its environment. During maturation, it needs a cool temperature and dry, sunny weather; even a slight dew, gloomy skies, or light rains shade the seeds. For seed germination, a temperature of 20–25 °C is needed, however for maturation, a temperature of 30–35 °C is needed. In the Indian states of Rajasthan, Madhya Pradesh, Gujarat, and Haryana, isabgol is cultivated during the Rabi season. The first week of November is the best time to sow. In the months of February and March, the crop is harvested. Two months after the crop is sown, blooming starts, and in February or March the crop is ready for harvest. The normal crop time for isabgol in Rajasthan and all of India is 110–130 days.

Table 4.1 High yielding varieties of Plantago

Variety	Source of availability	Yield (kg/hectare)
Jawahar isabgol (MIB 4)	MAP unit, College of Horticulture, RVSKVV, Mandsaur, Madhya Pradesh	1300–1500
Niharika	CIMAP, Lucknow, Uttar Pradesh	1000–1200
Haryana isabgol 5	MAP unit, CCS Haryana University of Agriculture, Hisar, Haryana	1000–1200
Gujarat isabgol 3 (GI 3)	Spices Research Station, Jagudan, S.D. Agriculture University, Sardarkhrushi Nagar, Gujarat	1300
Gujarat isabgol 2 (GI 2)	MAP unit, Anand Agricultural University, Anand, Gujarat	900–1000
Gujarat isabgol 1 (GI 1)	MAP unit, Anand Agricultural University, Anand, Gujarat	800–900

4.4 Mineral and Nutrient Composition

Plantago's flavonoids, phenols, alkaloids, phenolic chemicals, and terpenoids perform a variety of important pharmacological actions. It contains 6.5% tannin, anvrtyn, emulsions, and the glycoside aucubin. They also contain polysaccharides, fatty acids, and iridoid glycosides. The major element in Plantago is thought to be vitamin C, which has therapeutic benefits. It contains two to six percent mucilage, which is made up of four polysaccharides. Pectins, heterozeid, diastase, and colouring agents are all present. They contain carboxylic acids and salicylic acids in amounts greater than 1%. Minerals include potassium and zinc. They also include silicic acid and saponin. The seeds of this plant include glotinic elements as well as plantenolic acid. Siccinic acid is made up of the amino acids' adenine, choline, and aeocoeine.

A substantial component of the psyllium husk is composed of hemicellulose, which has a xylan backbone and also contains sections of arabinose, rhamnose, and galacturonic acid (arabinoxylans). Plantago species are very effective at producing a wide variety of secondary bioactive metabolites, including iridoids, phenols, polysaccharides, sterols, alkaloids, and cumatines, due to the seed's lignin, cellulose, and hemicellulose, which make up 35% of the soluble and 65% of the insoluble polysaccharides. Because of its propensity to form a gel in an aqueous solution, psyllium is classed as a mucilaginous fibre. This property stems from its role as the endosperm of the *P.* ovata seed, where it acts to retain water and keep the seed from drying out.

Seeds contain a variety of bases, carbohydrates, sterols, and proteins. Psyllium seeds have more than 30% hydrocolloidal polysaccharide (mucilage) in the external seed coat, as well as fixed oils, tannins, aucubin glycosides (iridoid), sugars, sterols, and proteins, and can be used to complement diets and treat human ailments. Rhamnose and galactose are added for good measure, although the majority of its ingredients are xylose, arabinose, and galacturonic acid. Psyllium husk contains 84.98% carbs, 0.94% ash, and 6.83% protein. An Osborne fractionation based on solubility

Table 4.2 Chemical composition of *Plantago ovata*

Nutrient composition	Leaves	Seeds
Crude fibre	6.8 ± 1.4	48.5 ± 9.8
Crude fats	–	43.2 ± 6.3
Crude proteins	21.87 ± 4.7	21.87 ± 4.7
Total proteins	9.3 ± 1.5	11.8 ± 0.9
Total ash	9.3 ± 2.3	8.6 ± 1.9
Total carbohydrate	15.9 ± 1.3	8.4 ± 2.5

yielded 35.8% albumin, 23.9% globulin, and 11.7% prolamin. Table 4.2 lists the chemical components of *P. ovata.*

Potassium was found to have the highest concentration of all the minerals examined (687 mg/100 g), while selenium had the lowest concentration (18 g/100 g). It was discovered that the contents of iron, zinc, copper, and manganese were 6.75, 3.15, 2.39, and 1.06 mg/100 g, respectively. The mineral content of psyllium seeds was reported by scientist as mg/kg for Fe (0.175), Mn (0.231), Zn (0.063), Cu (0.021), and K (0.105) (Gilani et al. 1998). The polysaccharide of *P. ovata* has a backbone chain of D-xylopyranosyl residues that are (1–4) connected. Although occasionally a xylopyranosyl side chain is seen at position 2, side chains (trisaccharide branches) are typically found at position 3 (Sagar et al. 2020). Table 4.3 provides an analysis of the carbohydrate profile of *P. ovata* seeds.

Among the water-soluble vitamins tested, niacin (4.18 mg/100 g) had the highest content, followed by total folates (32.19 g/100 g), pantothenic acid (0.87 mg/100 g), Vitamin B6 (0.25 mg/100 g), and riboflavin (0.13 mg/100 g). Tocotrienol and tocotrienol concentration were assessed to be 6.0 and 5.0 g/100 g in the vitamin E isomers studied, respectively. The overall carotenoids concentration was 183.86 g/ 100 mg, although carotenoids such as zeaxanthin, lutein, and lycopene have only been identified in trace levels (Sagar et al. 2020).

Psyllium seeds have higher amounts of ash (3%) and fat (6.7%) than other nutrients, including protein (17%), dietary fibre (25%), and total carbs (49%) according to some researchers (Romero-Baranzini et al. 2006). On the other hand, a different study found comparable dietary fibre content with high carbohydrate content (87.4%) and

Table 4.3 Carbohydrate composition of Plantago husk

Carbohydrates	Amount (%)
Arabinoxylan	60
Xylose	56.72
Arabinose	21.96
Galactose	3.76
Rhamnose	1.5
Glucose	0.64
Mannose	0.4

low protein (2.93%) and fat (0.3%) contents (Farahnaky et al. 2010). The different kinds, ambient conditions, and soil composition where they were cultivated may be responsible for the results' variability.

4.5 Health Benefits and Pharmacological Properties

Isabgol is a widely used medicinal drug for the treatment of hypercholesterolemia, colon cancer, diabetes, irritable bowel syndrome, ulcerative colitis, inflammatory bowel disease, constipation, and diarrhoea. Since psyllium husk is a mild natural laxative, it makes digestion easier. When submerged in water, it transforms into a gelatinous, sticky substance that absorbs water to carry out its usual functions. It is believed that psyllium husk's dietary fibre can operate as a bulking agent to aid with weight control and fat loss. Increasing satiety, lowering calorie intake, ingestion rate, and nutrient absorption are important considerations in this context.

4.5.1 Anti-Diarrheal and Anti-constipation Activity

P. ovata's antidiarrheal and anti-constipating properties were investigated, and it was found that muscarinic and 5-HT receptor activation enabled mice to experience laxative effects from a crude extract of *P. ovata* at concentrations of 100–300 mg/kg (Mehmood et al. 2011).

4.5.2 Anti-inflammatory Effects

When tested in HLA-B27 transgenic rats, psyllium was found to reduce inflammatory mediators connected to intestinal inflammatory mechanisms such as NO, leukotriene B4, and TNF, suggesting its usage as an intestine anti-inflammatory therapy (Hannan et al. 2006).

4.5.3 Effect on Autonomic Gastrointestinal Dysfunction

Psyllium husk can be used to treat Parkinson's disease that is accompanied with autonomic gastrointestinal dysfunction caused by anticholinergic medications. When used with levodopa/carbidopa, *P. ovata* husk improves levodopa pharmacokinetics, resulting in a more constant plasma concentration, reducing the 'wearing out' phenomena, and delaying the start of dyskinesia (Theuwissen and Mensink 2008).

4.5.4 Hypoglycaemic Activity

With the exception of inducing sucrose inflow into the large intestine, an aqueous extract of *P. ovata* husk has been shown to reduce postprandial glucose in blood and slow small intestinal absorption, implying that *P. ovata* may be an important source of active phytochemical components that may open up new treatment options for diabetes (Hannan et al. 2006).

4.5.5 Hypocholesterolemic Activity

Ispaghula (7 g/day for 6 months) reduced LDL cholesterol levels by 8.7% (4.1–0.42 mM/L) when combined with diet. Total cholesterol was reduced by 7.7–8.9%. *P. ovata* has been shown to significantly cut total cholesterol and LDL cholesterol in both rats and humans (Singh and Kumar 2008).

4.6 Application in Food Industry

Due to the related health benefits, the FDA has allowed the use of food items comprising psyllium husk. Due to the tasty flavour and improved storage stability, consumers like fibre supplements. For hundreds of years, traditional Iranian medicines have prescribed psyllium seeds. Foods enriched with *P. ovata* mucilage gum may have better customer acceptance due to its pharmacological properties. Consumer goods like high-fibre breakfast cereals frequently contain it. Psyllium supplements come in a variety of forms, including granules, powder, wafers, and capsules in addition to being a component of fibre formulations. With particular regard to baked goods, yoghurt, and beverages, functional and nutraceutical foods comprising fibre are being produced and evaluated for their efficacy (Perrigue et al. 2009). Psyllium husk is used to make a variety of bakery products, and research suggests that up to 50% of it can be substituted without negatively affecting the product's quality.

Due to the remarkable capacity of the mucilage found in seed husks to act as a thickening, the seed husk finds a range of uses in the food sector. In addition to being used as a fundamental stabiliser in ice cream, it is also a component of chocolate and other food items. In order to increase the food's bulk and increase its fibre content, psyllium can be used as a thickener, binder, or substitute in the food and beverage industry. Breakfast cereals, instant noodles, instant drinks, ice cream, bread, biscuits, and other bakery products are examples of these products. Psyllium husk can also be added to flavoured or fresh fruit drinks to improve the mouthfeel, richness, and consistency. Psyllium is used in the food and beverage sectors as a binder and stabiliser to improve softness and body texture as well as to provide

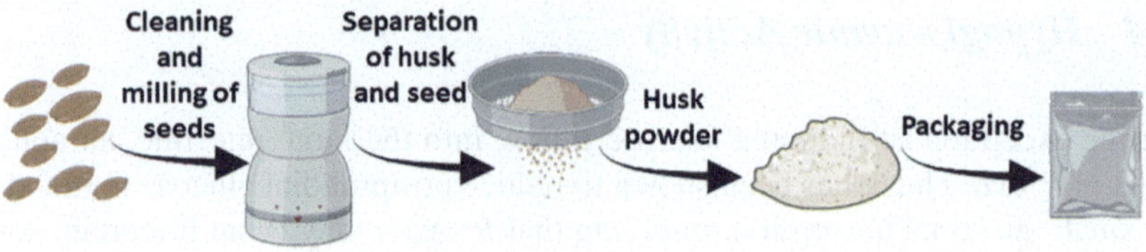

Fig. 4.3 Processing of *P. ovata* husk for marketable products (created with BioRender.com.)

strength (Turnbull and Thomas 1995). Processing of blond plantain is demonstrated in the Fig. 4.3.

Products containing psyllium can be a good substitute to increase functionality and create nutritious diets. The inclusion of psyllium in food products (such gluten-free goods, dairy, meat, and baked goods). Can provide the health claim for a product containing fibre, which is a significant competitive advantage. Additionally, it might make it possible for customers to consume enough fibre without upping their caloric consumption, as well as contribute to other health advantages including cholesterol control, glycaemic control, and satiety, among others. Psyllium is simple to incorporate into food recipes and has no discernible flavour altering effects. As a result, the use of psyllium in food items may help the food sector significantly in the prevention of chronic non-communicable ailments (Franco et al. 2020).

4.7 Role of Biotechnology

Biotechnological approaches have made the enhanced quality and production of *P. ovata* seeds on a large scale. The various biotechnological tools used in *P. ovata* improvement research have significantly helped to increase the competence of crop enhancement. Improvements made in terms of abiotic stress tolerance, expanded yield quality, disease, and pest resistance are the milestones in the increased *P. ovata* production. Various breeding methods such as hybridization, selections, induced mutations, tissue culture, and polyploidy have been used to genetic improvement of *Plantago* (Dhar et al. 2005). The three replicas produced by the Mutation breeding in *P. ovata* by gamma rays have significantly boosted the seed production of the medicinal herb by 200% (Sareen and Koul 1999). In addition to this, the molecular markers used in the genetic variability analysis and germplasm characterisations are discussed in this book chapter.

4.7.1 Germplasm Diversity and Conservation

Plant breeders develop new varieties of a plant using the unprocessed material called germplasm. And the germplasm of a plant species includes different types of plant varieties from natural hybrids wild varieties and wild relatives, breeding lines, mutant varieties, primitive cultivars, elite lines, obsolete varieties, and specifically developed intergenic hybrids. Agro-morphological characterization is considered as preliminary step in the analysis of germplasm diversity. These characters include plant height, number of spikelet, length, and weight of spike (Mohsenzadeh et al. 2022).

Accumulating a clear knowledge about the genetic variations among the populations of species is the initial step in an ideal conservation strategy. A clear idea of the genetic diversity of the species can assist the development of new improved varieties from available varieties. Even though India is known as the larger producer of *P. ovata* seeds in the world, very few attempts are made to understand the genetic diversity of the species. An experiment was done by a group of researchers to assess the genetic diversity of *P. ovata* by random amplified polymorphic DNA (RAPD) markers on 18 Indian-grown isabgol genotypes (Ashok et al. 2012). In a similar study, the genetic diversity of the *P. ovata* and some of its wild genotypes were analysed at the molecular level using AFLP and SSR markers and their phylogenetic relationship was determined and a genetic map was derived (Kotwal et al. 2013). Herliana et al. in 2023 has generated a genome assembly of *P. ovata* including gene annotations using de no assembly of PacBio CLR data and Hic data. They have generated 500 Mb size genome assembly consisting of 4 chromosomes. The produced genome can be used to improve the seed production and quality through breeding programs using selected traits of the species (Herliana et al. 2023).

Analysis of karyotype characters, haploid ideograms and metaphase chromosomes of 6 varieties of *P. ovata* was carried out using RAPD, SSIR and SCoT markers and made into 2 distinguished subgroups (Osman and Abdein 2019). Researchers from Pakistan have used callus DNA of *P. ovata* to analyse the somaclonal variations by RAPD markers. And they successfully explored the polymorphism among the callus samples (Mahmood et al. 2012). In a similar study done in Iran to analyse the genetic determinations of agronomic traits using SCoT markers have revealed the genome mapping of the *P. ovata* and resulted ecotypes were divided into 3 sub-groups (Rahimi et al. 2018). Similarly, genetic variability of 2 endangered species of *Plantago* in Portugal was studied using SSIR and RAPD molecular markers (Ferreira et al. 2013). Better understanding of somaclonal variation will help in the crop improvement. Ghorbanpour and Khadivi-Khub in 2015 have studied the somaclonal variations of 18 in vitro raised callus cultures of *Plantago* produced from explants such as leaf and root tips. And have grouped the samples into 2 subgroups on the basis of polymorphism explored in the somaclonal studies done using the ISSR markers (Ghorbanpour and Khadivi-Khub 2015). The information obtained from germplasm diversity studies can be used effectively in the conservation and crop strategies.

4.7.2 *Genetic Resources Conservation*

Approaches

The drastic changes in the climatic conditions, the outburst of new-fangled diseases and vermin, frequently occurring natural calamities, global warming, and the threats from globalization and industrialization announces the need of conservation of genetic resources of medicinal values. Conservation strategies should be implemented at different levels to save the diverse resources to save from endangering. The importance of conservation of high-quality seeds is highly demanding as the *P. ovata* undergo sexual propagation methods. In-situ, ex-situ, in-vitro conservation, and gene banks are the common methods to save the genetic diversity of a plant species.

The perennial herb *P. ovata* poses difficulties for in situ conservation strategies, which aim to safeguard species in their native ecological systems. However, ex-situ conservation offers a robust approach for safeguarding its genetic diversity. Seed banks provide a simple and cost-effective solution for storing dried seeds for extended periods (Rajpurohit and Jhang 2015). This technique is particularly useful for *P. ovata* seeds, which can tolerate desiccation.

Furthermore, field gene banks are ideally suited for this perennial species. Here, researchers can cultivate live *P. ovata* plants, ensuring the preservation of genetic variation (Reed et al. 2004). Additionally, cryopreservation offers a secure option for long-term storage of even those species whose seeds are difficult to bank in the traditional sense. However, this method necessitates specialized resources and expertise (Engelmann 1999). Finally, DNA banks serve as a valuable archive of genetic information for *P. ovata*. This information proves useful for research and potentially even future restoration efforts (Vijayan et al. 2018). In conclusion, these ex-situ conservation methods working in tandem can effectively safeguard the irreplaceable genetic heritage of *P. ovata*.

4.8 Molecular Breeding

4.8.1 *Molecular Marker-Assisted Breeding*

The crop improvement of *Plantago* species has focused mainly in the improvement of seed husk quality and yield as the seed husk of *P. ovata* has medicinal importance and it is widely used in medicinal food applications. The researches on breeding in *Plantago* has produced improved varieties of *Plantago* species. Similarly, improvement in the *Plantago* with better growth rate, seed yield, rate of genetic gain, sprouting, propagation efficiency, abiotic stress tolerance can be accomplished through breeding strategies using molecular markers (Vijayan et al. 2018). The molecular markers such as RAPD, ISSR, RFLP and SCoT are commonly used in genetic diversity studies and breeding approaches for crop improvement. Fougat et al. in 2014 have established molecular markers for crop development using a microsatellite enrichment approach using next generation sequencing and obtained SSRs of *P. ovata.* 249 SSRs were obtained from the 3447 contigs that were assembled during the study. Study has composed microsatellite markers that has showed the amplification of excepted size also showed cross amplification of other 6 species of *Plantago*. The study has aimed in the development of genetic resources which has showed amplification as well cross- amplification in *P. ovata* and other species respectively (Fougat et al. 2014). Similarly other molecular markers such as RAPD, RFLP, ISSR, AFLP are used in various researches in the development of new varieties and the genetic diversity assessment described in the Table 4.4.

4.8.2 Functional Genomics

The functional genomics studies in *P. ovata* were intended to explore the genetic diversity of the *P. ovata* plant species and to develop improved varieties of desired quality. In a transcriptomic analysis of *P. ovata* the transcriptome of the plant species ovary was sequenced using genome analyser and the mucilage biosynthesis pathway was characterized. Functional characterization and categorization of the *P. ovata* transcripts disclosed the existence of various functional genes associated with different biological processes such as mucilage biosynthesis, metabolic pathways, synthesis of anti-oxidants and secondary metabolites (Kotwal et al. 2016). A similar study has sequenced the complete plastome of *P. ovata* and that has compared with the related species of Plantaginaceae family. The study revealed high sequence similarity with analysed species. The investigation has disclosed 139 microsatellites with 71 non-coding regions, 32 forward, 17 palindromic repeats and 34 tandems. The genome sequence revealed the *matK* and *rbcL* gene generation of same phylogenetic signals with related species. The study uncovered the superior stability of *P. ovata* plastome genome over other Plantaginaceae species (Asaf et al. 2020). Early and late flowering genotypes' leaf transcriptomes *P. ovata* was studied in order to identify the genes and processes involved in flowering and maturation. Differentially expressed genes (DEGs) linked with blooming time and maturation were found. The study's findings can be effectively employed as the foundation for flowering time management breeding as part of crop development (Patel et al. 2021).

4.8.3 Bioinformatics

The bioinformatics tools and databases can be efficiently used the genome diversity analysis and the breeding programs developed at molecular level. The established databanks provide the information about the species such as functional illustrations of genes, gene structures, and transcriptome data. Illumina Hi-Seq platform (Illumina®,

Table 4.4 Molecular markers used for genetic characterization in *P. ovata*

Marker used	Purpose	Country	References
RAPD	Phylogenetic analysis and for breeding programmes to develop new crop varieties	India	Ashok et al. (2012)
AFLP, SSR	Interspecific and intraspecific genetic diversity analysis	India	Kotwal et al. (2013)
RAPD, SSR, SCoT	Perceptual mapping, cluster analysis based on cytological and molecular attributes of 6 species	Saudi Arabia	Osman and Abdein (2019)
RAPD	Analyse the somaclonal variations and assess the polymorphism among callus	Pakistan	Mahmood et al. (2012)
SCoT	Comprehensive genome-wide association mapping	Iran	Rahimi et al. (2018)
RAPD	Genetic variability of 5 cultivars	India	Das nee Pal and Raychaudhuri (2003)
RAPD	Genetic relationship in 10 *Plantago* species	India	Samantaray et al. (2010)
RAPD, ISSR	Genetic variability of 2 endangered *Plantago*	Portugal	Ferreira et al. (2013)
Isozymes	Isozyme diversity assessment	India	Kaswan et al. (2012)
ISSR	Somaclonal variations of 18 in-vitro raised callus samples	Iran	Ghorbanpour and Khadivi-Khub (2015)
ITS	Genetic variability of *P. ovata* of Iran by morphological and molecular means	Iran	Mohsenzadeh et al. (2022)
RAPD	Analysis of genetic and phytochemical difference of *Plantago*	Sweden	Zubair et al. (2012)
RAPD, SSR	Genetic diversity among 38 genotypes of *Plantago*	India	Kumar et al. (2014)
RAPD	Genetic variability of 15 *Plantago*	India	Vala et al. (2011)
RAPD	Genetic variability in 80 germplasm accessions of *Plantago*	India	Singh et al. (2009)
RAPD, ISSR	Genetic diversity of *Plantago* to develop crop improvement strategies	India	Kaswan et al. (2013)
RAPD	Molecular and morphological diversity evaluation of 22 *Plantago* populations	Iran	Vahabi et al. (2008)
PCR–RFLP	Phylogenetics *Plantago* on the basis of rps14 chloroplast gene	Pakistan	Saeed et al. (2011)
ISSR	Somaclonal variations 18 callus samples	Iran	Esmaeili et al. (2014)
RAPD	Genetic diversity in Plantago and wild allies	India	Kour et al. (2016)
SSR	Genetic diversity and development of genomic resources	India	Fougat et al. (2014)

San Diego, California, USA) was successfully used in the RNA library construction of *Plantago*. The NCBI basic local alignment search tool (BLASTX) is a popular tool for identifying gene annotations in *Plantago* species. FLOR-ID (flowering interactive database) and uniport databases are interactive databases of *Arabidopsis* flowering time gene networks. BLAST searches can be used to identify Plantago orthologous genera. Some of the most common databases used in bioinformatics include DDBJ, EMBL, and GenBank (Patel et al. 2021).

4.9 Genetic Engineering

Genetic engineering is an ideal method for crop improvement for the development of new varieties with desirable characters. Genetic engineering methods can be used for the development of new varieties with improved seed husk production, high tolerance level to abiotic stress, and improved flowering time and seed germination properties. Gene bombardment and the vector- mediated (E.g., *Agrobacterium tumefaciens*) gene transfer methods are the two common approaches of genetic engineering to advance varieties of plant with specific traits. To develop *P. ovata* varieties with improved salinity tolerance, the responses of *P. ovata* to salinity was analysed by in -vitro tissue culture method. Somaclonal variations of the tolerant genotypes were also studied. The finding of the study can be efficiently used for the development of salt tolerant varieties of *Plantago* through genetic engineering techniques (Safarnejad et al. 2017). Et al have done complete genome analysis of *Plantago* at chromosome level which identified 41,820 protein-coding genes, 411 non-coding RNAs, 108 ribosomal RNAs, and 1295 transfer RNAs. This genome can be used as the basis of breeding programs to get desired properties if the plant such as increase quality psyllium yield, reduce seed shattering and increased disease tolerance (Herliana et al. 2023).

4.10 Conventional Mutagenesis

Mutation breeding in *P. ovata* using gamma rays have produced three replicas of the *Plantago* and have reported to have superior seed production by the new varieties over the natural species (Sareen and Koul 1999). The gamma rays mediated mutations can produce positive and negative mutations. To develop the *Plantago* variety adaptable north Indian conditions, induced mutation breeding was initiated. *Plantago* dry seed were subjected to gamma irradiations and treated with 20% ethidium bromide- a chemical mutagen. The induced mutation has produced qualitative and quantitative varieties with various agronomic traits. Plant materials subjected irradiation more than 50 kR doses caused a reduction in seed germination rate. And 40–50 kR doses produced positive mutant strains. Niharika and FEA-5 are examples of mutant varieties (Lal and Sharma 2000).

4.11 Conclusion and Future Prospects

As a consequence of the discussion about the commercial and medical value of *P. ovata*, it was determined that the cultivation and advancement of this crop's culture methods on a wide scale is absolutely essential for meeting current needs. Thus, in the framework of building policy to link crop cultivation with the suppliers for industry, it may become a beneficial agricultural practise for the local populace and farmers in that region. Additionally, it was deduced from earlier research that isabgol seeds have the highest fibre levels when used as a natural source of fibre with a variety of health advantages. This is the primary driver of the scientific community's growing interest in conducting additional study in this area in an effort to discover fresher approaches to incorporating this fibre into nutritional food products that have health-protective effects. The inclusion of psyllium to a regular diet may help with concerns associated to food lacking in soluble fibre. Based on the information above, we may conclude that *P. ovata* has a potential function in industries for the development of marketable products and for the benefit of a healthy society. The food industry's creation of staple food products enriched with psyllium husk may contribute to a better supply of soluble fibre food products. Mutagenesis offers a potential avenue for future *P. ovata* improvement and conservation. By employing mutagens like EMS, scientists can induce targeted mutations in *P. ovata* genes. This could lead to the development of new varieties with enhanced characteristics, such as increased yield, improved stress tolerance, or resistance to disease. Furthermore, -induced mutations can be a valuable tool for inducing genetic variation and studying *P. ovata* genes and their associated functions. Government and research institutions can play an important role in sponsoring new research in this field and sharing findings with other public institutions in regard to these activities for economic development.

References

Asaf S, Khan AL, Lubna KA, Khan A, Khan G, Lee I-J, Al-Harrasi A (2020) Expanded inverted repeat region with large scale inversion in the first complete plastid genome sequence of *Plantago ovata*. Sci Rep 10(1):3881. https://doi.org/10.1038/s41598-020-60803-y

Ashok KR, Mukesh K, A S, K SB (2012) Genetic diversity analysis of the medicinal herb *Plantago ovata* (Forsk.). Afr J Biotechnol 11(92):15835–15842. https://doi.org/10.5897/AJB12.681

Dagar JC, Kumar Y, Tomar OS (2006) Cultivation of medicinal isabgol (*Plantago ovata*) in alkali soils in semiarid regions of Northern India. Land Degrad Dev 17(3):275–283. https://doi.org/10.1002/ldr.700

Das nee PM, Raychaudhuri SS (2003) Estimation of genetic variability in Plantago ovata cultivars. Biologia Plantarum 47:459–462. https://doi.org/10.1023/B:BIOP.0000023896.08612.f2

Dhar MK, Kaul S, Sareen S, Koul AK (2005) *Plantago ovata*: genetic diversity, cultivation, utilization and chemistry. Plant Genet Resour 3(2):252–263. https://doi.org/10.1079/PGR200582

Engelmann F (1999) Management of field and in vitro germplasm collections. Proceedings of a consultation meeting, 15–20 January 1996, CIAT Cali, Colombia

Esmaeili F, Shooshtari L, Ghorbanpour M, Etminan A (2014) Assessment of somaclonal variation in *Plantago major* using molecular markers. J Biodivers Environ Sci JBES 5(4):402–408

Farahnaky A, Askari H, Majzoobi M, Mesbahi G (2010) The impact of concentration, temperature and pH on dynamic rheology of psyllium gels. J Food Eng 100(2):294–301. https://doi.org/10.1016/j.jfoodeng.2010.04.012

Fernandez N, Lopez C, Díez R, Garcia JJ, Diez MJ, Sahagun A, Sierra M (2012) Drug interactions with the dietary fiber *Plantago ovata* husk. Expert Opin Drug Metab Toxicol 8(11):1377–1386. https://doi.org/10.1517/17425255.2012.716038

Ferreira V, Matos M, Correia S, Martins N, Gonçalves S, Romano A, Pinto-Carnide O (2013) Genetic diversity of two endemic and endangered Plantago species. Biochem Syst Ecol 51:37–44

Fischer MH, Yu N, Gray GR, Ralph J, Anderson L, Marlett JA (2004) The gel-forming polysaccharide of psyllium husk (*Plantago ovata* Forsk). Carbohydr Res 339(11):2009–2017. https://doi.org/10.1016/j.carres.2004.05.023

Fougat RS, Joshi C, Kulkarni K, Kumar S, Patel A, Sakure A, Mistry J (2014) Rapid development of microsatellite markers for *Plantago ovata* Forsk.: using next generation sequencing and their cross-species transferability. Agriculture 4(2):199–216

Fradinho P, Nunes MC, Raymundo A (2015) Developing consumer acceptable biscuits enriched with Psyllium fibre. J Food Sci Technol 52(8):4830–4840. https://doi.org/10.1007/s13197-014-1549-6

Franco EAN, Sanches-Silva A, Ribeiro-Santos R, De Melo NR (2020) Psyllium (*Plantago ovata* Forsk): From evidence of health benefits to its food application. Trends Food Sci Technol 96:166–175. https://doi.org/10.1016/j.tifs.2019.12.006

Ghorbanpour M, Khadivi-Khub A (2015) Somaclonal variation in callus samples of Plantago major using inter-simple sequence repeat marker. Caryologia 68(1):19–24

Gilani A, Aziz N, Khan MA, Khan S, Zaman V (1998) Laxative effect of ispaghula: physical or chemical effect? Phytother Res 12(S1):S63–S65. https://doi.org/10.1002/(SICI)1099-1573(1998)12:1+%3cS63::AID-PTR252%3e3.0.CO;2-7

Hannan JMA, Ali L, Khaleque J, Akhter M, Flatt PR, Abdel-Wahab YHA (2006) Aqueous extracts of husks of *Plantago ovata* reduce hyperglycaemia in type 1 and type 2 diabetes by inhibition of intestinal glucose absorption. Br J Nutr 96(01):131. https://doi.org/10.1079/BJN20061819

Herliana L, Schwerdt JG, Neumann TR, Severn-Ellis A, Phan JL, Cowley JM, Shirley NJ, Tucker MR, Bianco-Miotto T, Batley J, Watson-Haigh NS, Burton RA (2023) A chromosome-level genome assembly of *Plantago ovata*. Sci Rep 13(1):1528. https://doi.org/10.1038/s41598-022-25078-5

Hussain MA, Muhammad G, Jantan I, Bukhari SNA (2016) Psyllium Arabinoxylan: a versatile biomaterial for potential medicinal and pharmaceutical applications. Polym Rev 56(1):1–30. https://doi.org/10.1080/15583724.2015.1078351

Kaswan V, Joshi A, Maloo SR (2012) Assessment of isozyme diversity in the medicinal plant isabgol (Plantago ovata Forsk.). J Cell Tissue Res 12(3):3349

Kaswan V, Joshi A, Maloo SR (2013) Assessment of genetic diversity in Isabgol (Plantago ovata Forsk.) using random amplified polymorphic DNA (RAPD) and inter-simple sequence repeat (ISSR) markers for developing crop improvement strategies. Afr J Biotechnol 12(23)

Kotwal S, Dhar MK, Kour B, Raj K, Kaul S (2013) Molecular markers unravel intraspecific and interspecific genetic variability in *Plantago ovata* and some of its wild allies. J Genet 92:293–298

Kotwal S, Kaul S, Sharma P, Gupta M, Shankar R, Jain M, Dhar MK (2016) De novo transcriptome analysis of medicinally important Plantago ovata using RNA-Seq. PLoS ONE 11(3):e0150273

Kour B, Kotwal S, Dhar MK, Kaul S (2016) Genetic diversity analysis in *Plantago ovata* and some of its wild allies using RAPD markers. Russ Agric Sci 42:37–41

Kumar M, Fougat RS, Sharma AK, Kulkarni K, Mistry JG, Sakure AA, Kumar S (2014) Phenotypic and molecular characterization of selected species of 'Plantago' with emphasis on '*Plantago ovata*'. Aust J Crop Sci 8(12):1639–1647

Ladjevardi ZS, Gharibzahedi SMT, Mousavi M (2015) Development of a stable low-fat yogurt gel using functionality of psyllium (*Plantago ovata* Forsk) husk gum. Carbohydr Polym 125:272–280. https://doi.org/10.1016/j.carbpol.2015.02.051

Lal RK, Sharma JR (2000) Induction by gamma irradiation (60 Co), characterization and utilization of mutants for economic traits in Isabgol (Plantago ovata Forsk.). In: Proceedings of the DAE-BRNS symposium on the use of nuclear and molecular techniques in crop improvement, Mumbai, pp 272–278

Mahmood T, Jameel A, Abbasi BH, Munir F, Naqvi SMS (2012) In vitro callogenesis and detection of somaclonal variations in *Plantago ovata* L. J Crop Sci Biotechnol 15:289–295

Mehmood MH, Aziz N, Ghayur MN, Gilani A-H (2011) Pharmacological basis for the medicinal use of Psyllium Husk (Ispaghula) in constipation and diarrhea. Dig Dis Sci 56(5):1460–1471. https://doi.org/10.1007/s10620-010-1466-0

Mohsenzadeh S, Sheidai M, Koohdar F (2022) Molecular investigation of the intra-specific genetic variation in *Plantago ovata* Forssk.(Plantaginaceae): An insight into potential ancestral area distribution and probable time of dispersal versus vicariance events. Phytotaxa 559(2):157–166

Osman AKE, Abdein MAE-H (2019) Karyological and molecular studies between six species of *Plantago* in the Northern border region at Saudi Arabia. J Taibah Univ Sci 13(1):297–308. https://doi.org/10.1080/16583655.2019.1571400

Patel S, Pachhigar K, Ganvit R, Panchal RR, Ponnuchamy M, Kumar J, Reddy NRR (2021) Exploring flowering genes in isabgol (*Plantago ovata* Forsk.) through transcriptome analysis. Plant Mol Biol Report 39:192–211

Perrigue MM, Monsivais P, Drewnowski A (2009) Added soluble fiber enhances the satiating power of low-energy-density liquid yogurts. J Am Diet Assoc 109(11):1862–1868. https://doi.org/10.1016/j.jada.2009.08.018

Rahimi M, Nazari L, Kordrostami M, Safari P (2018) SCoT marker diversity among Iranian Plantago ecotypes and their possible association with agronomic traits. Sci Hortic 233:302–309

Rajpurohit D, Jhang T (2015) In situ and ex situ conservation of plant genetic resources and traditional knowledge. In: Salgotra RK, Gupta BB (eds) Plant genetic resources and traditional knowledge for food security. Springer, Singapore, pp 137–162

Reed BM, Engelmann F, Dulloo ME, Engels JMM (2004) Technical guidelines for the management of field and in vitro germplasm collections. Handbooks for Genebanks n.7, 106 p. ISBN: 978-92-9043-640-9

Romero-Baranzini AL, Rodriguez OG, Yanez-Farias GA, Barron-Hoyos JM, Rayas-Duarte P (2006) Chemical, physicochemical, and nutritional evaluation of Plantago (*Plantago ovata* Forsk). Cereal Chem J 83(4):358–362. https://doi.org/10.1094/CC-83-0358

Saeed S, Munir F, Naveed I, Raja GK, Mahmood T (2011) Phylogenetics of selected Plantago species on the basis of rps14 chloroplast gene. J Med Plants Res 5(19):4888–4891

Safarnejad A, Shoorvarzi M, Dalir M (2017) In vitro selection of Plantago ovata for NaCl tolerance. Iran J Rangel for Plant Breed Genet Res 25(2):289–301. https://doi.org/10.22092/ijrfpbgr.2017.113317

Sagar S, Goudar G, Sreedhar M, Panghal A, Sharma P (2020) Characterization of nutritional content and in vitro—antioxidant properties of Plantago ovata seeds. Int J Food Nutr Sci 9(2):27. https://doi.org/10.4103/IJFNS.IJFNS_27_20

Samantaray S, Dhagat UM, Maiti S (2010) Evaluation of genetic relationships in Plantago species using Random Amplified Polymorphic DNA (RAPD) markers. Plant Biotechnol 27(4):297–303

Sareen S, Koul AK (1999) Mutation breeding in improvement of *Plantago ovata* Forsk. Indian J Genet Plant Breed 59(03):337–344

Shah AR, Sharma P, Gour VS, Kothari SL, Dar KB, Ganie SA, Shah YR (2020) Antioxidant, nutritional, structural, thermal, and physico-chemical properties of Psyllium (*Plantago ovata*). Seeds 8(3):727–743. https://doi.org/10.12944/CRNFSJ.8.3.06

Singh B, Kumar S (2008) Synthesis and characterization of psyllium-NVP based drug delivery system through radiation crosslinking polymerization. Nucl Instrum Methods Phys Res Sect B Beam Interact Mater at 266(15):3417–3430. https://doi.org/10.1016/j.nimb.2008.04.022

Singh N, Lal RK, Shasany AK (2009) Phenotypic and RAPD diversity among 80 germplasm accessions of the medicinal plant isabgol (*Plantago ovata*, Plantaginaceae). Genet Mol Res 8(3):1273–1284

Talukder P, Talapatra S, Ghoshal N, Sen Raychaudhuri S (2016) Antioxidant activity and high-performance liquid chromatographic analysis of phenolic compounds during *in vitro* callus culture of *Plantago ovata* Forsk. and effect of exogenous additives on accumulation of phenolic compounds: Polyphenol accumulation during *in vitro* callus culture in *Plantago ovata* Forsk. J Sci Food Agric 96(1):232–244. https://doi.org/10.1002/jsfa.7086

Thakur VK, Thakur MK (2014) Recent trends in hydrogels based on psyllium polysaccharide: a review. J Clean Prod 82:1–15. https://doi.org/10.1016/j.jclepro.2014.06.066

Theuwissen E, Mensink RP (2008) Water-soluble dietary fibers and cardiovascular disease. Physiol Behav 94(2):285–292. https://doi.org/10.1016/j.physbeh.2008.01.001

Turnbull WH, Thomas HG (1995) The effect of a Plantago ovata seed containing preparation on appetite variables, nutrient and energy intake. Int J Obes Relat Metab Disord J Int Assoc Study Obes 19(5):338–342

Vahabi AA, Lotfi A, Solouki M, Bahrami S (2008) Molecular and morphological markers for the evaluation of diversity between *Plantago ovata* in Iran. Biotechnology 7(4):702–708

Vala AG, Fougat RS, Jadeja GC (2011) Genetic diversity of *Plantago ovata* Forsk. through RAPD markers. Electron J Plant Breed 2(4):592–596

Vijayan K, Ravikumar G, Tikader A (2018) Mulberry (Morus spp.) breeding for higher fruit production. Adv Plant Breed Strateg Fruits 3:89–130

Yadav N, Sharma V, Kapila S, Malik RK, Arora S (2016) Hypocholesterolaemic and prebiotic effect of partially hydrolysed psyllium husk supplemented yoghurt. J Funct Foods 24:351–358. https://doi.org/10.1016/j.jff.2016.04.028

Zubair M, Nybom H, Ahnlund M, Rumpunen K (2012) Detection of genetic and phytochemical differences between and within populations of *Plantago major* L. (plantain). Sci Hortic 136:9–16

Chapter 5
Breeding Methods for Caper Plant (*Capparis spinosa* L.)

Maryam Ahmadi and Hojjatollah Saeidi

Abstract *Capparis spinosa* (caper bush), is a xerophytic and perennial winter-deciduous species. This species grows naturally in Europe, Africa, Madagascar, Asia, Australia and Oceania from sea level to an altitude of 1900 m. Caper is called the Plant of the Millennium with diverse economic and medicinal values. This species is used as a medicinal plant contained numerous beneficial compounds, including alkaloids, glucosinolates, carotenoids, polyphenols, fatty acids, strEroids, terpenoids, vitamin C, tocopherols and minerals. This species is an important source of flavonoids such as quercetin, rutin and kaempferol. Immature flower buds, semi-mature fruits and young shoots with small leaves are used as food or condiment. Caper is propagated by seeds, stem cuttings and grafting. The growing market demand, difficulty in caper propagation and limited area under cultivation have led to overharvesting of the wild caper for domestic uses and trades that have made this species at risk of genetic erosion. Therefore, the conservation of genetic resources and developing new varieties of caper would require the improvement of current conventional and innovative biotechnological techniques. In this chapter, we represent information concerning taxonomic approaches, the origin and distribution, characterization and conservation of genetic resources, crop cultivation practices, recent developments in biotechnology and molecular biology as well as their application for crop improvement in association with conventional breeding methods of caper.

Keywords Medicinal plants · Breeding · Pharmaceutical compounds · Caper

M. Ahmadi · H. Saeidi (✉)
Department of Plant and Animal Biology, Faculty of Biological Science and Technology, University of Isfahan, Isfahan, Iran
e-mail: ho.saeidi@sci.ui.ac.ir

J. M. Al-Khayri et al. (eds.), *Biodiversity and Genetic Improvement of Medicinal and Aromatic Plants I*, Advances in Plant Breeding Strategies 9,
https://doi.org/10.1007/978-3-031-81846-2_5

5.1 Introduction

Capparis spinosa L., also known as caper bush, is a xerophytic and perennial winter-deciduous species. This species belongs to the genus *Capparis* Linnaeus (1753), family Capparaceae and order Brassicales (Angiosperm Phylogeny Group et al. 2016). It is widely distributed in Europe, Africa, Madagascar, Asia, Australia and Oceania (Fici 2014, 2015). Linnaeus (1753) described *C. spinosa* and since that time, the high level of polymorphism in morphological characters and the presence of intermediate phenotypes have led to longstanding taxonomic disagreement in the literature. Some authors believed in the broad species concept and combined all the variations into a single species, *C. spinosa*, with several intraspecific taxa (Jacobs 1965; Boissier 1867; Higton and Akeroyd 1991; Fici 2014, 2015) and the others followed the narrow species concept and classified this taxonomically complex group to several closely related species (Zohary 1960; Inocencio et al. 2006; Danin 2010). Boissier (1867) recognized one species with a number of varieties from the eastern Mediterranean to India. Zohary (1960) believed that hybridization in contact zones has blurred species boundaries in *C. spinosa* group. Thus, he described several species based on the floral characters in the Mediterranean region and West Asia for this group. Jacobs (1965) revised the taxonomy of the genus *Capparis* in South and Southeast Asia, Malaysia, Australia and Pacific. He considered *C. spinosa* as a monotypic species in sect. *Capparis*. In Europe, Heywood (1964) recognized two species, which were later combined into *C. spinosa*, with two subspecies by Higton and Akeroyd (1991). the finding of Inocencio et al. (2005) based on AFLP markers suggested that hybridization between *C. orientalis* Veill. and *C. sicula* Veill. has given rise to *C. spinosa*. Inocencio et al. (2006) classified this group from the Western to Central Asia, North Africa, and Europe into 10 species, some with several subspecies. According to Fici (2014, 2015) variability in the main diagnostic morphological characters of the *C. spinosa* group could be attributed to various factors such as phenotypic plasticity, hybridization events, selection of cultivated form and ecogeographic differentiation. Therefore, He included all the variations in one species, *C. spinosa*, represented by 6 subspecies and some varieties. Recent taxonomic treatments of *C. spinosa* group has been presented in Table 5.1, and Fig. 5.1 shows some taxa of the group growing in Iran.

5.1.1 Origin and Distribution

The caper bush is frequently found in coastal areas and islands (Ravindran et al. 2004). This species is distributed in Europe, Africa, Madagascar, Asia, Australia and Oceania (Fici 2014, 2015; Fig. 5.2). Different subspecies and varieties of this species grow mostly in Southern Europe, North Africa, the Middle East, Southwest Asia, India, Malaysia, and China, in a saline habitat, within a halophytic ecological community (Chedraoui et al. 2017). Caper might have originated in the dry regions

Table 5.1 Taxonomic investigations in *C. spinosa* in recent years

Taxonomic treatment	Source
1. ***C. cartilaginea* Decne** 2. ***C. sicula* Veill. subsp. *herbacea* (Willd.) Inocencio, D. Rivera, Obo´n & Alcaraz** *C. sicula* subsp. *leucophylla* (DC.) Inocencio, D. Rivera, Obo´n & Alcaraz *C. sicula* subsp. *mesopotamica* Inocencio, D. Rivera, Obo´n & Alcaraz *C. sicula* subsp. *sicula* *C. sicula* subsp. *herbacea* (Willd.) Inocencio, D. Rivera, Obo´n & Alcaraz *C. sicula* subsp. *leucophylla* (DC.) Inocencio, D. Rivera, Obo´n & Alcaraz *C. sicula* subsp. *mesopotamica* Inocencio, D. Rivera, Obo´n & Alcaraz *C. sicula* subsp. *sindiana* Inocencio, D. Rivera, Obo´n & Alcaraz 3. ***C. spinosa*** L. 4. ***C. zoharyi* Inocencio, D. Rivera, Obo´n & Alcaraz** 5. ***C. mucronifolia* Boiss. subsp. *mucronifolia*** *Capparis mucronifolia* Boiss. subsp. *rosanoviana* (B. Fedtsch.) Inocencio, D. Rivera, Obo´n & Alcaraz 6. ***C. parviflora* Boiss. subsp. *parviflora*** *C. parviflora* subsp. *kurdica* (Zohary) Inocencio, D. Rivera, Obo´n & Alcaraz *C. parviflora* Boiss. subsp. *sphaerocarpa* Inocencio, D. Rivera, Obo´n & Alcaraz 7. ***C. aegyptia*** Lam. 8. ***C. atlantica* Inocencio, D. Rivera, Obo´n & Alcaraz** 9. ***C. orientalis* Veill** 10. ***C. ovata* Desf. subsp. *ovata*** *C. ovata* Desf. subsp. myrtifolia Inocencio, D. Rivera, Obo´n & Alcaraz	Inocencio et al. (2006)
1. ***C. spinosa* subsp. *spinosa* var. *spinosa*** *C. spinosa* subsp. *spinosa* var. *atlantica* (Inocencio, D. Rivera, Obón & Alcaraz) Fici *C. spinosa* L. subsp. *spinosa* var. *parviflora* (Boiss.) Boiss. *C. spinosa* L. subsp. *spinosa* var. *canescens* Cosson *C. spinosa* L. subsp. *spinosa* var. *aegyptia* (Lam.) Boiss. *C. spinosa* L. subsp. *spinosa* var. *mucronifolia* (Boiss.) Hedge & Lamond *C. spinosa* subsp. *spinosa* var. *herbacea* (Willd.) Fici 2. ***C. spinosa* subsp. *rupestris*** var. *rupestris* *C. spinosa* subsp. *rupestris* var. *ovata* (Desf.) Fici *C. spinosa* subsp. *rupestris* var. *myrtifolia* (Inocencio, D. Rivera, Obón & Alcaraz) Fici 3. ***C. spinosa* L. subsp. *cartilaginea*** (Decne.) Maire & Weiller 4. ***C. spinosa* subsp. *cordifolia*** (Lam.) Fici 5. ***C. spinosa* subsp. *nummularia*** (DC.) Fici 6. ***C. spinosa* subsp. *himalayensis*** (Jafri) Fici	Fici (2014, 2015)

Fig. 5.1 Fruit and flower of *C. spinosa* var. *aegyptia* (**a**, **b**) habitat and fruit of *C. spinosa* subsp. *parviflora* (**c**, **d**) Vegetative and reproductive parts of *C. cartilaginea* (**e**, **f**; taxonomic treatment: Ahmadi et al. 2020)

of west or central Asia (Jacobs 1965; Zohary 1960). Analysis of molecular data revealed that Peninsular India is the origin center of *Capparis* and the genus has subsequently diversified to different parts of the old world (Maurya et al. 2023).

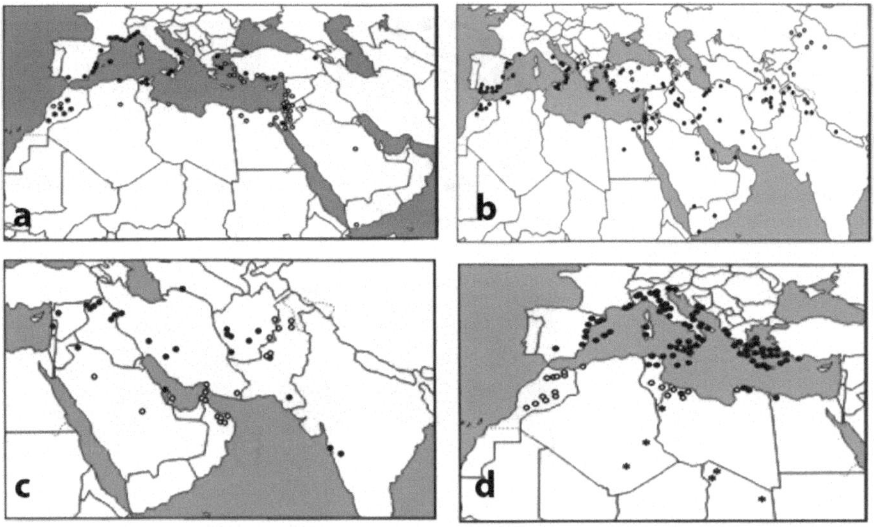

Fig. 5.2 Distribution map of varieties of *C. spinosa*: **a** subsp. *spinosa*: var. *spinosa* (•); var. *aegyptia* (○); var. *atlantica* (*), **b** subsp. *spinosa*: var. *canescens* (•); var. *herbacea* (○), **c** subsp. *spinosa*: var. *parviflora* (•); var. *mucronifolia* (○), **d** subsp. *rupestris*: var. *rupestris* (•); var. *ovata* (○); var. *myrtifolia* (*). *Map Source* Fici (2014)

5.1.2 Economic Importance

C. spinosa has diverse economic and medicinal properties. Immature flower buds, semi-mature fruits and young shoots with small leaves are consumed as food or condiment (Fig. 5.3). Flower buds are subjected to considerable trade, particularly at the international trade and they are used as condiment or in cosmetic industry. Commercial caper is immature flower buds that are pickled in vinegar or preserved in granular salt (El amri et al. 2019a). The leaves of caper are served in fish and meat dishes, pizza, pasta sauces, salads and in cheese-making (Ng et al. 2011; Sher and Alyemeni 2010). The processed caper berries are used as spices or condiments. Furthermore, they have sharp piquant flavor which makes them useful to include in hundreds of recipes such as a complement to pasta, salads and other foods or as an appetizer with olives, meet cheese and nuts (El armi et al. 2019b). Moreover, flower buds of *C. spinosa* are a good source of glucose, fructose, saccharose and have a potential value for animal feeding plant (Ulukapi et al. 2016). Due to its esthetic blossom and a sweet scented flower, *C. spinosa* can also be used as an ornamental in gardens and terraces exposed to sun (Chedraoui et al. 2017). *C. spinosa* is also a drought tolerant, fire resistant species and produces a great canopy which could be useful for the prevention of soil erosion and desertification in arid and semi-arid areas (Sakcali et al. 2008). In general, different parts of caper have a wide range of beneficial compounds such as alkaloids, glucosinolates, carotenoids, phenolic compounds, stereoids, terpenoids, fatty acids, vitamin C, tocopherols and minerals (Tilli et al.

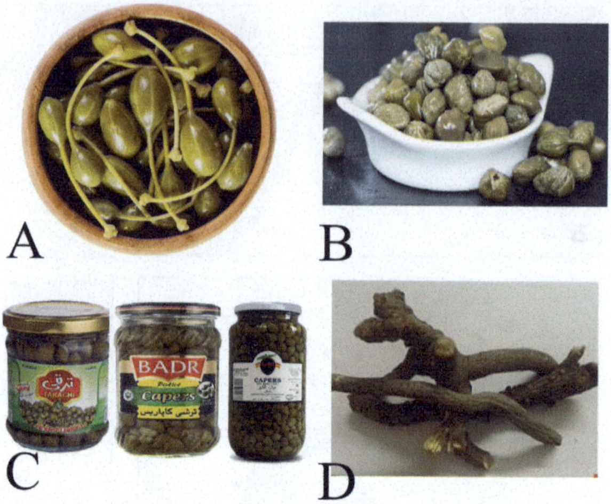

Fig. 5.3 **a** pickled caper berry. *Image Source* https://rasekhoon.net/article/show/1559857/; **b** pickled young flower buds. *Image Source* https://media.iscanews.ir/d/2022/10/22/3/571235.jpg; **c** commercial pickled. *Image Source* https://tabaye.ir/caper/ and **d** roots of caper. *Image Source* https://basalam.com/behsa/product/4550116

2009, 2010; Aliyazicioglu et al. 2013; Duman and Özcan 2014; Benzidane et al. 2020). Caper is used in traditional medicine to cure headache, malaria and hemorrhoids, toothache, fever, stomachache, convulsions, diabetes, anemia, menstruation, skin, kidney and liver diseases, rheumatism, ulcers and back pain (Shahrajabian et al. 2021; Annaz et al. 2022).Various pharmacological effects such as antioxidant, anti-inflammatory (Zhou et al. 2010; Moutia et al. 2016; Eid et al. 2023; Mahmoodpoor et al. 2023), anticancer and protective against the side effects of chemotherapeutic agents (Moghadamnia et al. 2019; Eid et al. 2023), antimicrobial (Gull et al. 2015; Mazarei et al. 2017; Al-Azawi et al. 2018; El-Shahaby et al. 2019; Benzidane et al. 2020) antiobesity (Lemhadri et al. 2007) and antidiabetic (Kazemian et al. 2015; Vahid et al. 2017; Farhangi et al. 2019; Assadi et al. 2021) have been reported for this plant.

5.1.3 Domestication, Selection and Early Improvements

Determining the time of Caper domestication is complicated, but the use of this species can be traced to ancient civilization. Caper seeds have been documented in archaeological contexts of the Mediterranean and west Asia (8000–8200 B.C.). Today, the center cultivation of this crop is Mediterranean countries. *C. spinosa* and *C. orientalis* are cultivated and *C. sicula* is harvested from wild populations for commercial purpose (Rivera et al. 2002).

5.2 Cultivation and Traditional Breeding

5.2.1 *Current Cultivation Practices*

C. spinosa is cultivated in Italy, Spain and Greece, Tunisia, Morocco and Egypt, Syria and Turkey (Sottile et al. 2021). This crop is traditionally propagated by seed, stem cutting and grafting. Seed propagation is the most common and grafting is the less accepted methods for the caper propagation. Seed propagation has its own problem due to the high degree of seed heterozygosity and the poor rate of seed germination (Gianguzzi et al. 2020). The seed propagation techniques are different, for example in Italy seeds are cultivated in small outdoors seedbeds, without need to disinfection, pre-germination treatment of seeds or regular irrigation. Seeds germinate 25–50 days after sowing, or sometimes after one or two years. In Spain, seeds are pre-treated by stratification in sand to break dormancy for nursery seed propagation, then they sow in raised beds from mid-spring and 1.5 or 2 g of pre-germinated seeds are used per linear meter. Seeds lots are collected in late summer to early autumn when ripen fruits are opened. Seeds are washed to separate from flesh and are dried before stratification. The full emergence of plantlets takes place 25–30 days after swing (Sottile et al. 2021). Vegetative propagation through stem cutting is also difficult due to low rate of rooting, but it is worth in avoiding high variation with respect to production and consistency of qualitative characteristics. Stem cuttings could be done using woody (hardwood and semi-hardwood) or herbaceous segments (Chedraoui et al. 2017; Benaoun et al. 2023).

5.2.2 *Current Agricultural Problems and Challenges*

Caper is considered as an orphan crop because researchers have paid a little attention to this crop, especially in terms of selecting the well-adapted, valuable and easily propagated candidate cultivars. There is any real cultivar for Caper and its cultivation has drastically decreased in Spain in recent years. On the other hand, caper is harvested from the wild in some countries which increases the risk of genetic erosion (Sottile et al. 2021).

5.2.3 *Crop Improvement*

Caper seed has both physical (seed coat) and physiological dormancy which has made it difficult to germinate and propagate through seedlings. The fruit weight, its position on the mother plant and maturity of the seed have also impact on the seed vigor (Chedraoui et al. 2017). various treatments such as mechanical scarification (Elazazi, 2016), cold stratification (Olmez et al. 2006) pre-treatment with sulfuric acid (H_2SO_4) following by soaking in either of potassium nitrate (KNO_3), GA3 or gamma ray treatment (Bhoyar et al. 2010), KNO_3 treatment (Fakhri et al. 2019) and manipulation of the environmental conditions (temperature and duration) (Basbag et al. 2009) has been reported for improvement seed germinating in *C. spinosa.*

Abiotic stresses are the main deterrent to sustainable crop production worldwide and the most critical stages of plant growth under stress conditions are seed germination and early seedling growth. Abiotic stresses such as drought stress can also affect seed germination and seedling growth of Caper. Seed priming has emerged as a promising crop stress management technique that can improve germination rate, shorten germination time, seedling emergence and improve plant establishment (Heydariyan et al. 2014; Anand 2014). Heydariyan et al. (2014) evaluated the effect of hormonal priming on germination and seedling growth of Caper under drought stress. The results of their survey revealed that priming improves the rate of germination, percentage of germination, seed vigor index, root length, seedling and shoot length under drought stress condition. Furthermore, priming with salicylic acid compared with gibberellic acid and ascorbic acid had a greater impact on germination characteristics of *C. spinosa.*

Micropropagation (i.e. in vitro clonal propagation of plants from small plant tissue) is a useful technology for the mass multiplication of species, which their regeneration through traditional methods is difficult (Shukla and Shukla 2024). A good number of studies have been carried out to develop and optimize micropropagation protocols for *C. spinosa* form different segments of this plant (Chalak and Elbitar 2006; Al-safadi and Elias 2011; Carra et al. 2012a; Elmaghrabi et al; 2017; Gianguzzi et al. 2020; Koufan et al. 2022).

5.2.4 Traditional Breeding Methodologies and Limitations

Thorny stipule is a disadvantageous character for harvesting of caper bush. Therefore, there is a strong selection against thorny stipules in caper cultivation (Gristina et al. 2014; Grimalt et al. 2022). On the other hand, the main desirable traits for cultivation of the caper bush are high yield, long stems, short internodes and high node fertility, round flower buds with dark green color, close non-hairy bracts and late opening, fruits with oval shape, light green pericarp and few seeds, easy stalk separation to simplify harvesting and post harvesting procedures, the ability of asexual propagation and resistance to biotic and abiotic stresses (Chedraoui et al. 2017).

5.2.5 Role of Biotechnology

Biotechnology provides new opportunities for the management and conservation of plant genetic resources using tissue, cell or organ culture and molecular biology. The use of tissue culture technique could overcome difficulties of conventional propagation in Caper such as rooting problems and improve mass production of this crop. A number of biotechnological studies were conducted in Caper include In vitro propagation using single nodal cuttings (Chalak and Elbitar 2006), nodal shoot segments (Rodriguez et al. 1990), plant regeneration from flower explant (Carra et al. 2012b), hypocotyl explants (Movafeghi et al. 2008) node explants (Sheikhi et al. 2020).

5.3 Germplasm Diversity and Conservation

Germplasm is a living source of information for all the genes exist in the respective plant, which can be preserved for a long time and restored whenever it is needed in the future (Bhatia 2015). In situ and ex situ are two conservation approaches for plant genetic resources and appropriate scheme for conservation can be determined based on the biological characteristics of the gene pool, infrastructure and human resources, the number of accessions in a particular collection and its geographical location (Rao and Riley 1994).

5.3.1 Germplasm Diversity

The assessment of genetic diversity within and among the plant populations can be done using various techniques such as (i) morphological, (ii) biochemical characterization/evaluation (allozyme) and (iii) DNA (or molecular) marker analysis especially single nucleotide polymorphism (SNPs) (Govindaraj et al. 2015). Morphological traits are sensitive to environmental conditions and their characterization is subjective in comparison to other methods (Bhandari et al. 2017). Therefore, molecular markers are more reliable for assessing genetic diversity because they are stable and not influenced by environmental factors. Chibani et al. (2017) used seven isozyme markers including isocitrate dehydrogenase (ICD), 6-phosphogluconate dehydrogenase (6-PGD), phosphoglucoisomerase (PGI), phosphoglucomutase (PGM), malate dehydrogenase (MDH), glutamate oxaloacetate transaminase (GOT) and Shikimate dehydrogenase (SKDH) to assess genetic structure among eight populations of *C. spinosa* for the first time in Tunisia. The results of their study indicated that some populations need prior conservation. On the other hand, DNA markers such as random amplified of polymorphic DNA (RAPD), inter simple sequence repeats (ISSR), amplified fragment length polymorphism (AFLP), single-nucleotide polymorphisms (SNPs), sequence-related amplified polymorphism (SRAP) and Start Codon Targeted Polymorphism (SCoT), sequence tag-simple sequence repeat (EST-SSR) and two cpDNA sequences (*rpS*12-*rpL*20, *ndh*F) were employed to assess the genetic diversity and systematic identification of taxa within *C. spinosa* group (Table 5.2). Najafian et al. (2021) constructed a phylogenetic tree based on the internal transcribed spacer (ITS) region of ribosomal DNA for species delimitation of *C.*

spinosa group in Iran. This is the first and only study that has used ITS as molecular marker in this group so far. Furthermore, the complete chloroplast genome of caper has also been reported recently (Alzahrani et al. 2021a, b). In addition, Mercati et al. (2019) presented the first genomic resource of *C. spinosa*, generated by transcriptomic approach and de novo assembly. Wang et al. (2022) also provided the whole-genome sequence of *C. spinosa* var. *herbacea* and their findings reported the whole genome duplication events, the enrichment of heat shock protein genes and the high proportion of tandem repeats in the genome of this variety. Genomic data provides valuable information for breeders to develop new cultivars with improved agronomic traits, such as higher productivity resistance to biotic and abiotic stresses, and to improve the nutritional value of food for the growing global population (Swarup et al. 2021). It is also useful to formulate effective conservation management strategies.

5.3.2 Genetic Resources Conservation Approaches

C. spinosa is mainly threatened by collection and excessive harvesting for household uses (medicinal and food) and for nationally or internationally trade because caper cultivation doesn't respond the growing market demand and most of the materials are collected from the wild populations. Caper is also threatened by overgrazing and the direct and indirect impact of anthropogenic activities such as leisure activities, tourism, infrastructure development, land clearing and management activities (Rankou et al. 2020).

5.3.2.1 In Situ Conservation

In situ conservation defines as the conservation of ecosystems and natural habitats, the maintenance and recovery of a viable species population in its natural environment and regarding the domesticated or cultivated species, in the place where they have developed their own characteristics (Rao and Riley 1994).

5.3.2.2 Ex Situ Conservation

Abo El-Fadl et al. (2021), Rankou et al. (2020) and Chibani et al. (2017) recommended micropropagation, re-introduction, seed collections, setting up specialized nurseries and national collections as 5 Ex situ conservation actions to protect the wild subpopulations of *C. spinosa* and its habitats.

Table 5.2 Molecular marker for genetic analysis of *C. spinosa* germplasm

S. no.	Marker type	Purpose	Country	References
1	AFLP	Determine the relationships among *Capparis* spp.	Spain, Morocco and Syria	Inocencio et al. (2005)
2	ISSR	Genetic diversity of *Capparis* plant (*Capparis* spp.)	Morocco	Saifi et al. (2011)
3	RAPD, ISSR	Genetic variability among the populations of *C. spinosa*	Cold arid desert of trans-Himalayas	Bhoyar et al. (2012)
4	RAPD	Genetic variation in natural populations of *Capparis*	Turkey	Özbek and Kara (2013)
5	IRAP, ISSR	Genetic diversity of some *Capparis* L. species	Syria	Al-Safadi et al. (2014)
6	ISSR	Taxonomic identification and evaluation the hybrid origin of intermediate phenotypes of *C. spinosa*	Italy	Gristina et al. (2014)
7	*rpS*12-*rpL*20, *ndh*F	Genetic diversity and differentiation in populations of *C. spinosa*	China	Wang et al. (2016)
8	AFLP	Genetic diversity and population structure	Tunisia	Aichi-Yousfi et al. (2016)
9	IRAP	Genetic diversity of *C. spinosa* group	Turkey	Çelik and Özcan (2016)
10	*rbc*L, *mat*K, ISSR	Genetic diversity and phylogeny of *Capparis* species	India	Tamboli et al. (2018)
11	ISSR	Genetic diversity and structure of *C. spinosa*	Iran	Ahmadi and Saeidi (2018)
12	ISSR, ITS	Species delimitation of *Capparis* species	Iran	Najafian et al. (2021)
13	SCoT, ISSR	Evaluation of molecular and morphological diversity	Iran	Mahmodi et al. (2022)
14	SNP	Genetic structure and evaluation of conservation strategies in *C. spinosa*	Arid Central Asia	Wang and Zhang (2022)
15	ISSR	Genetic diversity and structure of *C. spinosa*	Spain	Grimalt et al. (2022)

(continued)

Table 5.2 (continued)

S. no.	Marker type	Purpose	Country	References
16	SRAP	Assessment of mutagenesis agents on genetic variability of *C. spinosa* Genetic variation of *C. spinosa*	Egypt	Elkanany et al. (2023)
17	SRAP and ScoT ITS2, matK and trnL-trnF	Genetic diversity and structure of *C. spinosa*	Greece	Pegiou et al. (2023)
18	EST-SSR	Genetic differentiation of the *C. spinosa* group	Mediterranean Area	Melzi et al. (2024)

5.3.2.3 In Vitro Conservation

Attia et al. (2017) applied the slow growth technique for conservation of *C. spinosa* germplasm. Shoot tips and nodal buds of in vitro culture plants of *C. spinosa* were utilized as explant materials for in vitro conservation experiments. The slow growth media contain full and half strength of MS medium, sucrose and sorbitol with different combination and concentration to evaluate the plant response to different treatment of slow growth medium on the rate of survival and growth in this species. The highest rate of survival was noticed on MS medium +20 g/L sucrose and MS medium + 10 g/L sucrose + 10 g/L sorbitol, with an average of plant height (13.03 and 11.1 cm) for axillary buds and shoot tips, respectively. Their results indicated the suitable slow growth medium for in vitro conservation of *C. spinosa* was contained full strength MS with low concentrations of sucrose and sorbitol. the genetic stability of the conserved germplasm was also tested with RAPD markers and high genetic stability was obtained for this marker.

5.3.2.4 DNA Banks

Total genomic DNA of *C. spinosa* is commonly extracted with CTAB method (Doyle and Doyle 1987) or with a slight modification of this procedure. DNA samples are diluted in distilled water and preserved at −20 °C (Ahmadi and Saeidi 2018). A survey of important DNA banks for plant genetic resources conservation shows that DNA samples of *C. spinosa* group is kept in Royal Botanic Garden Kew DNA bank (https://dnabank.science.kew.org/) and the DNA bank of the Botanic Garden and Botanical Museum Berlin (https://www.bgbm.org/en/dna-bank).

5.3.3 *Cytogenetics*

It has been suggested that x = 7 is probably the primary basic chromosome number of the Capparaceae family and a number of secondary basic numbers are derived from it. Polyploidy and aneuploidy play an important role in the evolution and speciation of this family (Raghavan and Venkatasubban 1940). Intraspecific diversity of chromosome numbers is common in the genus *Capparis*, which may be due to aneuploidy, but misidentification of samples or incorrect counting may have been involved (Kers 2003; Subramania and Susheela 1988). 2n = 20, 2n = 24, 2n = 38 and 2n = 40 has been reported for *C. spinosa* group (Baquar et al. 1966; Magulaev 1976; Sandhu and Mann 1989; Fici et al. 1995; Al-Turki et al. 2000).

5.4 Molecular Breeding

There has been not much literature available on the molecular marker-assisted breeding, bioinformatics and genetic engineering in this crop.

5.4.1 *Functional Genomics*

Mercati et al (2019) developed the first transcriptome profile of *C. spinosa* subsp. *rupestris* from its vegetative tissue (leaf) and described a number of biological pathways of *C. spinosa* involved in abiotic stress tolerance and the production of bioactive compounds in detail. pathways such as purine, thiamine and phenylpropanoid biosynthesis, α-linolenic acid and lipid metabolism, genes involved in development and distribution of stomata and encoding for Stress Associated Proteins (SAPs) were studied. The first panel of codominant markers (EST-SSR) in caper was also developed. Alzahrani et al (2021a, b) reported the complete plastid genome of *C. spinosa* with 157,728 bp length. Two inverted repeats (IRs) of 26,337 bp have been divided by a large single-copy (LSC) region of 86,732 bp and a small single-copy (SSC) region of 18,322 bp. There are 136 genes in the plastid genome of *C. spinosa* that 116 of them are found in the single copy regions, while the rest are located in the inverted repeats region. The plastome also contained 80 protein coding genes, 31 tRNA genes and 4 rRNA genes. Moreover, Kianersi et al. (2020) identified and isolated the partial cDNA of four genes including 4-coumaroyl CoA ligase (4CL), flavonoid 3′-hydroxylase (F3′H), flavonol synthase (FLS) and flavonol-3-O-glucoside L-rhamnosyltransferase (RT) encoding rutin biosynthesis enzymes in *C. spinosa*, for the first time. They evaluated the effects of salicylic acid (SA) and methyl jasmonate (MeJA) on the accumulation of rutin and the expression of rutin biosynthetic pathway genes in this species. The results of their investigation revealed that MeJA and SA treatments enhance the rutin content remarkably in different tissues of *C. spinosa* and the expression patterns of genes involved in rutin biosynthesis are regulated by these chemical elicitors. The highest and the lowest amount of rutin were detected in young leaves and fruits, respectively. Transcriptome analysis will provide valuable information for breeding programs and the development of improved cultivars to sustainable food production in arid conditions. The chloroplast genome sequences are useful in species identification, population genetic and evolutionary studies.

5.5 Mutation Breeding

Mutation breeding is the method of using artificial mutagenesis to development of new biological cultivars, mainly through chemical or physical mutagenic agents (Ma et al. 2021; Penna et al. 2023). The use of physical mutagens (electromagnetic and particle radiation) is more common than chemical (such as ethyl methane sulfonate [EMS]) agents to artificially induce mutations (Beyaz and Yildiz 2017). Gamma radiation is one of the most frequently used physical mutagenic agents due to its ability to induce high genetic variability in plant species (Riviello-Flores et al. 2022; Bharath et al. 2024).

5.5.1 Conventional Mutagenesis

Bhoyar et al. (2010) treated caper seeds with different doses of gamma rays along with concentrated H_2SO_4 treatment for different durations. Maximum germination of 26% was observed at 30 Kr gamma rays. Bahmani et al. (2016) investigated the effect of different doses of gamma radiation on seed germination factors of Caper species. Although length and weight of the stem let, root weight and seed vigor increased at 100 Gray dose compared with the control, gamma irradiation had not major effect on the improvement of seed germination characteristics in *C. spinosa* var. *parviflora*.

5.5.2 *In Vitro* Mutagenesis and Selection

Al-Safadi and Elias (2011) studied the effect of different gamma ray doses (50, 100, and 150 Gy) on the seed germination of *C. spinosa*. The best results were obtained at 100 Gy dose of gamma rays after culturing in vitro and in the peat moss. The impact of irradiation on the growth of caper shoot was also investigated in vitro. irradiation of the shoots with 3 doses of gamma rays and subsequent culture on the MS medium without any growth regulators showed that the 10 Gy dose stimulated shoot growth. The same irradiation dose also increased the shoot length, shoot rooting percentage and the number of roots per plant. Elkanany et al. (2023) assessed the effect of different physical (gamma and ultraviolet rays) and chemical (caffeine and colchicine) mutagenesis agents on the appearance, chemical composition and genetic variability of *C. spinosa* and the results of their study revealed that ultraviolet rays had best performance among the studied agents and the number of leaves, shoots, rutin content as well as shoot length increased by using this agent for 4 h. Saleh et al. (2024) reported that the exposure of nodal explants of *C. spinosa* to gamma radiation with a dose of 0.1 Gy as well as UV-C rays accelerate the content of flavonoid and rutin compared to the control plants.

5.5.3 Molecular Analysis

Elkanany et al. (2023) utilized SRAP markers and SDS-protein electrophoresis to assess the influence of mutagenesis agents on genetic variability of *C. spinosa.* SRAP markers produced unique alleles for gamma ray 1.5 k gray, colchicine 50 mg/1 and caffeine 50 mg/1. Analyzing data obtained from SRAP markers indicated the highest and the lowest level of genetic similarity for treatment with colchicine (50 mg/l and 100 mg/ l) and Gamma ray 0.5 k gray compared to the control samples, respectively. The analysis of SDS-protein data confirmed the results of SRAP analysis, but the lowest level of similarity was also found for ultraviolet ray (2 and 4 h).

5.5.4 Enhanced Traits and Improved Cultivars

Little work has been done in mutational breeding of *C. spinosa* and they are focused on seed germination or seedling growth (Bhoyar et al. 2010; Al-Safadi and Elias 2011; Bahmani et al. 2016). Plant growth and seed vigor can be increased by using gamma irradiation, but these experimental studies haven't been utilized in a practical manner to develop new varieties of caper.

5.6 Hybridization

5.6.1 Conventional Hybridization

Several studies on *C. spinosa* group indicated the occurrence of hybridization between different taxa of this group in nature (Rivera et al. 2002; Chedraoui et al. 2017; Gristina et al. 2014; Ahmadi et al. 2020). According to Rivera et al. (2002) Hybridization could easily occur between *C. sicula* and C. *orientalis* and there are many intermediate phenotypes in culture which have led to a large selection of cultivated varieties, most of them within *C. spinosa.* The neighboring habitats and entomophilous pollination have given rise to high level of gene flow between cultivated and wild populations (Rivera et al. 2002). Rivera et al. (2002) described *C. spinosa* as a hybrid of *C. orientalis* and *C. sicula.* Their suggestion was later confirmed by Inocencio et al (2005) who applied AFLP markers to determine genetic relationship among *Capparis* species in Spain, Morocco and Syria. In their study *C. spinosa* occupied the intermediate position of *C. sicula* and *C. orientalis.* Gristina et al. (2014) characterized wild populations and cultivated forms of *C. spinosa* using ISSR markers and morphological data. Their investigation supported the hybrid origin of the intermediate phenotypes of *C. spinosa* subsp. *spinosa* and subsp. *rupestris* found in contact zones. Propagation by grafting could also offer an interesting perspective for developing caper hybrids (Chedraoui et al. 2017).

5.6.2 Hybrid Cultivars

Most capers and caper berries collected from wild and cultivated plants in producing countries is referred to *C. spinosa* s. str. that morphological and molecular evidence suggest the hybrid origin this species (Inocencio et al., 2005). Cultivated form are belonged to two subspecies of *C. spinosa,* namely subsp. *spinosa* and subsp. *rupestris.* Gristina et al (2014) and Grimalt et al. (2022) reported the hybrid origin of some cultivated forms between these two subspecies (Gristina et al. 2014; Grimalt et al. 2022). To the best of our knowledge, there is no experimental study for developing new cultivars of caper from hybridization and farmers often select cultivated forms based on highly appreciated traits from local populations.

5.7 Conclusion and Future Prospects

So far, a good number of studies have reported on the population genetic, in vitro propagation, traditional pharmacological effects and phytochemical evaluation of *C. spinosa. C. spinosa* has been included among the orphan crops because this species has received a little attention regarding the selection of well-adapted, valuable and easily propagated candidate cultivars. This crop has a recent history and it is considered as a semi-domesticated or a pre-domesticated crop. Based on the present knowledge on *C. spinosa,* researchers are not able to determine the characteristics of the genetic material of cultivated forms with any degree of certainty. Therefore, it is said that there is no caper cultivar. Individual plants that are used in production don't have any certain origin and they come from seedlings or from cuttings of harvested plants in the wild. The taxonomical confusion of the *C. spinosa* group and the lack of genomic resources have hampered breeding programs and the efficient utlization of this neglected crop (Mercati et al. 2019; Sottile et al. 2021; La Bella et al. 2021).

There is a need to focus on genetic improvement programs for the modernization of the caper culture and it can be started by paying attention to the huge genetic potential which exists in both wild and cultivated caper.

The world is experiencing an abrupt change in climate due to global warming. Excess CO_2 and high temperatures have an impact on soil salinization and agriculture production, leading to the issue of water and food security (Ullah et al. 2021). During the recent decades temperature has increased faster than the global average in the Mediterranean area and model projections confirm the future warming and drying of this region, with a likely rise of heat waves and dry spells (Lionello et al. 2014). Mediterranean agriculture has to cope with the new environmental conditions by cultivation of drought tolerant crop species. One option is selection and introduction of stress-tolerant cultivars of existing crops, but this process is time consuming and cost effective due to intensive research and field work. Promoting drought resistant crop species is an alternative option. *C. spinosa*, a xerophytic shrub, has been suggested as a potential crop to combat the threat of global warming in countries having Mediterranean climates (Chedraoui et al. 2017; Mercati et al, 2019). In fact, climate change will have a minor impact on potential distribution areas of *C. spinosa* and its distribution may increase in the arid, semi-arid, and Mediterranean climate regions in the future (Ashraf et al. 2018). Tolerance to drought and salt stress allow *C. spinosa* to adapt to different ecological amplitude and to withstand harsh environments. The ability of survival at high temperatures of summertime up to 40 °C, growing in nutrient-poor soils and deep penetration into the earth by having an extensive root system have made caper a well-adapted crop to drought stress conditions (Ravindran et al. 2004). Therefore, caper cultivation in the long-term agricultural strategy could be a good solution to combat climate change impacts in the future.

Development of genomics resources is a priority research need for taxonomic identification, the improvement and modernization of the caper cultivation and to provide a comprehensive view of the biology and evolutionary history of this species. Genetic variation of different genotypes for tolerance to drouth stress should be studied for screening germplasm and selecting cultivars with high resilience genotypes under drought stress condition. Selection of hybrids and mutation breeding can also reveal the high genetic potential of this crop. Identification of molecular marker -assisted breeding and the development of useful tools for genetic modification of caper are also necessary for promoting stress related breeding programs and to improve desirable traits for economic purposes. *C. spinosa* by having a rich source of bioactive compounds and different health benefit effects on human diseases is a suitable candidate for the discovery of new drugs. Nonetheless, these findings are still in the early stage and more experimental and clinical trials are required to evaluate the medicinal potential of caper in modern pharmaceutical science.

References

Abo El-Fadl R, Ahmed ME, Abd Elaziem TM (2021) Micropropagation, conservation, molecular and biochemical studies on *Capparis spinosa* var. *deserti*, a wild endangered plant in Egyptian flora. EJDR 71(2):209–243

Ahmadi M, Saeidi H, Mirtadzadini M (2020) Species delimitation in *Capparis spinosa* group (Capparaceae) in Iran: Multivariate morphometric analysis and ISSR markers. Phytotaxa 456(1):1–26

Aichi-Yousfi H, Bahri BA, Medini M (2016) Genetic diversity and population structure of six species of *Capparis* in Tunisia using AFLP markers. C R Biol 339:442–453

AL-Azawi AH, Ghaima KK, Salih HH (2018) Phytochemical, antibacterial and antioxidant activities of *Capparis spinosa* L. Cultivated in Iraq. Biosci Res 15(3):2611–2618

Aliyazicioglu R, Eyupoglu OE, Sahin H (2013) Phenolic components, antioxidant activity, and mineral analysis of *Capparis spinosa* L. Afr J Biotechnol 12(47):6643–6649

Al-Safadi B, Elias R (2011) Improvement of caper (*Capparis spinosa* L.) propagation using in vitro culture and gamma irradiation. Sci Hortic 127(3):290–297

Al-Safadi B, Faouri H, Elias R (2014) Genetic diversity of some *Capparis* L. species growing in Syria. Braz Arch Biol Technol 57(6):916–926

Al-Turki TA, Filfilan SA, Mehmood SF (2000) A cytological study of flowering plants from Saudi Arabia. Willdenowia, 339–358

Alzahrani D, Albokhari E, Yaradua S, Abba A(2021a) The complete plastome sequence for the medicinal species *Capparis spinosa* L. (Capparaceae). Gene Rep 23:101059

Alzahrani D, Albokhari E, Yaradua S, Abba A (2021b) The complete chloroplast genomes of two species from Capparaceae. Mitochondrial DNA B 6(2):601–603

Anand A (2014) Magnetopriming- an alternate strategy for crop stress management of field crops. In: Proceedings of the international conference on radiation biology: frontiers in radiobiology-immunomodulation, countermeasures and therapeutics: abstract book, souvenir and scientific programme

Annaz H, Sane Y, Bitchagno GTM et al (2022) Caper (*Capparis spinosa* L.): an updated review on its phytochemistry, nutritional value, traditional uses, and therapeutic potential. Front Pharmacol 13:878749

Angiosperm Phylogeny Group, Chase MW, Christenhusz MJ et al (2016) An update of the Angiosperm Phylogeny Group classification for the orders and families of flowering plants: APG IV. Bot J Linn Soc 181(1):1–20

Ashraf U, Chaudhry MN, Ahmad SR et al (2018) Impacts of climate change on *Capparis spinosa* L. based on ecological niche modeling. PeerJ 6:e5792

Assadi S, Shafiee SM, Akmali EM (2021) Antioxidative and antidiabetic effects of *Capparis spinosa* fruit extract on high-fat diet and low-dose streptozotocin-induced type 2 diabetic rats. Biomed Pharmacother 138:111391

Attia AO, Dessoky EDS, Al-Sodany YM, Ismail IA (2017) Ex situ preservation for some endemic and rare medicinal plants in Taif, KSA. Biotechnol Biotechnol Equip 31(5):912–920

Bahmani M, Yousefi S, Kartolinezhad D (2016) The effects of gamma radiation on seed germination and vigor of Caper (*Capparis spinosa* var. parviflora) medicinal plant Iran J Seed Res 3(1):15–26

Basbag M, Toncer O, Basbag S (2009) Effects of different temperatures and duration on germination of caper (*Capparis ovata*) seeds. J Environ Biol 30(4):621–624

Baquar SR, Husain AFAQ, Akhtar SHAMIM (1966) Meiotic chromosome numbers in some vascular plants of Indus delta II. Bot Not 119:24–32

Bhatia S (2015) Application of Plant Biotechnology. In: Bhatia S, Sharma K, Dahiya R, Tanmoy B (eds) Modern applications of plant biotechnology in pharmaceutical sciences. Academic Press

Benzidane N, Aichour R, Guettaf S et al (2020) Chemical investigation, the antibacterial and antifungal activity of different parts of *Capparis spinosa* extracts. J Drug Deliv 10(5):118–125

Beyaz R, Yildiz, M (2017) The use of gamma irradiation in plant mutation breeding. Plant Eng, 34–46

Bhandari HR, Bhanu AN, Srivastava K et al (2017) Assessment of genetic diversity in crop plants-an overview. Adv Plants Agric Res 7(3):279–286

Bharath RA, Prathmesh SP, Sarsu F, Suprasanna P (2024) (2024) Induced Mutagenesis using Gamma Rays: Biological Features and Applications in Crop Improvement. OBM Genetics 8(2):233. https://doi.org/10.21926/obm.genet.2402233

Bhoyar MS, Mishra GP, Singh R, Singh SB (2010) Effects of various dormancy breaking treatments on the germination of wild caper (*Capparis spinosa*) seeds from the cold arid desert of trans-Himalayas. Indian J Agric Sci 80(7):621

Bhoyar MS, Mishra GP, Naik PK, Murkute et al (2012) Genetic variability studies among natural populations of *Capparis spinosa* from cold arid desert of trans-Himalayas using DNA markers. Natl Acad Sci Lett 35(6):505–515

Boissier E (1867) Flora Orientalis sive enumeratio plantarum in Oriente a Graecia et Aegypto ad Indiae fines hucusque observatarum. Vol. 1, H. Georg, Basel, Genèv.

Benaoun H, Zouaoui R, Elaloui M et al (2023) Cuttings Vegetative Multiplication of Smooth Caper (*Capparis spinosa* Inermis). J Biotech Biores 4(4):594–600

Carra A, Del Signore MB, Sottile F et al (2012a) Potential use of new diphenylurea derivatives in micropropagation of *Capparis spinosa* L. J Plant Growth Regul 66:229–237

Carra A, Sajeva M, Abbate L et al (2012b) In vitro plant regeneration of caper (*Capparis spinosa* L.) from floral explants and genetic stability of regenerants. PCTOC 109(2):373–381

Çelik M, Özcan T (2016) Analysis of some *Capparis* L. accessions from Turkey based on IRAP and seed protein pattern. Internat J Environ Agri Res 2:43–55

Chalak L, Elbitar A (2006) Micropropagation of *Capparis spinosa* L. subsp. rupestris Sibth. & Sm. by nodal cuttings. Indian J Biotechnol 5:555–558

Chedraoui S, Abi-Rizk A, El-Beyrouthy M et al (2017) *Capparis spinosa* L. in a systematic review: a xerophilous species of multi values and promising potentialities for agrosystems under the threat of global warming. Front Plant Sci 8:1845

Chibani F, Skouri-Gargouri H, Salem AB et al (2017) Using genetic structure data and phylogenetic criteria in attributing prioritization scores for conservation of spontaneous *Capparis spinosa* L. populations from Tunisia. J Nat Conserv 37:96–105

Danin A (2010) Capparis in the East Mediterranean countries. Flora Mediterr 20:179–185

Doyle JJ, Doyle JL (1987) A rapid DNA isolation procedure for small quantities of fresh leaf tissue. Bull. Bot Soc Am 19:11–15

Duman E, Özcan MM (2014) Mineral contents of seed and seed oils of *Capparis* species growing wild in Turkey. Environ Monit Assess 186(1):239–245

Eid AM, Hawash M et al (2023) Exploring the potent anticancer, antimicrobial, and anti-inflammatory effects of *Capparis spinosa* oil nanoemulgel. Coatings 13(8):1441

El amri N, Errachidi F, Abdellatif Bour A et al (2019a) Morphological and nutritional properties of Moroccan *Capparis spinosa* Seeds. Sci World J, 1–8

El amri N, Errachidi F, Bour A, Chabir R (2019b) Characterization of Moroccan raw and processed caper berries. Mater Today: Proc 13:841–849

Elazazi, EM (2016) Effective methods to improvement *Capparis spinosa* L. (Caper) seeds germination by breaking seed dormancy in Qatar Gene Bank. In: Qatar Foundation Annual Research Conference Proceedings. Hamad bin Khalifa University Press (HBKU Press)

Elkanany EE, Serag AM, Abdelwahab MH et al (2023) Molecular evaluation of *Capparis spinosa* and its antibacterial effects on the activity of bacillus subtilis. JACB 14(11):153–164

Elmaghrabi AM, Abughnia E, Hamoud S (2017) In vitro propagation of the wild medicinal plant, caper (*Capparis spinosa* L.). Afr J Biotechnol, 1684–5315

El-Shahaby O, El-Zayat M, El-Fattah A, El-Hefny M (2019) Evaluation of the biological activity of *Capparis spinosa* var. *aegyptiaca* essential oils and fatty constituents as Anticipated Antioxidant and Antimicrobial Agents. Prog Chem Biochem Res 2(4):211–221

Fakhri F, Kaboli S, JafariA BP (2019) Methods for improving seed germination of *Capparis spinosa.* Iranian J Range Desert Res 2(75):388–399

Farhangi A, Mohammadi A, Zamani F et al (2019) Antidiabetics activity of water extract of *Capparis spinosa* on STZ-induced diabetics Rat. J Adv Pharm Res 9(S2):149
Fici S (2014) A taxonomic revision of the *Capparis spinosa* group (Capparaceae) from the Mediterranean to Central Asia. Phytotaxa 174(1):001–024
Fici S (2015) A taxonomic revision of the *Capparis spinosa* group (Capparaceae) from eastern Africa to Oceania. Phytotaxa 203(1):024–036
Fici S, Campo G, Chifari N, and Colombo P (1995) Biosystematic researches on the *Capparis spinosa* L. complex: preliminary anatomical and karyological data. Giornale Botanico Italiano, p 35
Gianguzzi V, Barone E, Sottile F (2020) In Vitro rooting of *Capparis spinosa* L. as affected by genotype and by the proliferation method adopted during the multiplication phase. Plants 9(3):398
Grimalt M., García-Martínez S, Carbonell P et al (2022) Relationships between chemical composition, antioxidant activity and genetic analysis with ISSR markers in flower buds of caper plants (*Capparis spinosa* L.) of two subspecies spinosa and rupestris of Spanish cultivars. Genet Resour Crop Evol 69(4):1451–1469
Gristina A, Fici S et al (2014) Hybridization in *Capparis spinosa* L. Molecular and morphological evidence from a Mediterranean island complex. Flora 209:733–741
Govindaraj M, Vetriventhan M, Srinivasan M (2015) Importance of genetic diversity assessment in crop plants and its recent advances: an overview of its analytical perspectives. Genet Res Int, 1–14
Gull T, Sultana B, Bhatti IA, Jamil A (2015) Antibacterial potential of *Capparis spinosa* and *Capparis decidua* extracts. Int J Agric Biol 17:727–733
Heydariyan M, Basirani N et al (2014) Effect of seed priming on germination and seedling growth of the caper (*Capparis spinosa*) under drought stress. Int j Adv Biol 2(8):2381–2389
Heywood VH (1964) *Capparis* L. In: Tutin TG, Heywood VH, Burges NA, Valentine DH et al (eds) Flora Europaea. Vol. 1, 1st ed. Cambridge University Press, Cambridge
Higton RN, Akeroyd JR (1991) Variation in *Capparis spinosa* L. in Europe. Bot J Linn Soc 106:104–112
Inocencio C, Cowan RS, Alcaraz F et al (2005) AFLP fingerprinting in *Capparis* subgenus *Capparis* related to the commercial sources of capers. Genet Resour Crop Evol 52(2):137–144
Inocencio C, Rivera D, Obón MC et al (2006) A systematic revision of *Capparis* section *Capparis* (Capparaceae). Ann Missouri Bot Gard 1:122–149
Jacobs M (1965) The genus *Capparis* (Capparaceae) from the Indus to the Pacific. Blumea 12(3):385–541
Kazemian M, Abad M, Haeri MR et al (2015) Anti-diabetic effect of *Capparis spinosa* L. root extract in diabetic rats. Avicenna J Phytomed 5(4):325–332
Kers LE (2003) Capparaceae. In: Kubitzki K, Bayer C (eds) Flowering plants: dicotyledons. The families and genera of Vascular plants. Vol. 5. Springer, Berlin, Heidelberg
Kianersi F, Abdollahi MR, Mirzaie-Asl A (2020) Identification and tissue-specific expression of rutin biosynthetic pathway genes in *Capparis spinosa* elicited with salicylic acid and methyl jasmonate. Sci Rep 10(1):1–5
Koufan M, Belkoura I, Mazri MA (2022) In vitro propagation of caper (*Capparis spinosa* L.): A review. Horticulturae 8(8):737
La Bella S, Rossini F, Licata M (2021) Four-year study on the bio-agronomic response of biotypes of *Capparis spinosa* L. on the island of Linosa (Italy). Agriculture 11(4):327
Linnaeus C (1753) Species plantarum, exhibentes plantas rite cognitas, ad genera relatas, cum differentiis specificis, nominibus trivialibus,synonymis selectis, locis natalibus, secundum systema sexuale digestas, Vol. 1, 1st ed. Impensis Laurentii Salvii, Holmiae
Lionello P, Abrantes F, Gacic M et al (2014) The climate of the Mediterranean region: research progress and climate change impacts. Reg Environ Change 14(5):1679–1684
Lemhadri A, Eddouks M, Sulpice T, Burcelin R (2007) Anti-hyperglycaemic and anti-obesity effects of *Capparis spinosa* and *Chamaemelum nobile* aqueous extracts in HFD mice. Am J Pharmacol Toxicol 2:106–110

Ma L, Kong F, Sun K, Wang T, Guo T (2021) From classical radiation to modern radiation: past, present, and future of radiation mutation breeding. Front Public Health 9:768071

Magulaev AJ (1976) The chromosome numbers of flowering plants of the Northern Caucasus (Part II). The Flora of the Northern Caucasus. 2:51–62

Mahmodi N, Sharifi-Sirchi GR, Cheghamirza K (2022) Evaluation of molecular and morphological diversity of caper (*Capparis spinosa* L.). Genet Resour Crop Evol 69(4):1–26

Mahmoodpoor F, Hosseini SH, Ahmadian E et al (2023) The histopathological evaluation of the effects of *Capparis spinosa* seed hydroalcoholic extract on Cisplatin-induced-nephrotoxicity in rats. J Adv Med Med Res 31(144):73–79

Mazarei F, Jooyandeh H, Noshad M, Hojjati M (2017) Polysaccharide of caper (*Capparis spinosa* L.) leaf: extraction optimization, antioxidant potential and antimicrobial activity. Int J Biol Macromol 95:224–231

Maurya S, Cornejo X, Lee C (2023) Molecular phylogenetic tools reveal the phytogeographic history of the genus *Capparis* L. and suggest its reclassification. PPEES 58:125720

Melzi Y, La Bella F, Gristina AS et al (2024) Genetic differentiation of the *Capparis spinosa* group in the Mediterranean area. PPEES 63:1–9

Mercati F, Fontana I, Gristina AS et al (2019) Transcriptome analysis and codominant markers development in caper, a drought tolerant orphan crop with medicinal value. Sci Rep 9(1):1–16

Moghadamnia Y, Kani SN, Ghasemi-Kasman M et al (2019) The anti-cancer effects of *Capparis spinosa* hydroalcoholic extract. Avicenna J Med Biotechnol 11(1):43–47

Moutia M, El Azhary K, Elouaddari A et al (2016) Capparis spinosa L. promotes anti-inflammatory response in vitro through the control of cytokine gene expression in human peripheral blood mononuclear cells. BMC Immunol 17(1):1–12

Movafeghi A, Habibi G, Ali AM (2008) Plant regeneration of Capparis spinosa L. using hypocotyl explants. Iranian journal of biology 21(2):289–297

Ng TB, Lam SK, Cheung RC et al (2011) Therapeutic use of caper (*Capparis spinosa*) seeds. In: Preedy VR, Watson RR (eds) Nuts and seeds in health and disease prevention. Academic Press, pp 279–284

Najafian S, Mehregan I, Iranbakhsh A, Assadi M, Fici S (2021) Species delimitation in *Capparis* (Capparaceae): Morphological and molecular. Genetika 53(2):609–627

Olmez Z, Gokturk A, Gulcu S (2006) Effects of cold stratification on germination rate and percentage of caper (*Capparis ovata* Desf.) seeds. J Environ Biol 27(4):667–670

Özbek Ö, Kara A (2013) Genetic variation in natural populations of *Capparis* from Turkey, as revealed by RAPD analysis. Plant Syst Evol 299(10):1911–1933

Pegiou S, Raptis P, Zafeiriou I et al (2023) Genetic diversity and structure of *Capparis spinosa* L. natural populations using morphological and molecular markers. J Appl Res Med Aromat Plants 34:100487

Penna, S., Shirani Bidabadi, S., Jain, S.M. (2023). Mutation breeding to promote sustainable agriculture and food security in the era of climate change. In: Penna S, Jain SM (eds) Mutation breeding for sustainable food production and climate resilience. Springer, Singapore, pp 1–23. https://doi.org/10.1007/978-981-16-9720-3_1

Raghavan TS, Venkatasubban KR (1940) Studies in the Capparidaceae. VIII. The cytology of Capparis zeylanica Linn. and related genera. Cytologia 11:319–331

Rankou H, M'Sou S, Diarra A, Ait Babahmad RA (2020) *Capparis spinosa*, Caper. The IUCN Red List of Threatened Species

Rao VR, Riley KW (1994) The use of biotechnology for conservation and use of plant genetic resources. Plant Genetic Resources Newsletter, 97

Ravindran PN, Pillai GS, Babu KN (2004) Under-utilized herbs and spices. In: Peter KV (ed) Handbook of herbs and spices, vol 2, pp 53–103

Rivera D, Inocencio C, Obon C et al (2002) Archaeobotany of capers (*Capparis*)(Capparaceae). Veg Hist Archaeobotany 11(4):295–314

Riviello-Flores MDLL, Cadena-Iñiguez J, Ruiz-Posadas LDM et al (2022) Use of gamma radiation for the genetic improvement of underutilized plant varieties. Plants 11(9):1161

Rodriguez R, Rey M, Cuozzo L, Ancora G (1990) In vitro propagation of caper (*Capparis* spinosa L.). Vitro Cell Dev Biol Plant 26(5):531–536
Saifi N, Ibijbijen J, Echchgadda G (2011) Genetic diversity of *Capparis* plant (*Capparis* ssp.) from north Morocco. JFAE 9(3&4):299–304
Saleh SS, M Serag El-Din, W, Youssef SM (2024) Enhancement Rutin Production from *Capparis spinosa* Plant by UV-C or Gamma Irradiation using In vitro Culture. EJOH 51(1):41–59
Sakcali MS, Bahadir H, Ozturk M (2008) Ecophysiology of *Capparis spinosa* L.: a plant suitable for combating desertification. Pak J Bot 40(4):1481–1486
Sandhu PS, Mann SK (1989) SOCGI plant chromosome number reports—VIII. CYTOL GENET 24:179–183
Shahrajabian MH, Sun W, Cheng Q (2021) Plant of the Millennium, Caper (*Capparis spinosa* L.), chemical composition and medicinal uses. Bull Natl Res Cent 45(1):1–9
Sheikhi M, Fahmideh L, Benakashani F, Solouki M (2020) Investigation of The Direct Regeneration of *Capparis spinosa* L. in MS Medium Containing Different Hormonal. J Crop Breed 12(34):54–61
Sher H, Alyemeni MN (2010) Ethnobotanical and pharmaceutical evaluation of *Capparis spinosa* L, validity of local folk and Unani system of medicine. J Med Plants Res 4(17):1751–1756
Sottile F, Caltagirone C, Peano C. et al (2021) Can the caper (*Capparis spinosa* L.) still be considered a difficult-to-propagate Crop? Horticulturae 7(9):316
Subramania D, Susheela G (1988) Cytotaxonomical studies of South Indian Capparidaceae. Cytologia 53:679–684
Swarup S, Cargill EJ, Crosby K et al (2021) Genetic diversity is indispensable for plant breeding to improve crops. Crop Sci 61(2):839–852
Tamboli AS, Yadav PB, Gothe AA (2018) Molecular phylogeny and genetic diversity of genus *Capparis* (Capparaceae) based on plastid DNA sequences and ISSR markers. Plant Syst Evol 304:205–217
Tlili N, Munne-Bosch SE, Nasri N et al (2009) Fatty acids, tocopherols and carotenoids from seeds of Tunisian caper "*Capparis spinosa*". J. Food Lipids 16(4):452–464
Tlili N, Khaldi A, Triki S, Munné-Bosch S (2010) Phenolic compounds and vitamin antioxidants of caper (*Capparis spinosa*). Plant Foods Hum Nutr 65(3):260–265
Ullah A, Bano A, Khan N (2021) Climate change and salinity effects on crops and chemical communication between plants and plant growth-promoting microorganisms under stress. Front Sustain Food Syst 5:618092
Ulukapi K, Özdemir B, Kulcan AA et al (2016) Evaluation of biochemical and dimensional properties of naturally grown *Capparis spinosa* var. *spinosa* and Capparis ovata var. palaestina. Int J Agric Innov Res 5(2):195–200
Vahid H, Rakhshandeh H, Ghorbani A (2017) Antidiabetic properties of *Capparis spinosa* L. and its components. Biomed Pharmacother 92:293–302
Wang Q, Zhang HX (2022) Population genetic structure and biodiversity conservation of a relict and medicinal subshrub *Capparis spinosa* in arid Central Asia. Diversity 14(146):1–16
Wang Q, Zhang ML, Yin LK (2016) Genetic diversity and population differentiation of *Capparis spinosa* (Capparaceae) in Northwestern China. Biochem Syst Ecol 66:1–7
Wang L, Fan L, Zhao Z et al (2022) The *Capparis spinosa* var. *herbacea* genome provides the first genomic instrument for a diversity and evolution study of the Capparaceae family. GigaSci 11:106
Zhou H, Jian R, Kang J et al (2010) Anti-inflammatory effects of caper (*Capparis spinosa* L.) fruit aqueous extract and the isolation of main phytochemicals. J Agric Food Chem 58(24):12717–12721
Zohary M (1960) The species of *Capparis* in the Mediterranean and the Near Eastern countries. Bull Res Council Israel 8D:49–64

Chapter 6
Breeding Strategies of *Isatis indigotica* (Chinese Woad)

Özlem Özbek

Abstract *Isatis indigotica* Fortune belongs to the genus *Isatis* L. in the Brassicaceae family, and widely grown in China. It produces indican and isatan B as the source of indigo dye, while terpenoids, lignans and indole alkaloids as effective metabolites against bacteria, viruses, and fungi. There is a high demand for plant secondary metabolites, but the production is in shortage. Although there is no any comprehensive conventional breeding study on *I. indigotica*, genetic engineering and tissue culture methods are promising. Biotechnological methods such as somatic hybridization and protoplast fusion methods enabled hybrid plants, which are impossible in conventional cross. Transgenik pest resistant tetraploid *I. indigotica* plants could be produced efficiently through *Agrobacterium tumefaciens* mediated transformation. The cloned tetraploid *I. indigotica* plants with genetic engineering could be used in conventional breeding studies as parental lines. The concentrations of effective secondary compounds such as indigo, indurubin, amino acids, anti-endotoxic and antiviral compounds were higher due to the larger organs of tetraploids, which obtained in tissue culture and displayed higher resistance to insects, cold, drought and diseases than their diploids. However, the demand for tetraploid *I. indigotica* as a medicinal plant has increased, but the demand cannot be met sufficiently. For future perspective, the molecular and conventional breeding methods might lead to improved varieties and increased production of the effective metabolites, which can be used in alternative medicine. Considering the characteristics of *I. indigotica*, the expansion and development of its production around the world will turn into a great health benefit for community.

Keywords *Isatis indigotica* · Chinese Woad · Indigo · Isatidis Radix · Isatidis Folium · and Isatidis Naturalis · Chinese Traditional Medicine

---

Ö. Özbek (✉)
Department of Molecular Biology and Genetics, Faculty of Arts and Science, Hitit University, Çorum, Türkiye
e-mail: ozlemozbek@hitit.edu.tr

J. M. Al-Khayri et al. (eds.), *Biodiversity and Genetic Improvement of Medicinal and Aromatic Plants I*, Advances in Plant Breeding Strategies 9,
https://doi.org/10.1007/978-3-031-81846-2_6

6.1 Introduction

Medicinal and aromatic plants have recently started to attract great attention, because the medicinally effective metabolites are used in alternative medicine in the treatment of various diseases caused by bacteria, viruses, and fungi (Bhowmick et al. 2024). Apart from that, due to their antimicrobial effects, extracts obtained from plants are used in packaging in the food industry, surface cleaning disinfectant in the cleaning industry, spice, and seasoning the meals. Some plants play an important role in the textile industry because they are used in the production of natural dyes such as indigo. In this chapter, *Isatis indigotica* Fortune ex Lindl, which is a medicinal plant in the production of indigo dye (Chinese woad or thein-cheing), will be discussed. Although *I. indigotica* is native to Central Asia, it also distributed in regions such as Europe, America, and somewhere else in the world. It is a difficult plant to cultivate agriculturally, hybrid plants are often found among closely related species within the genus *Isatis* L, which means the production of sterile plants and non-productive seeds. Developing crop plants with desired traits and overcoming hybrid sterility by using biotechnological methods including conventional and molecular breeding studies, genetic engineering and tissue culture methods in medicinal and aromatic plants will be among the desired goals to be achieved now and in the future. The high genetic diversity levels in the gene pools of their wild relatives of these species provide important contributions to the development of new varieties or lines, since they are used as parental lines in breeding programs. For this reason, determining the genetic diversity levels and population genetic structures of existing domesticated cultivars, wild and closely related species will provide an important database in future breeding programs.

The plant family Brassicaceae (Curicifereae) or Mustard family comprises 338 known genera and around 3709 species scattered around the world (Appel and Al-Shehbaz 2003; Al-Shehbaz et al. 2006). In Brassicaceae, the tribe Isatidae includes over 90 species in eight genera recognized by Hayek (1911) and Schulz (1936) as a subtribe of Arabidae and Lepidae respectively (Al-Shehbaz et al. 2006). According to molecular studies, a monophyletic origin was observed for *Isatis* L. and *Myagrum* L. (Beilstein et al. 2006). *Isatis* L. is a Eurasian genus, the distribution area extending from Central Asia to the Middle East, and the Mediterranean region includes around 79 species (Al-Shehbaz et al. 2006) (Fig. 6.1). However, its primary dispersal area is Irano-Turanian region where most of its species (around 90%) grow (Appel and Al-Shehbaz 2003; Davis 1965). The major woad cultivars are present in this genus and found in the regions throughout Europe and China (Wong et al. 2022).

The Brassicaceae family includes many economically important crop species such as ornamental plants, vegetables, medicinal and aromatic plants, industrial and edible oil sources, feed, and seasonings (Al-Shehbaz et al. 2006). Among these, some species in the genus *Isatis* L. belonging to the tribe Isatidae stand out with their medicinal aromatic content as well as their use in the production of indigo dye. *Isatis* L. species produces two types of indoxyl-forming substances in their leaves; these are indican (Schunk 1855) and isatan B (Beijerinck 1900). When these substances

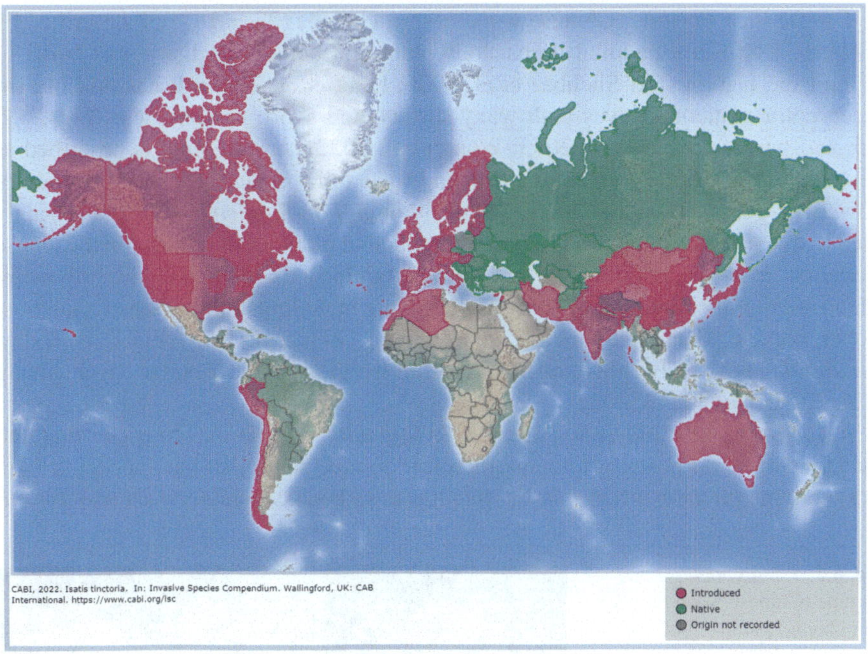

Fig. 6.1 The native and introduced worldwide distribution area of *I. tinctoria* L. Available at: *Isatis tinctoria* (dyer's woad) (cabi.org) with permission by the CABI Asset Reuse Licence

are exposed to air, they form a blue compound known as indigo (Epstein et al. 1967). The most commonly used species in the production of this dye are *Isatis tinctoria* (dyer's woad) and *Isatis indigotica* Fortune ex Lindl. (Chinese woad or tein-cheing). *I. tinctoria* is native to Central Asia (Spataro et al. 2007), and humans have transported it to southern and western Asia, Europe, North Africa, Australia, North America, and in South America Peru and Chile, (CABI 2022) (Fig. 6.1). *I. tinctoria* L. known widely as dyer's woad is a diploid (2n = 28) outbreeding, herbaceous biennial or shortly perennial (Ball 1964; Davis 1965; Darlington and Wyle 1955; Akeroid 2002; Al-Shehbaz et al. 2006; Warwick and Al-Shehbaz 2006; Spataro and Negri 2008a) plant species. It has pubescent or glabrous thin leaves oblong or oblong-lanceolate shape (Figs. 6.5 and 6.6) and the dimensions of the fruit are 10–20 × 2–5 mm, tapering from near apex to the base or oblong, glabrous or puberulent, wing thick (Davis 1965).

In ancient times, *I. tinctoria* was very actively cultivated in Europe, particularly England, France, Germany, Italy, and Switzerland between the twelfth and seventeenth centuries and was used in the production of indigo dye (Gilbert et al. 2002; Spataro and Negri 2008a). Since indigo was isolated easily from tropical indigo (i.e. *Indigofera* ssp.) than *I. tinctoria*, therefore it has been replaced by *I. tinctoria.* In the nineteenth century, with the use of synthetic cheap dyes, the cultivation of *I. tinctoria* was ceased and almost disappeared (Spataro and Negri 2008b).

I. indigotica Fortune (Chinese woad or tein-cheing) (2n = 14) first described by Fortune (1846) is a biennial herbaceous plant and belongs to the genus *Isatis* L. in the tribe Isatidae (Al-Shehbaz et al. 2006; Feng et al. 2021). It has dull bluish-green color glaucous leaves with waxy surface and short, upright stem (Figs. 6.2, 6.5, and 6.6) (Sun and Pang 2013; Wong et al. 2022). *I. indigotica* is described as having glabrous forms with oblong fruit and weak or broad auricles and yellow color flowers (Fig. 6.3) (Flora of China 2001). Although *I. indigotica* is not a plant widely used in dye production, it is an important medicinal plant used in traditional Chinese medicine. It is widely cultivated in China particularly distributed to the provinces, which are mainly continental-subtropical climate including Anhui, Gansu, Guizhou, Hebei, Jiangsu, Shaanxi, Shandong, and Zhejiang (Zou et al. 2005) (Fig. 6.4). Today, *I. indigotica* is actively cultivated in the Northern regions of the Yangtze River (Zhang et al. 2019; Wong et al. 2022).

I. tinctoria and *I. indigotica* are morphologically very similar to each other and used for similar purposes by humans, therefore some taxonomists have accepted them as synonyms for each other. *I. indigotica* has even been considered as a variety of *I.*

Fig. 6.2 *I. indigotica* Fortune (Chinese woad or tein-cheing) plants. Photo by Susan Dye Trade with permission

Fig. 6.3 The flowers of *I. indigotica* Fortune (Chinese woad or tein-cheing) Photo by Susan Dye Trade with permission

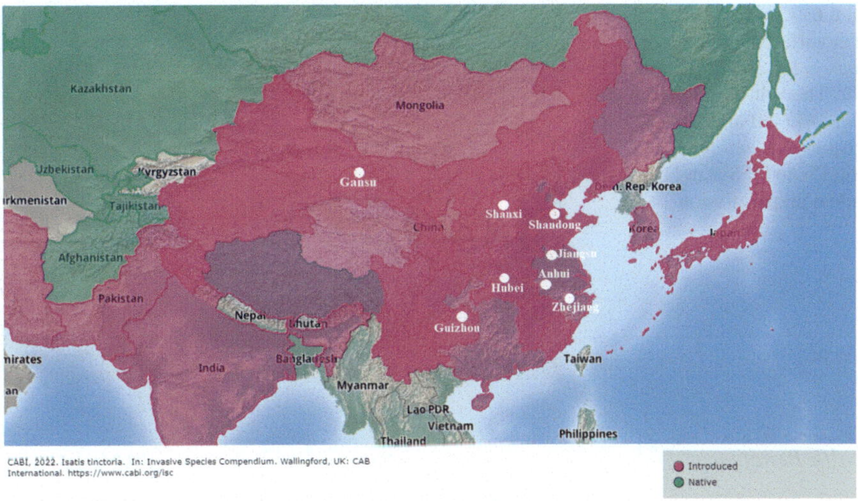

Fig. 6.4 The distribution area of *I. inctoria* L. introduced cultivar and the assigned provinces in China where the distribution area of *I. indigotica*. Available at: *Isatis tinctoria* (dyer's woad) (cabi.org) with permission by the CABI Asset Reuse Licence

Fig. 6.5 Comparison of the upper surfaces of leaves of *I. tinctoria* left and *I. indigotica* right. In *I. indigotica*, the veins and midrib are elevated, whereas in *I. tinctoria* they are sunken. Photo by Susan Dye Trade with permission

tinctoria (*I. indigotica* Fortune: *I. tinctoria* var. *indigotica* (Fortune) T. Y. Cheo and K. C. Kuan; *I. tinctoria* var., *yezoensis* (Ohwi) Ohwi; *I. yezoensis* Ohwi) in the Flora of China (2001). However, some taxonomists agreed that both are different species according to the studies on morphology, genetics, and physiology (Angelini et al. 2007; Danz et al. 2001; Sun and Pang 2013; Feng et al. 2021). Figures 6.5 and 6.6 are displaying the morphological differences between *I. tinctoria* and *I. Indigotica* leaves. The veins and midrib on the upper surface of the leave of *I. indigotica* are raised and sunken in the leave of *I. tinctoria* (Fig. 6.5). The veins are raised in *I. tinctoria* left and almost invisible in *I. indigotica* right on the lower surfaces of their leaves (Fig. 6.6).

Fig. 6.6 Comparison of the lower surfaces of leaves of *I. tinctoria* left and *I. indigotica* right. The veins are raised in *I. tinctoria* left and almost invisible in *I. indigotica* right. Photo by Susan Dye Trade with permission

Health Benefits

I. indigotica is a medicinal plant and widely cultivated in China. Especially the root and leaves of the plant are used in the production of three different products used in traditional Chinese medicine (TCM). These are Isatidis Radix, Isatidis Folium, and Isatidis Naturalis. Isatidis Radix is produced from the root of the plant and is known as Banlangen (Ban-Lan-Gen) which was used habitually in northern China and sometimes people call it Pei-Ban-Lan-Gen (Institute of Pharmaceutical Sciences 1993). The medicinal properties of Banlangen are its anti-inflammatory and antiviral effects. In addition, when further processed, Banglangen Keli is obtained. Both Banglengen and Banglangen Keli are used in epidemic diseases and flu treatment (Zhang et al. 2016a), seasonal febrile diseases, pestilence, mumps, eruptive diseases, inflammatory diseases with redness of skin, sore throat (Zhai and Guo 1987; Ho and Chang 2002; National Pharmacopeia Committee 2000), erysipelas, influenza, carbuncles, epidemic hepatitis and epidemic encephalitis B (Wu et al. 1997) and as an antipyretic (Huang et al. 1981) in China. Isatidis Folium is obtained from the dried leaves of the plant and is known as Daqingye, which is similar to Banglangen in pharmacological activities. Daqingye is widely used in the treatment of high fever, epidemic parotitis, and pharyngitis. Fufang Daqingye, which is consumed as a liquid from the mouth, is the main effective preparation of Daqingye Heji (Zhu et al. 2004). When the leaves of the plant is further processed yields the product named as indigo naturalis and is known as Qingdai, which is used to treat skin surface wounds, acne, erysipelas, and carbuncles (Chen et al. 2013).

The major compounds isolated from the dried leaves and roots of *I. indigotica* are terpenoids, lignans and indole alkaloids (Li et al. 2005; Deng et al. 2008; Chen et al. 2012; Zhou and Zhang 2013; Kang et al. 2020). Previous studies report that these compounds have antivirals (Lin et al. 2005; Sun et al. 2010), antibacterial (Xia et al. 2007), anti-inflammatory (Ho and Chang 2002; During et al. 2012) and antileukemia (Hoessel et al. 1999; Molina et al. 2001) properties. Shan et al. (2015) investigated the effects of arabinogalactan compound extracted from *I. indigotica* on H1N1 influenza and hepatitis B agents. Chen et al. (2016) analyzed the antiviral activities of four stereoisomers with antiviral activity on influenza virus A (H3N2), coxsackievirus B3 and HSV. They observed that these stereoisomers have selective antiviral effects and that this selective activity depends on the configurations of the molecules. Adjuvants are significant in vaccine studies because they increase vaccine immunogenicity and can stimulate the development of long-lasting immune responses. II-P-A-1 and II-P-2 polysaccharides with adjuvant activity isolated from the roots of *I. indigotica* were compared to the administration of the inactive rabies virus rCVS-II-G to the muscles of mice via the aluminum hydroxide, it was reported that both polysaccharides changed faster and caused higher virus neutralization assay (VNA) titers than aluminum hydroxide (Zhang et al. 2016a).

6.2 Current Cultivation Practices and Challenges

The increasing interest in organic natural dyes recently (Bechtold et al. 2002), various projects and scientific studies have started to be carried out to determine the necessary conditions for agricultural practices and production of *I. tinctoria*. Extensive studies conducted in Italy (Angelini et al. 2004, 2005, 2007; Campeol et al. 2006) and Spain (Sales et al. 2006). Good Agricultural Practices (GAP) proposes recommendations applicable to indigo-producing plants and specifically on-farm indigo extraction, addressing environmental, economic and social sustainability through on-farm production and post-production processes (John and Angelini 2009). *I. tinctoria* was brought to North America by European settlers and soon became an invasive weed (Farah et al. 1988; Monaco et al. 2005). Agricultural production of *Isatis* for the production of indigo dyes is regularly carried out on a commercial scale in at least three regions of Europe. These regions are Bleu de Pastel de Lectoure, Toulouse France (http://www.bleu-de-lectoure.com); Woad-incTM, Norfolk, United Kingdom (http://www.woadinc.co.uk); and La Campana, Montefiore sull'Aso, Italy (http://www.lacampana.it) (John and Angelini 2009).

Isatis spp. germinates towards the end of summer, spends the winter as rosette leaves, blooms the next spring, and produces seeds at the end of the following summer and the rosette leaves of the plant are used in the production of indigo (Sales et al. 2006). The rosette leaves can be harvested repeatedly, maximizing the yield from a crop. The regrowth of the cut leaves from the basal meristems, the growth conditions, the duration of the summer season and the permanent root system affect the yield. Depending on the plant's sowing date, climate and prevailing weather conditions, several consecutive harvests can be obtained from a plant, with previous studies reporting 7 harvests in Southern Spain, 4 in Italy, 3 in Germany, and 2 in the UK and Finland. Temperatures around 20–25 °C promote the development and expansion of leaves, while increasing temperature has a growth arresting effect (John and Angelini 2009). When compared in terms of sequential harvest yields, *I. tinctoria* in southern Spain was harvested seven times, while *I. indigotica* was harvested four times. This is because *I. indigotica* has had slightly less success in regrowth (Sales et al. 2006).

Many of the major pests that affect other plants in *Brassicaceae* also show their effects on *Isatis*. According to the results reported from Project SPINDIGO (2002), the main damaging pests were the striped flea beetle (*Phyllotreta undulata*) (Finland), the cabbage stem flea beetle (*Psylliodes chrysocephala*) (Italy) and the great white butterfly (*Pieris brassicae*) (Italy). The project in Spain revealed that total indigo production in *I. indigotica* and *I. tinctoria* plots increased along with N fertilization, nevertheless the leaf indigo content was not significantly influenced by this factor (Sales et al. 2006). Weeds are very effective in the initial stages of development in newly established *Isatis* plant fields, where effective weed control is necessary (John and Angelini 2009).

I. indigotica Fortune ex. Lindley has been widely grown in China for a long time (Spataro et al. 2007; Zhang et al. 2019; Wong et al. 2022). Robert Fortune and a French group researching the Chinese silk industry (Cardon 2007) brought it to Europe in the 1840s. Although it is similar to *I. tinctoria* in general features (Gilbert et al. 2002), it differs in morphological features. In terms of indigo production, *I. indigotica* has a higher yield potential than *I. tinctoria* (Angelini et al. 2007). According to previous studies, while high yield was obtained in Italy (Angelini et al. 2007), it was similar in Spain, and higher leaf yield was observed in *I. tinctoria* (Sales et al. 2006). As *I. indigotica* is adapted to warmer climates, it is more susceptible to vernelization triggers. When the temperature is low (< 5 °C), flowering begins in *I. indigotica* whereas, *I. tinctoria* requires the development of rosette leaves before flowering (Vetter and G. Angelini, unpublished: cited in John and Angelini 2009; Asghari et al. 1992). It is also more susceptible to the above-mentioned pests and mildew on *Isatis*.

6.3 Germplasm Biodiversity and Conservation

Genetic diversity is investigated using molecular markers, which scan either coding or non-coding DNA sequences of distinct individuals within a population or between populations of a species or interspecies (Vellend and Geber 2005; Rubenstein et al. 2005; Ellegren and Galtier 2016). Genetic diversity is an important issue in describing the structure of the populations within a species and predicting their futures. The genetic diversity that emerges in populations is actually the driving force of evolution (Vellend and Geber 2005).

Evaluation of the genetic diversity allows determining the rate of differentiation within and between the populations of a species. On the other hand, when we consider, for example, medicinal plants that are related or unrelated to each other may have different profiles in terms of biochemical content or biological active substance content and diversity, while they may be similar in terms of morphological characteristics. These plants can be used for similar therapeutic purposes, in which case their effects on treatment will be different from each other. In fact, among the secondary metabolites in plants, there will be those that have harmful toxic effects as well as beneficial ones. For this reason, it is necessary to identify the species, subspecies or lines of plants to be used for medicinal purposes. For the effective management of genetic resources, evaluating the genetic diversity in the gene pools of crop plants is of primary importance. The markers used from past to present in the evaluation of genetic diversity follow the order of morphological markers, biochemical markers and molecular markers.

Gilbert et al. (2002) investigated genetic diversity among the *Isatis* ssp. L. (*I. indigotica, I. tinctoria* and *I. glauca*) by using AFLP. The similarity percentage between *I. glauca* and *I. tinctoria* groups were observed as 55.3–59.6%, while the similarity between *I. indigotica* and other groups were ranged between 18.6 and 21.9%. Cluster analysis indicated that *I. tinctoria* and *I. glauca* groups were clustered in the same group and *I. indigotica* grouped as outgroup and compared to *I. indigotica, I. tinctoria* and *I. glauca* are more closely related to each other. The second internal transcribed spacer (ITS2) of ribosomal DNA, the chloroplast *matK* gene revealed that *I. indigotica* and *I. tinctoria* are two distinct species (Sun and Pang 2013: cited in Feng et al. 2021).

Moazzeni et al. (2010) revealed the sequence variation of internally transcribed spacer (ITS) regions and using the 5.8S gene of nuclear ribosomal DNA (nrDNA) to investigate the systematics of the genus Isatis (Brassicaceae). In the study, they sampled 20 Isatis species, three species of Pachypterygium, two of nine species of Sameraria, and twenty-eight taxa of Isatis and related genera of the *Isatis* L. tribe, including Boreava, Myagrum, and Tauscheria. Moazzeni et al., (2007) analyzed the seed coat microsculpture, and the results of the study based on phylogenetic tree and fruit characters based on molecular markers converged significantly. They also remind that making a genus distinction based only on the morphological characters of the fruit can lead to erroneous phylogenetic and taxonomic results.

DNA barcoding technology is a molecular method used to distinguish between different plant species (CBOL Plant Working Group 2009). This technique uses a standardized short DNA region (Chen et al. 2010; Shiran et al. 2014; Yang et al. 2012). Chen et al. (2014) investigated an accurate and effective identification system for the medicinal plant *I. indigotica* using three DNA barcode candidates (ITS2, trnL-F, and rbcL). When these barcodes were used to search for phylogenetic relationships, they showed that trnL-F and rbcL together could allocate *I. indigotica* to the correct genus in the phylogenetic tree, while the one created with rbcL could not distinguish *I. indigotica* from adulterans. However, they explained that the UPGMA tree of ITS2 correctly distinguishes *I. indigotica* from adulterants and they concluded that it has a more suitable and accurate DNA barcode than trnL-F and rbcL for identification of *I. indigotica* and its adulterans.

Ying (2012) displayed the genetic variation between one diploid and ten tetraploids of *I. indigotica* using ISSR molecular markers. The diploid and tetraploid genotypes shared some loci, and additional bands or missing bands were observed in some tetraploid genotypes. Conclusively, a distinctive genetic variation was determined between the diploid and tetraploid genotypes. It has been reported that 17 woad lines were analyzed to determine the production of indigo precursors indican (indoxyl β-D glucoside) and isatan B (indoxyl ketogluconate) in their leaves under the field conditions in Central Italy from 2001 to 2003. Significant variations in phytochemical and agronomic traits were observed according to the genotypes of *I.*

tinctoria and *I. indigotica* lines. The higher amount of Isatan B and indican, and leaf yield *per* harvest was reported in *I. indigotica* lines than *I. tinctoria* lines. *I. tinctoria* is well adapted to Mediterranean climate conditions with its high postharvest regrowth capacity and high biomass production, while *I. indigotica*, despite having higher indigo precursor content per leaf, showed greater sensitivity to climatic conditions and it produced 25% leaf yield per hectare per season (Angelini et al. 2007).

The genetic diversity was investigated using molecular markers in *I. tinctoria* L. and *Isatis glauca* Aucher ex Boiss. in previous studies. Fifteen *I. tinctoria* L. populations from Europe and central Asia were analyzed to determine the genetic variation and similarity levels among and within natural populations using AFLP and SAMPL molecular markers. According to the results they obtained, *I. tinctoria* L. originated from eastern Asia and moved western Asia and Europe in the meantime its gene pool differed from the original (Spataro et al. 2007). Rocha et al. (2011) assessed the genetic diversity in woad (*I. tinctoria* L.) accessions using ISSR markers. They observed a considerably high total genetic diversity as 0.3272 and ranged between 53.8% (Austria) and 73.1% (Belgium). A considerably high genetic diversity ($h = 0.23$) in subspecies of *Isatis glauca* Aucher ex Boiss. from Central Anatolia, in Turkey was revealed by using Amplified Fragment Length Polymorphism (AFLP) technique (Özbek et al. 2013). They concluded that Ankara province might be the area, where the species *I. glauca* Aucher ex Boiss. originated. Görgülü et al. (2013) analyzed morphological variation in four subspecies of *I. glauca* Aucher ex Boiss.; ssp. *glauca*, ssp. *galatica* Yıldırımlı, ssp. *sivasica* (Davis) Yıldırımlı and ssp. *iconia* (Boiss.) Davis, naturally growing in Turkey. The most variable morphological character was the number of branches and the least variable morphological character was fruit width. The *Pearson*'s correlation and multiple regression analyses revealed that eco-geographical factors had a substantial effect on morphological variation. For the chromosome number of *I. indigotica* Fortune ex Lindl, which has the accepted name as *I. tinctoria* L., four chromosome counts were submitted on the website of Chromosome Count Database (Wang et al. 1985a; Ge et al. 1988; Zhang et al. 1990; Xiao et al. 1995: at Chromosome Count Database website).

Kang et al. (2020) displayed the chromosomal scale genome assembly of *I. indigotica*, using single-molecule real-time long readings and high-throughput chromosome conformation capturing techniques. They revealed that 53.27% of the assembled sequences were repeat sequences, of which 34.67% were retrotransposons and 7.37% were DNA transposons. They found 30,323 genes of which 5973 were processed with alternative spliced transcripts in the plant genome.

6.4 Traditional and Molecular Breeding

I. indigotica used widely for therapeutic purposes in traditional Chinese medicine. However, the recent increase in demand for this plant and the shortage in production lead to breeding studies. A formal comprehensive conventional breeding program for *I. indigotica* has not been reported. Previous studies have generally been carried out (i) to determine the secondary metabolites of *I. indigotica* and the gene/genes involved in the synthesis of these molecules (ii) to determine the expression profiles of the genes (iii) to investigate the molecular mechanisms and functions of the genes (iv) the responses caused by the abiotic stress factors such as salt or water deficit and how these factors affect the yield.

For breeding studies, it is important to display the differences in terms of yield, bioactive components accumulation and stress tolerance among the polyploidy of a plant species such as tetraploid and diploid *I. Indigotica.* Xiao et al. (2020) showed that the transcription factor IiWRKY34, has large pleiotropic effects on an array of traits, containing biomass growth rates, lignan biosynthesis, as well as salt and drought stress tolerance, had higher expression levels in tetraploid than in diploid *I. indigotica.* In breeding studies, the differentially expressed genes like *IiWRKY34* in different polyploidy levels could be used to improve the *I. indigotica* varieties with higher yield, higher content of bioactive molecules and higher growth rate and resistant to biotic or abiotic stress factors for industrial production of bioactive components.

In plants, CDPKs are responsible sensors for Ca^{2+} flux in response to varieties of biotic and abiotic stress factors. The expression of a full-length cDNA clone (*IiCPK1*) encoding CDPK (Ca^{2+}-dependent protein kinases) was isolated from *I. indigotica* cDNA library. When tetraploid and diploid *I. indigotica* were stimulated in response to cold stress, tetraploids showed good fertility and exhibited enhanced resistance and higher yield based on Real-time PCR analyses. Tetraploids presented to be more responsive and adaptable. The *IiCPK*1 gene involved in adapting to the environmental stress (Pan et al. 2008).

Cao et al. (2021) displayed the expression profile of *IiWRKY* gene, which is a member of transcriptional regulators in plants and has important activity in regulating growth and development, particularly in diverse stress responses by using transcriptome data and quantitative real-time PCR analysis. Such studies will contribute to the understanding of the functional analysis of *IiWRKY* genes and provide a basis for further elucidation of the molecular mechanism of stress responses in this important plant species.

The growing conditions play a pivotal role in the production of effective components in plants as well in *I. indigotica.* The investigation of the genes involved in plant growth such as the *GRAS* genes a group of transcriptional regulators provides important knowledge about the gene structure, conserved motifs, about the order of domains (N-terminal), and phylogeny of the genes of interest. Zhang et al. (2016b) revealed that *I. indigotica* and *Arabidopsis* shared 43 orthologous gene pairs and interaction networks of these orthologous genes.

In breeding studies, to improve the content of bioactive molecules, it is important step to understand the catalytic steps in the general biosynthetic pathways of those molecules such as indole, terpenoid, and phenylpropanoid in *I. indigotica.* For this, identifying the unigenes, which involved in these pathways from the transcriptome, could provide important knowledge about the pool of candidate genes. The detailed analysis and characterization of the important pathways is expected to give a better insight into the diversity of chemical composition, synthetic properties, and regulatory mechanism working in *I. indigotica* (Chen et al. 2013).

Little is known about the regulatory pathways of the appropriate flowering and bolting times of the *I. indigotica*. These are important stage for the reproductive success in plants. A number of differentially expressed genes (DEG) and 91 potential bolting and flowering-related genes from the RNA-seq data were reported in *I. indigotica* (Bai et al. 2018). These kind of studies will contribute to understand the molecular mechanisms of bolting and flowering in *I.* indigotica as well in Brassicaceae plants.

Kumagai et al. (2016) investigated the effect of indirubin, one of the bioactive molecules in the Ban-Lan-Gen on the transcriptional activation of the *CYP3A4* gene, is responsible for the metabolism of a wide variety of xenobiotics. The overexpression of the human pregnane X receptor raised indirubin- and rifampicin-induced *CYP3A4* gene reporter activity. As a result, indirubin activated *CYP3A4* gene transcription through the activation of the human PXR.

Tang et al. (2020) analyzed the salt stress, which stimulates secondary metabolism in *I. indigotica* by molecular analysis. The salt stress, decreases the yield promotes the increase in epigoitrin and indigo production in root and leaf of Radix isatidis respectively. The expression pattern of the genes of interest determined with qPCR revealed the genes involved in the first step in biosynthesis of core short-chain aliphatic glucosinolates, which is extremely important in dealing with salt stress.

Water is the most important ingredient for photosynthesis, which affects the net photosynthesis rate, the biomass, yield, transpiration rate, and stomatal conductivity during the vegetative growth period in plants. Therefore, it is very important to determine the irrigation regimes particularly during the vegetative growth and fleshy root growth. Zhou et al. (2021) investigated the effects of the water resource scarcity during the vegetative growth and fleshy root growth in *I. indigotica* between 2017 and 2019 in the Hexi Oasis, China. They observed that the application of moderate and severe water deficiency resulted a reduction (6.74–17.74%) in the economic yield of woad. To achieve maximum yield, it has been suggested that a continuous slight deficit from the vegetative growth stage to the fleshy root growth stage and provision of adequate water during other growing periods would be the optimal irrigation regime.

6.5 Tissue Culture Applications

In this section, tissue culture studies and the results obtained in the production of *I. indigotica.* as a medicinal plant are considered. The aims of tissue culture studies in medicinal plants include rapid plant reproduction, plant regeneration with rapid and stable genetic constitution and increasing the concentration of effective secondary metabolites produced in the roots and leaves of plants, producing virus-free plants, haploid or polyploidy plants. *I. indigotica* autotetraploids obtained in tissue culture displayed higher resistance to insect, cold, drought, disease than their diploid parents

(Li et al. 2005). Qiao et al. (1995) produced a tetraploid giant plant with 2n = 28 chromosomes from *I. indigotica.* When the properties of tetraploid and diploid *I. indigotica* were compared, it was observed that the concentrations of effective compounds such as indigo, indurubin, amino acids, anti-endotoxic and antiviral compounds were higher due to the larger organs of tetraploids (Wang et al. 2000). For this reason, the demand for tetraploid *I. indigotica* as a medicinal plant has increased, but the demand cannot be met sufficiently. Zhang et al. (2003a) investigated in vitro regeneration systems from cotyledons in the tetraploid *I. indigotica.* In their studies, they examined culture medium type, different combinations of plant growth regulators, sucrose concentrations in the medium and illuminations as factors affecting regeneration. As a result, they concluded that the analyzed factors are important for an effective regeneration system from cotyledons and that this method is an effective propagation method for tetraploid *I. indigotica.* While the regeneration methods used in tissue culture studies give successful results in some plant species, they may tend to synthesize less stable genetically, biochemically and useful secondary metabolites at very low levels in some species. Xu et al. (2004) induced hairy roots by transformation of cotyledon and hypocotyl explants of *I. indigotica* via *Agrobacterium rhizogenes* strains A4, R1601, and ATCC15834. It was determined that the A4 strain was the most suitable for the transformation of cotyledons as explants. Capillary electrophoresis showed that the five organic acids [CPQ [3-(2-carboxyphenol)-4 (3H)-quinazolinone], syringic acid, salicylic acid, benzoic acid and 2-aminobenzoic acid] examined were found in natural roots, hairy roots, and liquid medium. As a result, it was determined that the main antiviral compound CPQ, was found in higher amounts in the hairy roots and the five analyzed organic acids could be transferred from hairy roots to liquid medium. Hu et al. (1999) regenerated mesophyll protoplasts obtained by enzyme digestion from an accession of *I. indigotica* in tissue culture. They reported the proportion of shoots developing from callus cultures as 26.6%. As a result, they planted 29 plantlets with 2–3 broad leaves in the soil in the greenhouse and reported that 22 of them grew normally and reached maturity. Tao et al. (2011) obtained haploids with seven chromosomes in an anther culture and haploid induction study in *I. indigotica.* They displayed that cold pretreatment could increase the induction rate of callus, after which transferring calluses to differentiation medium and rooting medium resulted successive haploids.

6.6 Genetic Engineering

I. indigotica produces secondary metabolites terpenoids, lignans, and indole alkaloids and uses some of them to defend itself against biotic pathogens (Li et al. 2005; Deng et al. 2008; Chen et al. 2012; Zhou and Zhang 2013; Kang et al. 2020). Some of these metabolites have been used in the treatment of various diseases from traditional Chinese medicine. Further studies were conducted to identify the genes and the genetic mechanisms controlling the synthesis of these molecules. The main scope of these studies is to identify the gene/genes encoding these compounds with medical and aromatic importance, and to identify the gene/genes responsible in the metabolic pathways where their syntheses take place. The development of high yielding, disease resistant, and able to adapt to wider range of environmental conditions commercial varieties have become a necessity. Therefore, while the genes themselves were studied directly in the molecular studies, some studies were conducted on the factors involved in the expression mechanisms of the genes responsible for the synthesis of secondary metabolites.

Agrobacterium tumefaciens mediated *Pinella Ternata Agglutinin* (*PTA*) gene was transferred to tetraploid *I. indigotica* to create a variety with anti-pest. The transgenik explant displayed up to 95% regeneration rate. The explants obtained from cotyledons, hypocotyls of tetraploid *I. indigotica* play important role in establishing a highly efficient plant regeneration system, and *A. tumefaciens* mediated transgenic transformation system (Zhang et al. 2003a). PCR and Southern Blot identified the integration of *PTA* gene in the genome of tetraploid *I. indigotica*, while RT-PCR screened the normal transcription of the gene. Transgenik pest resistant tetraploid *I. indigotica* plants could be produced efficiently through *A. tumefaciens* mediated transformation (Zhang et al. 2003b). *A. tumefaciens* LBA4404 mediated two insect resistance genes *Bacillus thuringiensis* (Bt) crystal protein gene *CryIA*(c) and cowpea trypsin inhibitor gene *CpTI* were transferred into tetraploid *I. indigotica* to improve the resistance to moths. Co-transgenic rate of the genes of interest was determined as 16.67% and co-transformation of *Bt* and *CpTI* genes is an efficient strategy to develop new tetraploid *I. indigotica* varieties resistant to moths (Ding et al. 2009). The cloned genetic engineering tetraploid *I. indigotica* plants could be used in conventional breeding studies as parental lines.

The giant organs and increased concentration of secondary metabolites in breeding studies are the features that make autopolyploids attractive in medicinal and agricultural plants. The studies were conducted to produce autopolyploids of *I. indigotica* ($2n = 2x = 14$) and made comparisons in terms of phenotypic and transcriptomic changes with their diploid parents. Depending on the increase in genome dosage, increases in organ size and secondary metabolites yield (Zhou et al. 2015).

6.7 Hybridization

Extensive genomic resources in the Brassicaceae family provide a basis for hybridization in the Brassica genus. Hybrid individuals formed in symmetrical crosses between *Brassica napus* and *Lesquerella fendleri* cannot self-fertilize and are sterile, while asymmetric hybrids between the same parents are fertile because they can self-fertilize (Skarzhinskaya et al. 1996).

I. indigotica has more potential to produce indigo dye than *I. tinctoria* (Angelini et al. 2007), it is less used in dye production due to it has a low seed generator rate and is more sensitive to environmental conditions. *I. indigotica* has resistance to tobacco mosaic virus (TMV) (Wang and Wan 1988) and stem rot (*Sclerotinia sclerotiorum*) (Zhao et al. 1994), the most serious rapeseed disease in China. For this reason, *I.*

indigotica was used as one of the fusion parents in the protoplast fusion method studies (Tu et al. 2008).

In a previous study, somatic hybrid plants were obtained by polyethylene glycol-derived symmetric fusions of mesophyll protoplasts of tribes *Raphanus sativus* (2n = 18, *RR*) and *Brassica rapa* spp. *chinensis* (2n = 20, *AA*) and *I. indigotica* (2n = 14, I I). When the hybrid individuals obtained from both crosses were analyzed by cytogenetic methods and molecular methods, it was determined that chromosome pairings and bivalent numbers according to chromosome numbers were as expected, but meiotic disorders were observed in the second division stage of meiosis. It was recommended that some hybrids obtained from the crosses made according to the results can be used for the genetic improvement of crops, dye and medicinal plants (Tu et al. 2008).

In order to expand the gene pool of *B. napus* and *I. indigotica*, both symmetrical and asymmetrical somatic hybridizations of these two species were tested, out of 237 calluses, only one symmetric hybrid (S2) and five asymmetric hybrids (As1, As4, As6, As7 and As12) were able to be transferred to the field and kept. While some morphological differences were observed in these hybrids, they were found to have very low pollen fertility (Du et al. 2009). It was determined that the plants obtained by asymmetric hybridization were mixoploid (2n = 48–62). When the As1 hybrid was crossed with *B. napus* morphologically had the characteristics of *B. napus* and the individuals produced without self-pollination or pollination with *B. napus* were able to produce seeds. In these hybrid breeding studies between *B. napus* and *I. indigotica*, the importance of parental contributions for genomic analysis and breeding studies is emphasized (Du et al. 2009).

Cytoplasmic male sterility occurs in hybrid plants formed between *Brassica napus* and *I. indigotica* (Chinese woad). The RF (fertility-restoring) gene(s) of *I. indigotica* are carried by an efficient monosomic alien addition line (MAAL) of rapeseed. Self-pollinated strains of MAAL were extensively selected and analyzed to screen for rapeseed type plants (2n = 38) with good male fertility and to produce double haploid (DH) lines by microspore culture. The restorer obtained in the study contained 38 rapeseed chromosomes, while no intact woad chromosomes. In addition, the restorer is very well developed in terms of pollen and seed productivity and low glucosinolate content. Successful cultivation of the restorer for Inap CMS made this new pollination control system suitable for the production of hybrid rapeseed plants (Li et al. 2019).

Somatic hybridization enables unique gene combinations of protoplast fusion. While the measurable success of somatic hybridization in terms of cultivar release is still limited, the potential for use in plant breeding is still great. Therefore, somatic hybridization will enable the production of new plant varieties in the future. Somatic hybridization studies between radish (*Raphanus sativa*) and Chinese woad (*I. indigotica*) may contribute to increasing the effective bioactive content of medicinal properties by improving the root structure of *I. indigotica*. In this way, new varieties can be developed and the production of the plants with medical and economic importance can be increased (Tu et al. 2008).

6.8 Conclusions and Prospects

I. indigotica Fort is an herb that is used extensively in traditional medicine in China as a medicinal and aromatic plant. Apart from this feature, it is no longer used in large-scale indigo production although it has a production potential even more than *I. tinctoria* L. in terms of indigo dye production, but it is more sensitive to environmental conditions and less seedling. Although *I. indigotica* is widely cultivated in China, there are some projects that were performed in Europe, and there are some farmers growing small-scale dye production. *I. indigotica* is a diploid ($2n = 2x = 14$) plant, however, researchers state that when compared to the diploid ancestor of tetraploid *I. indigotica* plants obtained by using colchicine in tissue culture studies, giant organs and biomass production are higher, and the amount of secondary metabolite production with antimicrobial effect is higher. Looking at previous studies, diploid *I. indigotica*, it is noteworthy that extensive conventional breeding studies have not been carried out in the ($2n = 2x = 14$) species. Conventional breeding studies take a lot of time, increase the budget, and labor intensive, there are problems in producing and keeping plants with stable genetic structure. On the other hand, important steps have been taken to increase the production amount of *I. indigotica* in the future with polyploidy production technology and biotechnology. Transgenic *I. indigotica* plants resistant to pests or weeds have been produced by genetic engineering studies. It has been reported that transgenic plants have a highly efficient regeneration ability in tissue culture media. In plants produced with transgenic technology, one or a few genes can be transferred at a time, and problems can be encountered about how they integrate into the host genome and how they are expressed in the new cellular environment. In addition to the suspicion and concern among the public against transgenic products in the market, studies are restricted by legal regulations and ethical problems are also encountered. On the other hand, the development of new technological methods in biotechnology, especially protoplast fusion techniques used in somatic hybridization methods have shown very promising results. With these methods, gene transfer for disease resistance, abiotic stress resistance, herbicide resistance and many other quality traits increases the improvement of crop plants, even if it is not possible with a conventional breeding program. Thus, traditional breeding can be facilitated by producing new lines to be used as elite breeding materials with superior properties in terms of yield, quality, and durability. Commercial varieties can be developed for large-scale agricultural productions that are compatible with wide environmental conditions, and solutions can be produced to meet the increasing demand for *I. indigotica* plant recently.

Acknowledgements I would like to thank Susan Dye Trade (Susan and Ashley Walker), who granted the photos of *I. tinctoria* and *I. indigotica* used in this chapter.

Appendix: Research Institutes

Institute name	Specialization research activities	Contact information and website
The Bleu de Lectoure	The cultivation of the plant The extraction and natural production (chemical free) of pure pigment Manufacturing and dyeing 100% natural	https://www.bleu-de-lectoure.com/en/
European Comission	The project: SPINDIGO (Sustainable Production of Plant-derived Indigo) to introduce indigo-producing crops into European agriculture	https://cordis.europa.eu/project/id/QLK5-CT-2000-30962
The Jade Institute	Advanced professional education in acupuncture and Chinese herbal medicine	https://www.jadeinstitute.com/
Susan Dye Trade	Growing Chinese Woad	55 Walsworth Road, Hitchin, Hertfordshire, UK SG4 9SX +44 1462 422814 ashleywalker@naturesrainbow.co.uk https://www.naturesrainbow.co.uk/2021/03/growing-chinese-woad-isatis-indigotica/
University of Bristol	A project: A journey into the blue Growing and production of indigo	http://www.bristol.ac.uk/news/2005/1119522957.html
The Woad Centre	Wood-inc is the trading name of a Norfolk farmer who has diversified into natural indigo pigment extraction following five years growing woad for the MAFF-LINK (INDINK) and Spindigo projects	http://www.thewoadcentre.co.uk/about-us/

References

Akeroid JR (2002) *Isatis* Linnaeus. In: Cullen J, Alexander JCM, Brady A, Brickell CD, Green PS, Heywood VH, Jørgensen PM, Jury SL, Knees SG, et al (eds) The European Garden Flora, vol. IV. Cambridge University Press, Cambridge, UK, p 135

Al-Shehbaz IA, Beilstein MA, Kellog EA (2006) Systematics and phylogeny of the Brassicaceae (Cruciferae): an overview. Plant Syst Evol 259:89–120

Angelini LG, Tozzi S, Nassi O, Di Nasso N (2004) Variation in indicant content, indigo yield and crop production among three *Polygonum tinctorium* Ait lines under Mediterranean climate. J Agric Food Chem 52(25):7541–7547

Angelini LG, John P, Tozzi S, Vandenburg H (2005) Extraction of indigo from *Isatis tinctoria* L. and *Polygonum tinctorium* Ait. as a basis for large-scale production. In: Pascual-Villalobos

MJ, Nakayama FS, Bailey CA, Correal E, Schloman WW Jr (eds), Proceedings of Annual Meeting of the Association for the Advancement of Industrial Crops: International Conference on Industrial Crops and Rural Development, Murcia, Spain, 17–21 September 2005, pp 521–534, IMIDA (Istituto Marciano de Investigation Desarollo Agrarioy Alimentario) Selegrafica, S L Murcia. ISBN 84-689-3363-5

Angelini LG, Tozzi S, Di Nasso N, Nassi O (2007) Differences in leaf yield and indigo precursors production in woad (*Isatis tinctoria* L.) and Chinese woad (*Isatis indigotica* Fort.) genotypes. Field Crop Res 101:285–295

Appel O, Al-Shehbaz IA (2003) Cruciferae. In: Kubitzki K, Bayer C (eds) The families and genera of vascular plants. Springer, Berlin, pp 75–174

Asghari JB, Evans JO, Dewey SA (1992) Low temperature vernalization forces Dyer's woad (*Isatis tinctoria* L.) flowering. Western Society of Weed Science. Research Progress Report, 6:6–7 [73665]

Bai Y, Zhou Y, Tang X, Wang Y (2018) De Novo transcriptome assembly of *Isatis indigotica* at reproductive stages and identification of candidate genes associated with flowering pathways. J Amer Soc Hort Sci 143(1):56–66

Ball PW (1964) *Isatis* L. In Flora Europaea; Tutin TG, Heywood VH, Burges NA, Valentine DH, Walters SM, Webb DA (1964) Vol. 1. *Lycopodiaceae* to *Platanaceae*, xxxii -f 464 p 5 maps. Cambridge University Press

Beijerinck MW (1900) Further researches on the formation of indigo from the woad. Proc K Akademie Van Wetenschappen TE Amsterdam 3:101–116

Beilstein MA, Al-Shehbaz IA, Kellogg EA (2006) Brassicaceae phylogeny and trichome evolution. Amer J Bot 93(4):607–619

Bhowmick S, Singh T, Chauhan PS (2024) Harnessing the power of aromatic and medicinal plants for natural product innovation. In: Kumar L, Bharadvaja N, Singh R, Anand R (eds) Medicinal and aromatic plants. Sustainable landscape planning and natural resources management. Springer, Cham, pp 211–222. https://doi.org/10.1007/978-3-031-60117-0_15

Campeol E, Angelini LG, Tozzi S, Bertolacci M (2006) Seasonal variation of indigo precursors in *Isatis tinctoria* L. and *Polygonum tinctorium* Ait. as affected by water deficit. Environ Exp Bot 58:223–233

Cardon D (2007) Natural Dyes: Sources, Tradition. Archetype Publications, London, Technology and Science

Chen SL, Yao H, Han JP, Liu C, Song JY, Shi LC, Zhu YJ, Ma XY, Gao T, Pang XH, Luo K, Li Y, Li XW, Jia XC, Lin YL, Leon C (2010) Validation of the ITS2 region as a novel DNA barcode for identifying medicinal plant species. PLoS ONE 1:e8613

Chen M et al (2012) Alkaloids from the root of *Isatis indigotica*. J Nat Prod 75:1167–1176

Chen J, Dong X, Li Q, Zhou X, Gao S, Chen R et al (2013) Biosynthesis of the active compounds of *Isatis indigotica* based on transcriptome sequencing and metabolites profiling. BMC Genomics 14(1):857

Chen S, Zhu Z, Ma H, Yang J, Guo Q (2014) DNA barcodes for discriminating the medicinal plant *Isatis indigotica* Fort. (Cruciferae) and its adulterants. Biochem Syst Ecol 57:287–292

Chen MH, Lin S, Wang YN, Zhu CG, Li YH, Jiang JD, Shi JG (2016) Antiviral stereoisomers of 3,5-bis(2-hydroxybut-3-en-1-yl)-1,2,4-thiadiazole from the roots of *Isatis indigotica*. Chinese Chem Let 27(5):643–648

Chromosome Count Database (retrieved on 05.05.2024). https://taux.evolseq.net/CCDB_web/countsByMatchedName/Isatis%20indigotica%20Fortune%20ex%20Lindl./

Danz H, Stoyanova S, Wippich P, Brattstrom A, Hamburger M (2001) Identification and isolation of the cyclooxygenase-2 inhibitory principle in *Isatis tinctoria*. Planta Med 67:411–416

Darlington CD, Wyle AP (1955) Chromosome atlas of flowering plants George Allen & Unwin LTD, London, p 38

Davis PH (1965) *Isatis*. In: Davis PH (ed) Flora of Turkey and the East Aegean Islands, Edinburgh University Press, Edinburgh, 1, pp 1–567

Deng X, Gao G, Zheng S, Li F (2008) Qualitative and quantitative analysis of flavonoids in the leaves of *Isatis indigatica* Fort. by ultra-performance liquid chromatography with PDA and electrospray ionization tandem mass spectrometry detection. J Pharm Biomed 48:562–567

Ding RX, Xiao Y, Wang K, Chen WS, Zhang HM, Zhang L (2009) Enhanced resistance of transgenic tetraploid *Isatis indigotica* by *Bt* and *CpTI* to moths. Chin Tradit Herb Drugs 40:621–624 (Chin)

Du X, Ge X, Yao X, Zhao Z, Li Z (2009) Production and cytogenetic characterization of intertribal somatic hybrids between *Brassica napus* and *Isatis indigotica* and backcross progenies. Plant Cell Rep 28:1105–1113

During A, Debouche C, Raas T, Larondelle Y (2012) Among plant lignans, pinoresinol has the strongest anti-inflammatory properties in human intestinal Caco-2 cells. J Nutr 142:1798–1805

Ellegren H, Galtier N (2016) Determinants of genetic diversity. Nat Rev Genet 17:422–433

Epstein E, Nabors MW, Stowe BB (1967) The origin of indigo in woad. Nature 216:547–549

Farah K, Tanaka A, West N (1988) Autecology and population biology of dyer's woad (*Isatis tinctoria*). Weed Sci 36:186–193

Feng J, Huang D, Yang Y, Chen J, Qiu S, Lv Z, Ma X, Li Y, Li R, Xiao Y, Chen W (2021) *Isatis indigotica*: from (ethno) botany, biochemistry to synthetic biology. Mol Horticulture 1:17. https://doi.org/10.1186/s43897-021-00021-w

Flora of China 13 (2001) *Isatis* Linnaeus, Sp. Pl. 2: 670. 1753, 8: 35–37

Fortune R (1846) The notice of the Tein-Ching or Chinese indigo. J R Hortic Soc 1:269–271

Ge CJYKL, Wan P, Li YX, Jiang FH (1988) Observations on the chromosome numbers of medicinal plants from Shandong Province (V). J Shandong Coll Traditional Chin Med 12:55–57

Gilbert K, Garton S, Karam M, Arnold G, Karp A, Edwards K, Cooke D, Barker J (2002) A high degree of genetic diversity is revealed in *Isatis* spp. (dyer's woad) by amplified fragment length polymorphism (AFLP). Theor Appl Genet 104:1150–1156

Görgülü E, Özbek Ö, Yıldırımlı Ş (2013) Determination of morphological variation in subspecies of *Isatis glauca* Aucher ex Boiss. from Central Anatolia. Plant Syst Evol 299:827–840

Hayek A (1911) Entwurf eines Cruciferensystems auf phylogenetischer Grundlage. Beih Bot Centralbl 27:127–335

Ho YL, Chang YS (2002) Studies on the antinociceptive, anti-inflammatory and antipyretic effects of *Isatis indigotica* root. Phytomedicine 9:419–424

Hoessel R et al (1999) Indirubin, the active constituent of a Chinese antileukaemia medicine, inhibits cyclin-dependent kinases. Nat Cell Biol 1:60–67

Hu Q, Andersen SB, Hansen LN (1999) Plant regeneration from mesophyll protoplasts in *Isatis indigotica*. Plant Cell Tissue Organ Cult 55:155–157

Huang QS, Yoshihira K, Natori S (1981) Isolation of 2-hydroxy-3-butenyl thiocyanate, epigotrin and adenosine from 'Banlangen', *Isatis indigotica* root. Planta Med 42:308–310

John P, Angelini LG (2009) Indigo - agricultural aspects. In: Bechtold T, Mussak R (eds) Handbook of Natural Colorants. Wiley, New York, pp 75–103

Kang M, Wu H, Yang Q, Huang L, Hu Q, Ma T, Li Z, Liu J (2020) A chromosome-scale genome assembly of *Isatis indigotica*, an important medicinal plant used in traditional Chinese medicine. An Isatis Genome. Hortic Res 7:18

Kumagai T, Yusuke Aratsu, Ryosuke Sugawara, Takamitsu Sasaki, Shinichi Miyairi, Kiyoshi Nagata (2016) Indirubin, a component of Ban-Lan-Gen, activates CYP3A4 gene transcription through the human pregnane X receptor. Drug Metab Pharmacokinet, 31, 139e145

Li B et al (2005) Phenylpropanoids isolated from tetraploid roots of *Isatis indigotica*. Chin Tradit Herbal Drugs 36:326–328

Lin CW et al (2005) Anti-SARS coronavirus 3C-like protease effects of *Isatis indigotica* root and plant-derived phenolic compounds. Antivir Res 68:36–42

Li P, Kang L, Wang A, Cui C, Jiang L, Guo S, Ge X, Li Z (2019) Development of a Fertility Restorer for inap CMS (*Isatis indigotica*) *Brassica napus* Through Genetic Introgression of One Alien Addition. Front Plant Sci 10:257

Moazzeni H, Zarrea S, Al-Shehbaz IA, Mummenhoff K (2007) Seed-coat microsculpturing and its systematic application in *Isatis* (Brassicaceae) and allied genera in Iran. Flora 202:447–454

Moazzeni H, Zarre S, Al-Shehbaz IA, Mummenhoff K (2010) Phylogeny of *Isatis* (Brassicaceae) and allied genera based on ITS sequences of nuclear ribosomal DNA and morphological characters. Flora 205:337–343

Molina P et al (2001) Inhibition of leukocyte functions by the alkaloid isaindigotone from *Isatis indigotica* and some new synthetic derivatives. J Nat Prod 64:1297–1300

Monaco T, Johnson D, Creech J (2005) Morphological and physiological responses of the invasive weed *Isatis tinctoria* to contrasting light, soil-nitrogen and water. Weed Res 45:460–466

National Pharmacopeia Committee (2000) The Pharmacopeia of PRC (Part 1, 2000th edn. Chemical Industry Press, Peking, China, pp 163–164

Özbek Ö, Görgülü E, Yıldırımlı Ş (2013) Genetic diversity in populations of *Isatis glauca* Aucher ex Boiss. ssp. from Central Anatolia in Turkey, as revealed by AFLP analysis. Bot Stud, 54(1):48. https://doi.org/10.1186/1999-3110-54-48.

Pan X, Xiao Y, Wang Z, Zhang L, Tang K (2008) Tetraploids *Isatis indigotica* are more responsive and adaptable to stresses than the diploid progenitor based on changes in expression patterns of a cold inducible Ii CPK1. Biologia (Section Cellular and Molecular Biology) 63(4):535–541

Qiao CZ, Dai FB, Cui X (1995) Pharmacognostical studies on the two-ploidy level of indigo woad (*Isatis indigotica*). Chin Tradit Herb Drugs 26:423–429

Rocha L, Martins S, Carnide V, Braga F, Carvalho C (2011) Genetic diversity in woad (*Isatis tinctoria* L.) accessions detected by ISSR markers. Plant Genet. Resour 9(2); 210–213

Rubenstein KD, Heisey P, Shoemaker R, Sullivan J, Frisvold G (2005) Crop Genetic Resources, an Economical Appraisal/EIB-2. A Report from the Economic Service/ USDA, [internet] Available from: http://www.ers.usda.gov

Sales E, Kanhonou R, Baixauli C, Giner A, Cooke D, Gilbert K, Arrillaga I, Segura J, Ros R (2006) Sowing date, transplanting, plant density and nitrogen fertilization affect indigo production from *Isatis* species in a Mediterranean region of Spain Industrial Crops and Products, 23: 29–39

Schulz OE (1936) Cruciferae In: Engler A, Harms H (eds) Die natu¨ rlichen Pflanzenfamilien, vol 17B Verlag von Wilhelm Engelmann, Leipzig, p 227–658

Schunk E (1855) On the formation of indigo blue (part I). Phil Mag J Sci Series 4(10):74–95 Science and Technology Press, Beijing, p 191

Shan J, Chunzhi Zhao, Qian Li, Ting Zhu, Jinwei Ren, Haixia Li, Junhua Wu, Hao Ma, Wensheng Qu, Yuxia Wang (2015) An arabinogalactan from *Isatis indigotica* and its adjuvant effects on H1N1 influenza and hepatitis B antigens. J Funct Foods 18 (Part A):631–642

Shiran B, Kiani S, Sehgal D, Hafizi A, Ul-Hassan T, Chaudhary M, Raina SN (2014) Internal transcribed spacer sequences of nuclear ribosomal DNA resolving complex taxonomic history in the genus *Vicia* L. Genet Resour Crop Evol 61, 909e925

Skarzhinskaya M, Landgren M, Glimelius K (1996) Production of Intertribal Somatic Hybrids between *Brassica napus* L. and *Lesquerella fendleri* (Gray) Wats. Theor Appl Genet 93(8):1242–1250

Spataro G, Taviani P, Negri V (2007) Genetic Variation and Population Structure in a Eurasian Collection of *Isatis tinctoria* L. Genet Resour Crop Evol 54:573–584. https://doi.org/10.1007/s10722-006-0014-4

Spataro G, Negri V (2008a) Adaptability and variation in *Isatis tinctoria* L.: a new crop for Europe. Euphytica 163:89–102

Spataro G, Negri V (2008b) Assessment of the reproductive system of *Isatis tinctoria* L. Euphytica 159:229–231. https://doi.org/10.1007/s10681-007-9479-2

Sun ZY, Pang XH (2013) Discussion on the botanical origin of Isatidis radix and Isatidis folium based on DNA barcoding. Yao Xue Xue Bao 48(12):1850–1855 Chinese PMID: 24689245

Sun DD, Dong WW, Li X, Zhang HQ (2010) Indole alkaloids from the roots of *Isatis ingigotica* and their antiherpes simplex virus type 2 (HSV-2) activity *in vitro*. Chem Nat Compd 46:763–766

Tang X-Q, Tan H-W, Shi S-L, Wang L-L, Wang F-Q, Wang K-C, Yang J (2020) Characterization of Transcriptome Expression: The Response of *Isatis indigotica* to Salt Stress. Russ J Plant Physiol 67(6):1135–1143

Tao L, Yuan LW, Wang ZZ (2011) Anther culture, and haploid induction from *Isatis indigotica* (Cruciferae). Plant Diversity Resour 33(2):225–228

Tu Y, Sun J, Liu Y, Ge X, Zhao Z, Yao X, Li Z (2008) Production and characterization of intertribal somatic hybrids of *Raphanus sativus* and *Brassica rapa* with dye and medicinal plant *Isatis indigotica*. Plant Cell Rep 27:873–883. https://doi.org/10.1007/s00299-008-0513-1

Xia X, Xiao J, Shi GWD (2007) Function research of resistance to Salmonella Typhimurium infection using Banlangen polysaccharide. Med J Wuhan Univ 28:348–350

Xiao CH, Zhao JW, He FX (1995) Studies on karyotypes of three species in Cruciferae. J. Wuhan Bot. Res. 13(3):283–285. https://doi.org/10.1016/S1673-8527(08)60146-0J

Xiao Y, Feng J, Li Q, Zhou Y, Bu Q, Zhou J, Tan H, Yang Y, Zhang L, Chen W (2020) IiWRKY34 positively regulates yield, lignin biosynthesis and stress tolerance in *Isatis indigotica*. Acta Pharm Sin BB 10(12):2417e2432

Xu T, Lei Zhang, Xiaofen Sun, Hanming Zhang and Kexuan Tang (2004) Production and analysis of organic acids in hairy-root cultures of *Isatis indigotica* Fort (indigo woad). Biotechnol Appl Biochem 39: 123–128 (Printed in Great Britain)

Vellend M, Geber MA (2005) Connections between species diversity and genetic diversity. Ecol Let 8(7):767–781

Wang Y, Qiao CZ, Liu S, Zhang H (2000) Evaluation on antiendotoxic action and antiviral action in vitro of tetraploid *Isatis indigotica*. Chin J Chin Mat Medica 25:327–329

Wang QY, Wan XB (1988) A TMV-resistant material-Banlangen (*Isatis indigotica* Fort.) Acta Agri Boreali-Sin 3:92–95

Warwick SI, Al-Shehbaz IA (2006) Brassicaceae: chromosome number and index database on CD-Rom. Plant Syst Evol 259(2–4):237–248

Wong LW, Goh CBS, Tan JBL (2022) A Systemic Review for Ethnopharmacological Studies on *Isatis indigotica* Fortune: Bioactive Compounds and their Therapeutic Insights. Am J Chin Med 50(1):161–207

Wu XY, Qin GW, Cheung KK, Cheng KF (1997) New alkaloids from *Isatis indigotica*. Tetrahedron 53(39):13323–13328

Yang JB, Wang YP, Moeller M, Gao LM, Wu D (2012) Applying plant DNA barcodes to identify species of Parnassia (Parnassiaceae). Mol Ecol Resour 12, 267e275

Ying Zi D (2012) Analysis of genetic variation between diploids and their homologous tetraploids of Isatis indigotica Fort. by ISSR. Acta Botanica Boreali-Occidentalia Sinica 32(8):1534–1538

Zhai SY, Guo JZ (1987) Treatment of 100 cases of acute and chronic dacryocystitis by irrigating the lacrimal passage with Banlangen solution. Chung I Tsa Chih 28(3):216

Zhang YH, Guan Q, Zhang Y, Mao G (1990) A report on the chromosome numbers of some medicinal plants. China J Chin Mat Med 15:19

Zhang L, Xu T, Sun X, Zhang H, Tang T, Tang K (2003a) Factors influencing shoot regeneration from cotyledons of tetraploid isatis indigotica fort. In Vitro Cell Dev Biol—Plant 39:459–462

Zhang L, Xu TF, Tan QM, Wang ZN, Ding RX, Tang KX et al (2003b) Agrobacterium tumefaciens mediated transgenic tetraploid *Isatis indigotica* carrying anti-insect gene from *Pinella ternata* agglutinin. Chin Tradit Herb Drugs (Chin) 34:258–261

Zhang W, Zheng X, Cheng N, Gai W, Xue X, Wang Y, Gao Y, Shan J, Yang S, Xia X (2016a) *Isatis indigotica* root polysaccharides as adjuvants for an inactivated rabies virus vaccine. Int J Biol Macromol 87:7–15

Zhang L, Li Q, Chen J, Chen W (2016b) Computational identification and systematic classification of novel GRAS genes in *Isatis indigotica*. Chin J Nat Med 14(3):161–176. https://doi.org/10.1016/S1875-5364(16)30013-9

Zhang D, Li J, Ruan D, Chen Z, Zhu W, Shi Y, Chen K, Li Y, Wang R (2019) Lignans from *Isatis indigotica* roots and their inhibitory effects on nitric oxide production. Fitoterapia 137:104189

Zhao HJ, Huang YJ, Wang YY (1994) A study on intergeneric hybridization *Brassica napus* L. and *Isatis indigotica* Fort Sci Agri Sin, 27:89–91

Zhou W, Zhang XY (2013) Research progress of Chinese herbal medicine Radix isatidis (Banlangen). Am J Chin Med 41:743–764

Zou P, Hong Y, Koh HL (2005) Chemical fingerprinting of *Isatis indigotica* root by RP-HPLC and hierarchical clustering analysis. J Pharm Biomed Anal 38:514–520

Zhu S, Chai X, Kuang R (2004) Pharmacodynamics study on Fufangdaqinye Mixture. Chin Tradit Patent Med 26(11):909–912

Zhou C, Zhang H, Li F, Wang Z, Wang Y (2021) Photosynthetic Characteristics and Yield Response of *Isatis indigotica* to Regulated Deficit Irrigation in a Cold and Arid Environment. Water 13(24):3510

Chapter 7
Advances in Common Hop (*Humulus lupulus* L.) Breeding Strategies

Khaled F. M. Salem and Amira A. Ibrahim

Abstract Hop (*Humulus lupulus* L.), commonly known as common hop, is a member of the Chenopodiaceae family and is characterized by its distinctive flowers. The Cannabaceae family, also known as the hemp family, is a small group of angiosperms that flower. This family includes approximately 170 species spread across 11 genera, such as *Cannabis* (hemp), *Humulus* (hops), and *Celtis* (hackberries). This crop is important both for its essential qualities and economic value. *H. lupulus*, commonly known as hops is mainly used in beer production because of its bitter taste and fragrant scent. Currently, the brewing sector consumes up to 98% of the global hop harvest. However, there are numerous alternative applications, for hops that have been recognized since ancient times. The strobili, cone-shaped female structures, are the most commonly utilized part of the hop plant but there is also interest in other tissues of the plant. There are multiple hop species with diverse chromosome counts And breeders are working to improve hop cultivars for various attributes like resistance to stressors, quality and yield. By using contemporary genetic methodologies in traditional breeding programs, breeders can achieve the attainment of these objectives. This chapter provides a comprehensive account of hop germplasm biodiversity and conservation, as well as the objectives and stages of hop breeding programs and conventional breeding methods. Furthermore, the chapter discusses contemporary techniques of plant breeding, including marker-assisted breeding and genetic modification.

Keywords Biodiversity · Biotechnology · Genetic improvement · Genetic map · Hop (*Humulus lupulus* L.) · Modern breeding · Traditional breeding

---

K. F. M. Salem (✉)
Department of Biology, Scientific Departments, Shaqra University, Qwaieah, Saudi Arabia
e-mail: khaled.salem@gebri.usc.edu.eg; khaledfathi@su.edu.sa

Department of Plant Biotechnology, Genetic Engineering and Biotechnology Research Institute (GEBRI), University of Sadat City, Sadat City, Egypt

A. A. Ibrahim
Botany and Microbiology Department, Faculty of Science, Arish University, Al-Arish, Egypt
e-mail: amiranasreldeen@sci.aru.edu.eg

J. M. Al-Khayri et al. (eds.), *Biodiversity and Genetic Improvement of Medicinal and Aromatic Plants I*, Advances in Plant Breeding Strategies 9,
https://doi.org/10.1007/978-3-031-81846-2_7

7.1 Introduction

Hop (*Humulus lupulus* L.) is a herbaceous perennial vine that belongs to the family Cannabaceae and is one of the three species of Humulus. The genus Humulus is believed to have originated in China, as all three of its species, including *H. yunnanensis* Hu. and *H. japonicus* Siebold & Zucc., are found in this region (Alonso-Esteban et al. 2019; Murakami et al. 2006). Currently, *H. lupulus* is distributed in temperate regions worldwide. The botanical specimen exhibits a dextral growth habit, whereby it ascends in a clockwise orientation along any viable supporting structure, facilitated by its robust trichomes that face downward. It has the potential to reach a vertical height of 10-18 meters. The leaves exhibit either an opposite or alternate arrangement and possess three to five lobes with serrated margins. The extensive root system requires deep, nutrient-rich soil with good drainage. According to Amoriello et al. (2020) and Van Cleemput et al. (2009), it has been observed that horizontal roots have the propensity to generate a substantial quantity of delicate rootlets.

H. lupulus is a dioecious plant species, meaning that male and female flowers are typically present on separate individuals, although monoecious specimens may occasionally be observed. The floral structures are small and rely on wind-mediated pollination. Female botanical specimens exhibit strobili (hops) which are cone-shaped structures measuring 2.5–5 cm in length. These structures emerge shortly after the pollination process. The floral structures are comprised of bracts that overlap each other and are attached to a central axis, displaying a yellowish-green coloration. Each bract envelops a small fruit, classified as achene, which is a desiccated fruit that does not open upon reaching maturity. The lupulin glands located at the base of the bracts in mature strobila contain a variety of chemicals. These chemicals are stored in glandular trichomes and have potential applications in several industries. One of the most widely recognized industries that utilize these chemicals is the brewing industry, although other industries also make use of them.

The seeds of *H. lupulus* are known for their small size, tough texture, and impermeability to water. Additionally, they have a long period of dormancy and a slow germination process. Female plants may take up to three years to start flowering and producing cones. Most cultivated varieties do not have seeds because they can affect the taste of beer when crushed during brewing (Almaguer et al. 2014; Liberatore et al. 2018; Wendakoon et al. 2018). Recent research has shown a trend toward using sex-linked DNA markers to identify sex early in seedlings (Liberatore et al. 2018; Rodolf et al. 2018). While primarily used in beer production, *H. lupulus* has other potential uses. This chapter will cover the chemical and pharmacological properties of *H. lupulus*, its traditional and modern uses and its future potential. It will also address genetic resources, breeding efforts, and the challenges of climate change on hop production. Our goal is to provide a comprehensive scientific review of *H. lupulus* biodiversity, conservation, breeding programs, and traditional and modern breeding techniques. We hope to offer valuable information and new perspectives to enhance future research, cultivation, and utilization of this plant.

7.1.1 Origin and Distribution

The hop plant, a perennial species, native to a significant portion of the Northern Hemisphere has its roots in Europe. It exhibits optimal growth and yield from April to September, requiring abundant sunlight during the fruiting period and high levels of rainfall throughout the growing season. The primary crop producing regions in the Northern Hemisphere include the United States, Germany, the Czech Republic, and England. Currently, the United States produces approximately 50 million tonnes of hops annually. During the 19th century, Central New York was one of the most productive hop farming regions in the United States, thanks to its favorable farming conditions and crop geographic tendencies (Awasthi et al. 2021; Karabín et al. 2016; Van Cleemput et al. 2009).

The extensive cultivation of the hop plant is intricately linked with the production of beer, as is the case with a significant portion of its historical background. Hops were initially cultivated in Bavaria in the year 736 A.D. and subsequently gained widespread usage as an additive in beverages during the medieval era in Europe. The organized cultivation of the bitter hop plant witnessed a surge in tandem with the rise in global beer production. In response to alterations in global transportation, supplementary stages were incorporated to safeguard the quality of beer during an extended journey. The crop was introduced to America during English colonization in the year 1629, as noted by Glazier (2007).

H. lupulus L., commonly known as hops, belongs to the family Cannabaceae and is a flowering plant classified under the class Dicotyledonae and order Rosales. The Cannabaceae family, commonly known as the hemp family, is a small group of angiosperms that exhibit the characteristics of flowering (https://www.itis.gov). This family encompasses around 170 species spread across approximately 11 genera, including *Cannabis* (hemp), *Humulus* (hops), and *Celtis* (hackberries). Celtis is the largest genus, including about 100 species (Stevens 2001). The Cannabaceae family is a part of the Rosales order. Plants in the family have an upright growth habit and produce flowers without petals, with dried fruits containing a single seed. According to Kenneth and colleagues (2002), hemp (*Cannabis*) and hops (*Humulus*) are commercially important species. Despite their common evolutionary lineage, plants in this family do not share many characteristics. Some, like Celtis, have tree-like growth habits, while others, like Cannabis, are herbaceous (Fig. 7.1).

Domain: EukaryotaKingdom: Plantae (Plants)
Subkingdom Viridiplantae – green plants
Infrakingdom: Streptophyta (land plants)
Superdivision: Embryophyta
Division: Tracheophyta (vascular plants, tracheophytes)
Subdivision: Spermatophytina (spermatophytes, seed plants, phanérogames)
Class: Magnoliopsida (Dicotyledons)
Superorder: Rosanae

(continued)

(continued)

Domain: EukaryotaKingdom: Plantae (Plants)
Order: Rosales
Family: Cannabaceae (Hemp)
Genus: *Humulus* L. (Hop)
Species: *Humulus lupulus* L.
Common names: Common hops, hops

The hop plant holds immense cultural importance, having been utilized as a medicinal herb for numerous centuries and as a crucial constituent in the brewing of beer for both its flavor-enhancing and preservative properties (Hamel et al. 1975; Neve 2012). The hop plant is a sizable, vine-like, dioecious species belonging to the Rosid taxonomy. The *Humulus* genus comprises three distinct species, namely *H. japonicus*, *H. lupulus* and *H. yunnanensis*. Among these, *H. japonicus* and *H. lupulus* have been identified as potential sources of compounds possessing advantageous pharmaceutical properties (Sung et al. 2015). The current knowledge regarding *H. yunnanensis* is limited and there exists a possibility of its extinction, despite attempts made to locate a surviving specimen (Boutain 2014). *Humulus* exhibits three common sex chromosome arrangements, namely *H. lupulus* (2n = 18 + XY), *H. lupulus* var. cordifolius (2n + 16 + X1X2 Y1Y2), and *H. japonicus* (2n = 14 + XY1Y2) (Ono 1955). The genome assembly of *H. lupulus* var. lupulus is potentially more manageable due to its simplicity in comparison to other configurations. The aforementioned configurations offer a compelling foundation for investigating the evolution of sex chromosomes in plants, and several research endeavors have already been dedicated to this topic (Hill et al. 2016; Karlov et al. 2003).

The hop plant grown for commercial purposes is typically diploid, with a chromosome count of 2n = 2x = 20, consisting of 9 pairs of autosomal bivalents and a pair of sex chromosomes. Female hop plants have XX sex chromosomes, while male hop plants have heteromorphic X and Y sex chromosomes. It is worth noting that polyploids have been created in hop plants, as reported by Haunold in 1971. Previous research has identified at least six distinct categories of sex chromosomes in *H. Lupulus*, as documented by Ainsworth in 1999. The determination of sex expression depends on the balance between X chromosomes and autosomes (Hill et al. 2016; Parker and Clark 1991; Shephard et al. 2000). Loose panicles produce male flowers with a perianth of five yellowish-green sepals and five anthers on short filaments. The anthers release pollen through dehiscence, which is dispersed by the wind.

In the realm of commercial hop production, the presence of male hop plants is considered undesirable due to the potential for seeded cones to introduce an unwanted off-flavor and increase cone weight without improving beer flavor. Haunold (1971) conducted a study that demonstrated the use of triploid male hops can boost cone yields without significantly affecting seed sets, as the dehisced pollen grains are sterile. Female hop flowers consist of bracts and bracteoles that enlarge to form a strobilus, commonly known as a cone when mature, which is the main product of interest. The process of pollination and fertilization leads to the creation of an achene

Fig. 7.1 As an example of hop's phenological development, below are some of the stages: **a** shoot emergence, **b** leaf development, **c** blooming, **d** cone development and **e** cone maturity. *Source* Rossini et al. (2021)

at the base of the bracteoles, which are protected by bracts, as shown in Neve's 1991 research.

7.1.2 Economic Importance

The hop plant, scientifically known as *H. lupulus* L., is a significant component in the beer production process. The female hop plant produces flowers that develop into 'cones', which are the main ingredient in brewing beer, adding bitterness, flavor, and aroma. Throughout history, hops have been used in medicinal practices as a herbal remedy for various health conditions. Biendl and Pinzl documented in historical records from the Middle Ages the anti-inflammatory and antimicrobial properties of hops (Fig. 7.3). The use of hops in beer production has been beneficial due to its bitterness, aroma, taste and antimicrobial properties leading to the replacement of gruit, a blend of wild herbs and spices, in the brewing process (Astray et al. 2020; Neve 1991; Ibrahim et al. 2023a, 2023b). The earliest mention of hops being used in brewing beer dates back to the 9th century when Abbot Adalhard of the Benedictine Monastery of Corbie in northeastern France noted that his fellow monks added hops to their ale. By the 11th century, hopped beer consumption had become widespread in France, as indicated by various sources (Neve 1991).

The commercial cultivation of hops in Europe began in 1524. In 1629, the Massachusetts Company introduced the first cultivated hops in the United States, sourced from Europe. Early American colonists also utilized Native American hop varieties, as noted by Schwartz (1973). In addition to being used in beer production, hop-derived prenylflavonoids are being explored as potential therapeutic agents for human consumption due to their significant physiological effects. The compounds currently being investigated include Xanthohumol (XN) and its derivatives, such as iso-xanthohumol (IXN) and the phytoestrogen 8-prenylnaringenin (8PN). The conversion of XN into IXN and 8PN occurs during beer production and through in vivo metabolism facilitated by liver microsomes and intestinal bacteria. This process has been extensively studied and documented (Nikolic et al. 2006; Possemiers et al. 2008; Possemiers and Verstraete 2009). Research has shown that Xanthohumol has the potential to inhibit various types of cancer (Diederich and Noworyta 2012; Gallo et al. 2016; Gerhäuser 2005; Liu et al. 2015; Stevens and Page 2004) and also possesses therapeutic properties for conditions such as metabolic syndrome, inflammation, and bone disease.

The possibility of synthesizing XN has been investigated, although the process is complex and the overall yield is relatively low, as reported by Khupse and Erhardt in 2007. The primary method for obtaining XN continues to be through extraction, isolation, and purification from female hop inflorescences. Iso-xanthohumol has shown a wide range of effects, including anti-tumor, anti-inflammatory, antioxidant, and antiangiogenic properties (Allsopp et al. 2013; Krajnović et al. 2016; Negrão et al. 2010; Serwe et al. 2012). The phytoestrogen properties of the hop compound 8PN have been confirmed through various in vitro studies (Kitaoka et al. 1998; Milligan et al. 1999; 2002; Schaefer et al. 2003), establishing it as one of the most potent phytoestrogens. The constituents in question are primarily sourced from beer, although the quantities present in typical commercial beer are insufficient to confer therapeutic benefits to human health (Liu et al. 2015). Several research studies have assessed the potential of hop β-acids derived from dehydrated hop cones as an innovative feed supplement for antimicrobial applications. According to Cornelison et al. (2006), incorporating Teamaker hops into grill chicken diets at a rate of 0.5 pounds per tonne could potentially enhance the growth rate and feed utilization in the absence of antibiotics. The hop variety known as Teamaker is characterized by a low concentration of alpha acids and is believed to possess greater acceptability among domesticated animals. The study conducted by Bortoluzzi et al. (2014) exhibited the favorable impacts of hops supplementation on the productive performance of broilers. The optimal outcomes were observed when hops β-acids were added at a concentration of 30 mg/kg. Chilliard (2009) conducted a study to examine the impact of a beta acid extract derived from hops on the ruminal metabolism of Angus cross steers (Bos taurus). The findings of this study indicate that the extracts of hops did not exhibit any noteworthy impact on ruminal fermentation or diet digestibility. The utilization of spent hops as an environmentally sustainable repellent for insect pests that affect stored food has been explored in previous studies (Bedini et al. 2015; DeGrandi-Hoffman et al. 2012). The therapeutic properties of hops have opened up the potential for its utilization beyond the scope of brewing, extending to medicinal and culinary applications (Fig. 7.2).

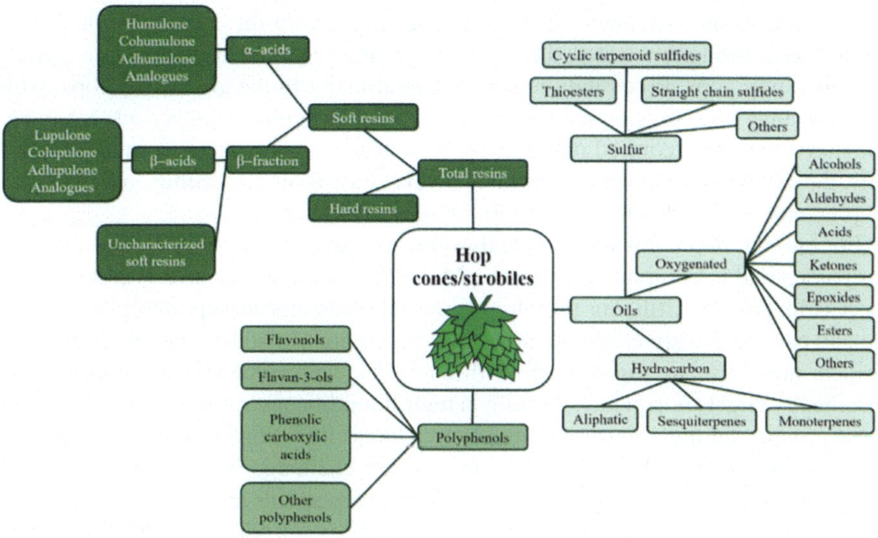

Fig. 7.2 Economic importance of hops. *Source* Astray et al. (2020)

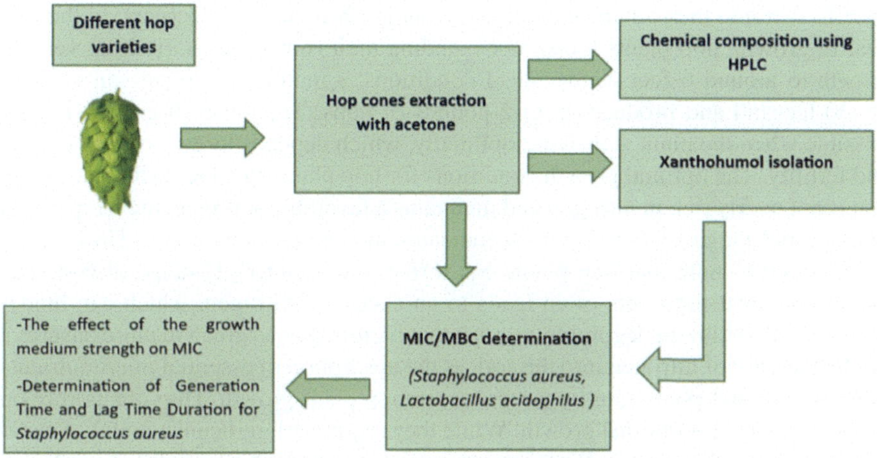

Fig. 7.3 Antimicrobial activities of hops varieties. *Source* Kolenc et al. (2023)

7.2 Current Cultivation Practices and Challenges

Hop farming can achieve optimal productivity and efficiency by meeting specific conditions, similar to any other crop. Optimal temperature conditions during the growth period, typically spanning from June to August, are characterized by an average range of 50–66 °F. The provision of sunlight is of equal significance as plants require approximately 15–27 hours per day within the aforementioned time frame. Considering this factor, the summer season is deemed the appropriate period for cultivating hops. Precipitation is an essential factor in the growth of crops, with an ideal range of 2.5–22.4 inches of rainfall during the cultivation period. Adequate moisture content is required in the soil while ensuring appropriate drainage. Preferably, the soil should not exhibit elevated levels of alkalinity or salinity, as suggested by previous studies (Carter et al. 1990; Donner et al. 2020).

For optimal hop cultivation, it is advisable to start planting after the risk of frost has passed and to complete planting by May 18th. A weed-free area is essential for successful growth. Utilizing thorough tiling methods and incorporating straw into the soil can effectively eliminate unwanted vegetation. In addition to weed control, farmers must also be vigilant about diseases. Downy mildew is a pathogen that specifically targets hop plants, thriving in high-humidity environments and hindering shoot growth. Farmers must act quickly to address infections to prevent crop loss, removing any diseased shoots. Pests are another challenge in gardening with hop aphids and spider mites being common culprits. Hop aphids are a particular issue in our region, but can be managed with insecticide or by introducing ladybugs (Carter et al. 1990).

Hops are native to temperate regions in the Northern Hemisphere and are grown worldwide between latitudes 30 and 50. The plant has a root system that can reach over 12 feet into the soil and can remain healthy for more than 25 years. In the first year of growth, hop plants focus on expanding their root systems, limiting vertical growth to around 6 feet. Under ideal conditions, a mature hop plant can grow to 25–30 feet tall and produce up to 2 pounds of hops. This level of growth is only possible when the plant is thriving optimally, which depends largely on soil quality and fertility. The optimal growth conditions for hop plants involve well-drained soil that is not overly rich in nitrogen and maintains a favorable soil structure. Hop plants prefer a pH range of 6–6.2, which indicates moderately acidic soil. However, it is important to note that hop plants can thrive in a broader pH range of 6–8. The use of chemical fertilizers often leads to an excess of nitrogen, which can hinder plant growth. Utilizing leguminous crops that fix nitrogen as green manure ensures a gradual release of nitrogen into the soil, as the absorption of essential micronutrients like calcium and potassium is necessary for hop plant growth. Hop plants rely on proper drainage for optimal growth. While they require a significant amount of water during the growth period, stagnant water around the plant's base can lead to unfavorable outcomes. This increases the risk of plant diseases, presenting challenges for management (Kneen 2001; 2003; Simieli et al. 2021). A common strategy employed by hop farmers is to plant their hop plants in raised mounds and leave inter-row spaces unused. This helps prevent water accumulation near the plant's base, which could impede its development (Fig. 7.3).

The growth of the hop plant is significantly influenced by temperature. This plant is a perennial species that goes dormant during the winter season. The aerial roots, located in the canopy region of the plants typically remain underground until the first blossoms of the spring season have bloomed. It takes a minimum of 120 days to produce a mature yield of flowers. The vine continues to grow even after the flowers have been harvested, accumulating enough energy reserves for the winter season until the first frost in autumn. The perennial root stock known as a crown produces hops annually. In addition to the underground roots and aboveground aerial vines, the top of the plant also produces rhizomes. Rhizomes are subterranean stems that resemble roots and have many buds. The use of rhizomes in vegetative propagation is a common practice to guarantee genetic uniformity among all plants of a specific variety (Kneen 2001, 2003; Simieli et al. 2021).

It is advisable to start planting rhizomes after the last frost to avoid rotting. For best growth plant the rhizomes with their buds facing up, about 1–2 inches deep the soil. If planting different varieties, space them 5 feet apart, but, if they are the same variety, 3 feet apart is enough. Providing a support system is crucial for plant growth. In the United States, commercial hop farms use an 18-foot trellis system, but smaller gardens can use shorter support structures like tall poles tied together with twine. Montell (1994) suggests a support structure of at least six feet, but this may cause the plant to become bushy due to its vine-like growth. Once vines reach about one foot long, select 2–3 strong vines and coil them clockwise around a string or pole. This method is used to train the vines to grow vertically along the support system, with the primary objective being to avoid entanglement. Regular pruning is crucial to eliminate lateral branches and enhance air circulation, minimizing the risk of disease. It is recommended to trim the lower parts of the plant every few weeks. The strobila is exclusively found in the uppermost regions of the hop plant. During the initial growth phase, the hop plant allocates a significant portion of its resources towards expanding its root structure. Consequently, promoting this growth will require recurrent brief watering sessions. Applying mulch to the soil surrounding the plant facilitates the absorption of surplus moisture and serves as a mechanism for managing weed growth. During the harvest season, the most convenient method of gathering hops is to lower the vines. The timing of the harvest is dependent on climate variations. The goal is to harvest the hops when the lupulin has the highest aroma levels and the strobila is beginning to dry out. This ensures that the bitterness is preserved and the cones remain intact during processing. After the harvest, it is advisable to trim the vines and bury the healthy basal vines to encourage the growth of rhizomes for the following spring. In the first winter, the hop plant is expected to expand its root system significantly, while only producing a few rhizomes. Once the plant reaches maturity and develops a full root system, an agriculturalist can expect a yield of about 20–30 sturdy rhizomes per plant.

7.3 Genetic Improvement

Since 1763, brewers have noticed that the amount of hops needed to achieve a certain level of bitterness varies slightly among different hop varieties. Additionally, beers brewed with different hop varieties have noticeable differences in shelf life. In the late1800s, European brewers observed that certain imported hop varieties had distinct differences in bitterness compared to local hops, due to the increasing international trade in hops. Improved chemical and analytical techniques quickly identified this as a result of the resinous component, with Oregon Clusters from the United States showing the highest levels of bitterness. Furthermore, the introduction of these imported varieties led to unique preservation qualities being imparted to the beer. In modern times, bitterness and shelf-life have become essential factors in beer production, leading brewers to search for hop varieties that can provide increased bitterness or longer shelf-life. At the same time, growers have been working on developing cultivars. According to Donner et al. (2020), the lack of effective chemical solutions for managing pests and diseases has led to significant fluctuations in crop yields from season to season, depending on the severity of infestations and weather conditions. The production of hop crops has been marked by variability and unpredictability, to the point where it has been associated with gambling. Growers have sought to improve the resistance of hops against pests and diseases. In response to requests from brewers and growers, and leveraging advancements in genetics, hop breeding programs have been established in most hop-producing nations.

The hop plant is highly sensitive to both day length and soil conditions in its local environment. As a result, hop varieties of ten do not perform well when grown outside of their original selection area. While the exchange of hop varieties between growers in different countries has been common practice, hop improvement programs are most successful when localized within the country of cultivation. Salmon's Wye program, which began in 1906, set the groundwork for hop breeding throughout the entirety of the twentieth century. Subsequent programs followed his methodologies, goals and cultivars. In the United States, the hop breeding program started in 1908 but was temporarily halted before resuming in Oregon in 1931. Denmark's Carlsberg laboratory also established a program in 1911. In Japan, hop breeding began in Hokkaido in 1912 and was later incorporated into the operations of the Sapporo Brewery. In Zatec, Czech Republic, hop breeding was initiated, under the guidance of Osvald. The primary focus of the program was a clonal selection of the Saazer variety. In response to the requirement for downy mildew-resistant varieties in the Bavarian Hallertau region, a program was initiated in 1926 in Hüll, Germany. Following the conclusion of World War II, various publicly funded programs were initiated in several countries, including Slovenia, Ukraine, Poland, China, India, and New Zealand. Furthermore, brewing enterprises initiated initiatives in Japan, Australia, and South Africa, and additional initiatives in the United States of America. While the impetus behind the initiation of each program may have varied, there exist several overarching goals that each program has endeavored to attain.

7.3.1 Breeding for High-Resin Content

Hop breeding has achieved a consistent rise in the resin concentration of hop cones, with a particular emphasis on the alpha-acid content, over the 20th century. Around 1900, only a limited number of varieties under cultivation exhibited an alpha-acid content surpassing 4%. However, towards the conclusion of the century, a greater selection of varieties became accessible, yielding a higher alpha-acid content of 16–18%. The augmentation in question is frequently regarded as the primary accomplishment of hop breeding initiatives throughout the twentieth century. The pursuit of achieving elevated levels of alpha-acid content remains a prominent objective in contemporary initiatives. Salmon was the first to enhance resin content through breeding by introducing wild North American hops into his breeding program. He recognized that the American hops had higher resin levels than European hops, despite having a less appealing aroma. Using wild accessions in a breeding program was a groundbreaking idea, as most breeding programs at the time only used commercial varieties as parents. Salmon believed that it was possible to develop a variant with both high resin content and a superior aroma. The first successful cultivars from work were Brewer's Gold (1934) and Bullion (1938), which had a unique American aroma and high alpha-acid content of around 8%. Northern Brewer introduced in 1944, was a hybrid of Golding and an American male plant, blending high resin concentration with desirable 'European' fragrance. These three cultivars were widely grown in Europe, the United States, and England.

Salmon's three cultivars have been the basis for high-alpha breeding programs worldwide, with only two bitter-type cultivars introduced that do not have a direct lineage to Salmon's varieties. The exceptions, Talisman and Comet were obtained from wild American breeding stock. It can be estimated that over 50% of the current global hop acreage consists of cultivars originating from Salmon's three primary varieties. The alpha-acid content of hop varieties around the world has consistently increased since the introduction of Northern Brewer. Various hop varieties, such as Pride of Ringwood in Australia (released in 1953), Atlas and Apollo (also known as the 'Super-Styrians') in Slovenia (released in 1971), Wye Northdown in the UK (released in 1971), and Galena and Eroica in the USA (released in 1979 and 1982 respectively), have surpassed the alpha-acid content of Northern Brewer.

The Wye Target, introduced in 1972 and originating from Wye, was the first variety to consistently surpass 10% alpha-acid in commercial production. As a result, it quickly gained dominance in the United Kingdom. The emergence of these bitter hops, with over two times the resin concentration of conventional strains, led to the commercialization of alpha-acid as a tradable commodity, with hops being marketed based on their alpha-acid weight in kilograms. Due to economies of scale, the United States became the primary producer of alpha acid. Successful alpha-type varieties were further developed by programs at Oregon and Washington State Universities. Breeding for higher alpha acid contents has become increasingly important due to significant advancements in hop processing since the mid-1980s. Currently, over 85% of the global hop harvest, undergoes processing into pellets, extracts, or other downstream commodities. These processes reduce the volumetric mass of unprocessed hop cones, making it easier to implement novel packaging methodologies that enhance and prolong shelf life. The economic viability of these operations is mainly determined by the alpha-acid content of the hops. The latest generation of high-alpha cultivars, introduced between 1990 and 2000 and commonly known as 'super-alpha' varieties, have emerged as a result of this trend. These cultivars have alpha-acid levels exceeding 15%. Some of these cultivars include Admiral from

Wye, Taurus and Magnum from Germany, Eastern Gold and Kitimadori from Japan, Pacific Gem from New Zealand, Victoria and Opal from Australia, and Columbus, Tomahawk, Zeus, Warrior and Millennium from the United States.

Increasing the alpha-acid concentration has been the main factor in improving hop cultivation efficiency. Plant breeding has led to significantly higher yields in many other crops. Before the late 1980s, newer hop varieties within a country did not show significant increases in yield compared to pre-1900 varieties selected by growers. The increase in alpha-acid content has a direct increase in productivity, measured by alpha-acid yield. This practice has mainly benefited individuals in the brewing industry, especially those who base purchasing decisions on alpha-acid price, regarding the cultivar or source. The higher alpha-acid levels have also reduced the acreage needed for production, as noted by growers. Advancements in analytical capabilities, particularly through the HPLC methodologies, have helped identify resin composition and quantify overall content (Verzele and De Keukeleire 1991). This has led to a prioritization focus on resin quality. During wort boiling or hop processing, six distinct isomers of alpha-acids are produced, each contributing a unique bitterness type. Brewers often prefer alpha-acid from hop varieties with low cohumulone content, as they find cohumulone bitterness unpleasant based on experience with different hop varieties. Most breeding programs prioritize selecting hop varieties with high alpha-acid content and low cohumulone. According to Murakami's research in 1999, the trait in question exhibits a high degree of heritability and there has been a significant decrease in cohumulone levels. Compared to the older bitter-type variety Brewers Gold, which has a cohumulone content of 45%, more recent varieties like Magnum exhibit a lower cohumulone content of only 25%. Since 1980, there has been a growing utilization of hop processing, and the economic significance of these procedures has led to a heightened interest in maintaining the stability of alpha-acid as a desirable quality parameter. Typically, it is deemed advantageous for high alpha-acid cultivars to possess favorable storage characteristics to minimize alpha-acid degradation during the initial months of storage before processing. Upon reviewing the progress made in enhancing the storage stability of various types, it is evident that the advancements in the overall alpha-acid content have typically occurred before any improvements in storage.

7.3.2 *Breeding for Aroma*

The development of novel aroma cultivars has exhibited comparatively lower levels of triumph in contrast to the alpha cultivars. Numerous brewers have persisted in utilizing conventional hop varieties to confer fragrance and hop flavor characteristics to their beer. Regrettably, brewers had to forgo numerous conventional strains after the implementation of stationary mechanical harvesters within the period 1945–1965. The antiquated cultivars tended to possess voluminous cones, frequently growing in clusters, which could be effortlessly harvested in multiple quantities. The cones in question were deemed unsuitable for use in the machinery due to their tendency to fracture or their inability to traverse the inclined belts utilized in the separation of leaves from cones, as noted by Burgess (1964) and Desalle and Tattersall (2019). The extant traditional cultivars, while coveted by brewers, frequently exhibited deficiencies from the perspective of the cultivator.

Breeders have endeavored to enhance the agronomic characteristics of particular preferred conventional cultivars while preserving a fragrance that closely approximates the original. The utilization of pedigree breeding techniques notwithstanding, the intricate and subjective nature of this undertaking, coupled with the dearth of scientific understanding regarding the role of hop aroma in beer aroma and flavor, renders the production of a seedling that approximates the fragrance of the maternal variety a formidable challenge. The adoption of novel varieties has been observed to occur primarily in situations where the cultivation of conventional varieties has been significantly jeopardized, frequently due to the emergence of a novel pathogen (Havill 2023; Young et al. 2023). The emergence of wilt disease, which is attributed to the pathogen *Verticillium alboatrum*, was first observed in the United Kingdom during the 1920s and subsequently identified in continental Europe during the 1950s. Chemical methods for managing this soilborne ailment are currently unavailable, and the existence of highly infectious strains may result in the demise of the hop plant and the subsequent loss of yield. The susceptibility of all traditional varieties has posed a significant threat to the sustained production of Fuggles and Goldings in the United Kingdom, as well as Hallertau Mittelfrüh in Germany. Production has persisted under conditions of minimal infection through the implementation of isolation measures and stringent phytosanitary protocols. In cases where the disease is prevalent in the soil, the cultivation of resistant varieties is the sole method for sustaining production. Salmon's introduction of wild germplasm into the USA was fortuitous as it brought with it resistance to wilt disease. Epistolary correspondences from Salmon addressed to hop cultivators in the region reveal that American-bred male specimens were also employed as pollinators in Kent farms. In 1911, a cultivator reported the emergence of a noteworthy hybrid within the Bates Brewer cultivar. The individual in question created a limited region on their agricultural property located in Beltring, Kent, utilizing the identifier 1147. The farm underwent a change in ownership when it was acquired by the Whitbread brewery. In the early 1930s, the farm experienced a significant outbreak of wilt. The hybrid strain 1147 exhibited evident resistance and was subsequently renamed as Whitbreads Golding Variety (WGV), which is presently cultivated as an aroma-type substitute for Goldings, with the added advantage of being resistant to wilt.

A comprehensive investigation of Wilt disease was conducted at East Malling, Kent. An individual conducted a field screening of advanced selections derived from the Wye program. These selections were often obtained by crossing wild USA hops with highly regarded European aroma hop varieties (Havill 2023; Keyworth 1942). Several resistant selections were detected and subsequently established as aroma-type cultivars, such as Early Choice (1948) and Bramling Cross (1951), both offspring of Goldings that are still being cultivated today. A collaborative initiative between Wye and East Malling aimed to develop a wilt-resistance strain of Fuggles. This effort resulted in the introduction of multiple variants, with Progress (1967) emerging as the most successful. Production is currently ongoing.

7.3.3 Breeding for High Yield

Before the late 1980s, endeavors to enhance hop yield through augmenting hop plant vitality for increased cone numbers or weights had yielded unsatisfactory outcomes. There is a correlation between enhanced self-shading and heightened vigor, while crop yields have not exhibited any significant improvement. The creation of triploid cultivars has presented a heightened capacity for yield augmentation, which has been successfully achieved in certain initiatives. Specifically, a number of the recently introduced aromatic cultivars referred to earlier are triploid. Illustrative instances comprise Willamette, Blisk, and the entire Slovenian C-series. In contrast to diploid hybridized varieties, which typically exhibit an equal genetic contribution from each parent, triploid varieties display a genetic makeup that is twice as much from one parent. The method, originally devised for hops in Japan, was implemented on various conventional European aroma hops, as documented by Dark in 1953 and Ono in 1948. This approach is highly conducive to aroma breeding due to the potential for greater retention of desired flavor components in the female parent. The experimentation with triploid varieties in England proved to be ineffective due to the heightened vitality of said plants, which resulted in shading, complications with disease, postponed maturation, and diminished crop output. In areas with higher light intensities and elevated wirework, triploid varieties have demonstrated a significant increase in crop yields. All cultivars produced in New Zealand after 1970 have been triploid, including the latest aroma cultivars like Hallertau Aroma (1988) and Pacific Hallertau (1994), as well as alpha cultivars like Pacific Gem (1987) and Pacific Sunrise (2000). The Millennium cultivar in the United States, classified as an alpha-type, is also a triploid variety.

7.3.4 Breeding for Pest and Disease Resistance

Before the onset of the twentieth century, powdery mildew was the only fungal disease of hop that had been documented, and it was prevalent across Europe. Before the late 1990s, the issue was not considered significant enough to warrant the implementation of a breeding initiative, except in England. Coincidentally, Salmon was an expert in mycology and worked as a mycologist at Wye College, rather than as a plant cultivator. Salmon introduced several wild accessions to the program, particularly from Italy and the USA, which contained resistance genes. A specific type of resistance, known as "blister" (Salmon 1920), was the focus of their research. Salmon identified resistance to powdery mildew in two distinct cultivars, Pride of Kent and Early Choice. To date, it has been observed that five out of the seven resistance genes for powdery mildew in hops, originated from Salmon's accessions, as noted by Darby in 2001. Neve, who succeeded Salmon, implemented a method of glasshouse screening for seedlings to assess their resistance and then incorporated this trait into all the varieties under his supervision. In 1971, the initial powdery mildew-resistant cultivar, Wye Challenger, was introduced by Neve. This cultivar had the 'blister' resistance, which provided immunity at Wye. However, during propagation for commercial distribution in East Anglia, the plant was subjected to diverse pathogenic strains, resulting in the emergence of a highly infectious variant. The resistance of Wye Target (1972) was due to the expression of the R2 gene and the absence of any virulent strains of the pathogen at the time. Wye Target showed significant resistance to the disease in the United Kingdom for about two decades. The Wye Target cultivar was used as a resistance source in the Hüll program to breed new varieties. Following the release of Magnum in 1993, which was highly vulnerable to powdery mildew, there was a greater emphasis on breeding for resistance to this disease. The R2 resistance, originating from Wye Target, was incorporated into the Merkur (2000) and Herkules (2005) varieties and continues to be effective in continental Europe. However, its effectiveness has ceased in all regions of the United Kingdom.

The development of aroma-type varieties has been reported to involve breeding for resistance to wilt disease and downy mildew. However, it should be noted that this trait is also observed in bitter types. The screening and selection of seedlings have become customary practices in various programs, and the integration of resistance to locally virulent strains of the aforementioned diseases is a standard feature in all newly developed varieties (Darby 2005a, b; Havill 2023; Young et al. 2023). The genetic association between resistance to UK strains of wilt disease and unfavorable agronomic and quality characteristics has been observed. The incorporation of wilt resistance has resulted in limitations to the productivity of the breeding program at Wye. The Wye Target cultivar encountered agronomic challenges during its farm trial phase, such as suboptimal climbing proficiency and diminutive cone size in certain years. Despite its poor agronomy, the aforementioned variety proved to be the sole type capable of enduring the more potent strains of wilt disease that emerged in the Weald of Kent during the 1970s and in the West Midland hop-growing regions during the 1980s. As a result, it emerged as the primary variety in the United Kingdom.

7.3.5 Breeding for Dwarf Traits

Currently, the market is completely globalized and free of any limitations. The dismantling of marketing boards and regulated national economies, which were common throughout most of the 20th century, has resulted in growers engaging in global competition. Hop farms in the United Kingdom are typically around 23 hectares in size, which is too extensive to be managed solely by family labor, as is common in many parts of continental Europe. However, these farms do not benefit from the economies of scale that are present in larger farms in the United States. The imperative to curtail labor expenses and mitigate reliance on the accessibility of appropriate labor has been a propelling factor in Britain's pursuit of trial and error with uncomplicated cultivation systems, specifically low-trellis production, to curtail production expenses and augment competitiveness.

The low-trellis system involves growing hops as a continuous hedge that is typically 2.3–3 m in height. These systems are known for being easy to build, cost-effective, and suitable for mechanization, which, reduces the need for labor. They are also more effective in controlling pests and diseases when sprayed. However, hope varieties that are typically grown on high trellis systems, which are at least 5 meters tall, are not suitable for low trellis systems, leading to lower yields. In 1977, Neve began a breeding program at Wye to develop new dwarf varieties. The goal of the program was to cultivate novel dwarf cultivars that exhibit reduced height and vigor while maintaining a yield equivalent to that of traditional cultivars.

The program's inception can be attributed to Salmon, as evidenced by its historical origins. The individual's records from 1911 document his observations of stunted flora within his cultivated garden. The author depicted a particular entity possessing laterals of moderate length that are densely arranged and highly productive, yet lacking in any explicit potential. The user's text implies that the plant in question was productive, but not well-suited to the prevailing agricultural practices of the era, as evidenced by the author's emphasis on its "direct" nature. The traits associated with dwarfism were observed in the untamed US accessions utilized in the researcher's program, as well as in the hybrids produced to study the resistance to powdery mildew known as 'blister'. The unique amalgamation of objectives in this particular program may account for the conspicuous presence of dwarf plants at Wye, which has not been documented in other programs worldwide until recently.

In 1977, Neve procured open-pollinated seeds from dwarf plants, which were subsequently employed as progenitors in 1982. Sufficient data was obtained from the examination of the growth patterns of segregating offspring resulting from these crosses. As a result, a program aimed at producing commercial dwarf cultivars for the United Kingdom was initiated in 1985 by Darby (Darby 1999a, 1999b; Darby and Farris 1991). In 1996, the initial dwarf varieties, namely First Gold, Herald, and Pioneer were officially recorded.

According to Barth (2005), breeding programs in Germany, Slovenia, Poland, and the USA have initiated the development of dwarf varieties in response to the increasing farm size and labor costs in these regions. The adaptability of daylength—The hop plant exhibits characteristics of a short-day plant. Upon reaching physiological receptivity, the plant's flowering initiation is triggered by daylength that falls below a critical threshold of approximately 15.5 h. It can be observed that hop cultivation is predominantly concentrated within the latitudinal range of 35°–55°. In a majority of nations beyond this specified range, the cultivation of hops is not financially viable, thereby necessitating the importation of this primary ingredient by breweries. Several breeding programs in South Africa, Mexico, and Columbia, as well as variety trialing programs in India, Argentina, and Kenya, have endeavored to identify and select varieties that are well-suited to significantly reduced daylengths.

The National Party's triumph in South Africa's 1948 general election and the subsequent implementation of new legislation in the early 1950s resulted in heightened political and commercial seclusion of the nation. As a reaction, South African Breweries Hop Farms Ltd., formerly known as Union Hop Growers Ltd., initiated a breeding initiative in 1956 by utilizing seeds from Fuggle and Hallertau breeding lines that were imported from Wye. The objective was to cultivate a cultivar that was more efficient and better suited to the local environment compared to the Golden Cluster variety that was being cultivated during that period. During the initial years of the 1960s, Golden Cluster was afflicted by black root rot, which led to a decline in production. However, the introduction of Pride of Ringwood from Australia in 1968 enabled the continuation of production. The initial cultivar surfaced from the breeding initiative in 1972. The predominant cultivar in the South African brewing industry is Southern Brewer, which originated as a Fuggle seedling and rapidly proliferated in the region. Despite being better adapted, the crop still necessitated longer daylight hours to attain economic production. This predicament was resolved by implementing artificial lighting techniques, and further breeding efforts were undertaken to cultivate strains that could endure shorter daylengths. The reintegration of South Africa into the global community during the early 1990s significantly expedited the necessity for such diversities (Brits 2001). Thankfully, several options had progressed to the advanced evaluation phases, and in 1992, Outeniqua and Southern Promise were introduced as the initial cultivars capable of being cultivated without supplementary lighting. Southern Promise exhibited a high level of adaptability to the absence of winter chill that is commonly observed in the hop-growing region of South Africa. A breeding program that is currently in operation has been ongoing, and in the year 2001, Southern Star, which is a seedling of Outeniqua with high-alpha properties, was introduced.

The hop breeders who were at the forefront of the industry during the early 1900s could not have foreseen the emergence of Verticillium wilt or downy mildew, nor could they have anticipated the advancements in hop cultivation or brewing methodologies that transpired throughout the twentieth century. The potential future obstacles that hop breeders may encounter remain uncertain and likely unforeseeable. As previously mentioned, numerous advancements in the field of hop breeding have been attributed to scientific advancements, particularly in the areas of genetics and chemical analysis. A significant portion of contemporary scientific advancements are occurring within the realm of molecular biology. Several techniques and concepts have the potential to significantly enhance comprehension of the biology, chemistry, and genetics of hops, thereby facilitating the emergence of novel advancements in hop breeding. Thus far, no novel cultivars have been generated utilizing said methodologies. Nonetheless, once these techniques are adequately refined, breeders may be capable of promptly adapting to alterations in the preferences of cultivators and brewers, thereby diminishing the duration required for the development of a new cultivar. Havill (2023) posits that the integration of desirable traits, such as pest or disease resistance, to minimize the application of pesticides, can be achieved with greater ease. Additionally, it may be feasible to effectuate particular modifications in extant varieties.

It is plausible that novel traits will be pursued and established. In ongoing breeding initiatives, researchers are investigating the potential for improvement in various components of hops, such as the antioxidant xanthohumol, the phytoestrogen properties, and the antimicrobial beta-acid content. In addition, there is a current application of molecular techniques to investigate the diversity present in untamed hop germplasm reservoirs. Salmon's pioneering utilization of wild germplasm in a hop breeding initiative remains pertinent even after a century, particularly in the era of molecular genetics, as per scholarly sources (Darby and Walker 2002; Haas and Barsoumian 1994; Murakami et al. 2006).

7.4 Germplasm Biodiversity and Conservation

7.4.1 Germplasm Diversity

Studies on molecular genetic diversity have focused on characterizing the genetic variability within hop, encompassing germplasm in local breeding programs and global collections. In certain instances, scholars have assessed the genetic variability of female hop assemblages to distinguish between closely affiliated lineages or to establish highly related cohorts or clusters of hop entries. Previous studies have endeavored to ascertain the heterotic crossing groups between male and female lines (Townsend and Henning 2005) or to determine the specific males that exhibit distant or close kinship with significant hop varieties (Henning et al. 2004). The ongoing research conducted by the author's group aims to elucidate the genetic variability among distinct male and female *H. lupulus* var. lupulus accessions and wild American counterparts, namely, *H. lupuloides* and *H. neomexicanus*. Various methodologies have been employed to assess the genetic variability among accessions, each exhibiting subtle distinctions and differential capabilities. Recent research conducted by Henning et al. (2004) utilized chemical and morphological characteristics to distinguish between various hop accessions. The present study aimed to classify a total of 129 accessions, comprising wild North American female accessions and cultivated hop lines from various regions across the globe. The classification was based on three primary groups, namely wild North American, European, and hybrids.

The European cohort was divided into two subgroups, known as the "English" and "Continental European" groups. Similarly, the Hybrids cohort was split into five distinct subgroups, each corresponding to a specific geographic origin. Abbott and Fedele (1994) were the first to publish a report utilizing molecular techniques to differentiate hop lines through the use of RAPDs. However, no assessments of genetic diversity were provided. The first investigations into the diversity of hop involved the utilization of RAPD analyses on 24 hop accessions, as conducted by Pillay and Kenny in 1996. The authors have documented that a total of nine primers out of 60 generated polymorphic bands. Furthermore, they have noted a limited degree of genetic diversity among the chosen hop accessions in their investigation. The present study investigated the segregation patterns of three primers, namely A11, A17, and C9, across five distinct families. The results indicated that the observed segregation ratios were consistent with the principles of Mendelian inheritance.

In their study, Sustar-Vozlic and Javornik (1999) used the PCR technique on a sample of 65 cultivars and identified two predominant clusters aligning with European and American hop cultivars. The European cohort was subsequently partitioned into five subgroups corresponding to distinct regions of geographic adaptation. In addition, the present investigation used the composition of essential oil in a distinct cluster analysis to affirm that the clustering analysis based on RAPD was in agreement with the analysis of essential oil. In 2000, Murakami conducted an analysis using RAPD on 51 cultivars from around the world. The results revealed six distinct groups among the cultivated hops. Pillay and Kenny (1996b) conducted RFLP analysis on rDNA samples obtained from both cultivated and wild hop accessions. Their findings showed the presence of three distinct genotypes (A, B, C) when XhoI restriction endonuclease was used. The study revealed that all accessions exhibited genotype A, while genotype B was exclusively in North American cultivars and genotype C was only detected in wild North American lines. The proposition was put forth that the utilization of genotype C may serve as a means of categorizing indigenous North American lineages. The technique employed in this study did not allow for differentiation at a fine level due to the limited number of molecular markers utilized. The genetic diversity (GD) among eight hop accessions was characterized in the initial AFLP studies conducted by Hartl and Seefelder in 1998, revealing a low level of genetic diversity. Three of the varieties, 'Saazer', 'Tettnanger', and 'Spalter', were indistinguishable.

The cultivars Hallertauer Magnum' and 'Wye Target' exhibited the highest genetic diversity, as evidenced by a GD value of 0.11. Subsequent research conducted by Seefelder and colleagues (2000) examined the genetic variability of 90 hop accessions (86 male and 4 female) and revealed a higher degree of polymorphism compared to their earlier findings. In the present study, it was observed that the cultivars 'Saazer', 'Tettnanger' and 'Spalter' could not be differentiated from each other. However, certain pairs such as 'Columbus' and 'Saazer' exhibited a genetic distance of up to 0.83. The findings of this research indicate the existence of two primary clusters, with an inter-cluster average genetic distance of 0.55. Cluster one comprised solely of aroma hop lines derived from Europe, whereas cluster two was composed of European germplasm that had been genetically modified with genes from American hop accessions. Several sub-groups were observed within each primary cluster. According to the authors, the clusters obtained through cluster analysis were deemed to agree with established pedigrees, with accuracy being determined through qualitative assessment. While the genetic diversity (GD) among female hop accessions is significant for germplasm preservation, its practical breeding applications are limited. The hop plant is a dioecious crop species that necessitates pollination from male hop accessions to produce offspring. In their research, Henning and colleagues (2004) undertook an investigation that encompassed a significant sample of male and female accessions. The study aimed to provide a characterization of 19 primary hop cultivars that are either grown or utilized for breeding purposes in the United States, as well as

82 male accessions that represent male hop lines in the hop collection of the USDA-ARS. The publication did not endeavor to ascertain genetic clusters. However, it did record certain male lineages as either genetically similar or genetically dissimilar to each female cultivar. This was determined through the use of AFLP fingerprints and the coefficient of coancestry, which was estimated from pedigrees. A noteworthy association was discovered between the coefficients of coancestry values and the genetic distance for male-female pairs that were either closely related or distantly related/unrelated. Nonetheless, this association did not hold for all conceivable male-female pairings. According to the authors, the significance of this study lies in the identification of male lines that have the potential to optimize heterosis.

Townsend and Henning (2005) analyzed a sample of 80 males and 26 females, in which they identified clusters that may exhibit heterotic effects between male and female lineages. The study conducted by the researchers revealed a substantial degree of polymorphism concerning AFLP markers, with 490 out of 550 loci exhibiting polymorphic characteristics. The researchers inferred that this high level of polymorphism could be attributed to the incorporation of diverse male and female genotypes. Townsend and Henning (2005) reported two main clusters in their molecular diversity analysis of hops, which were the European types and the wild American-European hybrids. This finding is consistent with previous studies on the molecular diversity of hops. The two principal clusters were subsequently differentiated into thirteen subgroups, comprising two exclusively female groups, nine exclusively male groups, and two groups with a mix of sexes. Multiple studies have documented the utilization of microsatellites for distinguishing between hop genotypes, gauging genetic variability, and categorizing genotypes into clusters based on their similarities.

In a study conducted by Brady et al. (1996), the efficacy of microsatellites and RAPDs in distinguishing hop cultivars was compared. The study revealed that the prevalent dinucleotide repeats in hops were (GA)n and (GT)n. The levels of genetic diversity in hops were not defined. In 2001, Jakse and colleagues conducted a comprehensive microsatellite analysis to assess genetic diversity. The study involved 41 genotypes, selected to represent female accessions from various regions across the globe. The authors conducted a comparative analysis of discrimination outcomes between microsatellites and AFLP. Their findings revealed that microsatellites demonstrated utility in grouping genetically related genotypes, while AFLP effectively clustered closely related genotypes and facilitated the separation of two geographically distinct hop clusters. Microsatellites have been utilized in various other applications, including the identification of particular hop cultivars from samples with unknown origins. In the field of varietal differentiation, Jakse et al. (2002) and Hadonou et al. (2004) have published three noteworthy works that discuss novel microsatellites and their development methods. These publications have been deemed useful for the aforementioned purpose. In their study, Jakse et al. (2004) conducted a thorough assessment of microsatellites in a sample of 124 female hop accessions, both wild and cultivated. The evaluation was carried out using four microsatellite loci. The present study successfully identified a total of 63 distinct alleles, with an average of 15.7 alleles per locus. The wild hop accessions from

North America and Europe exhibited the greatest number of alleles per locus among the different geographical groups.

The germplasm of wild North American accessions exhibited a substantial count of exclusive alleles, which may imply a significant degree of genetic diversity, as per the authors' proposition. In their study, Cerenak et al. (2004) employed a two-stage microsatellite technique to distinguish among 63 hop accessions. These accessions comprised clonal selections from other lines that were part of the investigation, as well as colchicine-induced tetraploids of female accessions that were also included in the study. In this study, it was observed that microsatellites were unable to distinguish between clonally-related lines and diploid±tetraploid pairs of accessions. Peredo et al. (2005) conducted a study comparing the efficacy of capillary electrophoresis and autoradiography in detecting and evaluating microsatellites. Their findings suggest that capillary electrophoresis may offer certain advantages over autoradiography in this regard. The researchers noted that the employment of fluorescence-labeled microsatellites in conjunction with capillary electrophoresis and fluorescence detection resulted in a greater degree of band resolution compared to the utilization of radioactively labeled microsatellites and visual examination of autoradiographs. The aforementioned studies enable the formulation of overarching observations regarding the comparative advantages of microsatellites and AFLP markers in the context of genetic diversity analyses. In contrast to certain literature (Peredo et al. 2005; Cerenak et al. 2004), it can be argued that microsatellites do not possess universal superiority in hop diversity studies.

7.4.2 Genetic Resources Conservation Approaches

7.4.2.1 In Vitro Conservation

The Humulus genus is native to Europe, Asia, and North America, with its probable place of origin being China, where all three species of the genus are found (Barth et al. 1994; Small 1978a, b). *H. japonicus* Siebold and Zucc. is indigenous to Japan, Taiwan, and China, whereas *H. yunnanensis* Hu is endemic to the elevated regions of the Yunan Province in China. *H. lupulus* L. is indigenous to China, Europe, and North America. It can be observed that all commercially utilized hops have been derived from the plant species Humulus lupulus. The majority of cultivars were initially obtained from untamed European hop selections, however, a considerable number of modern cultivars have their ancestry rooted in North American germplasm (Haunold et al. 1993).

Perennial rhizomes are a common method of maintaining hops germplasm collections in fields. The susceptibility of these plants to various hazards such as diseases, insects, and environmental stresses poses a significant threat. Furthermore, it has been observed that viral infections have the potential to aggregate within a field collection and subsequently spread to other locations through vegetative propagation, as reported by Young et al. (2023). The National Clonal Germplasm Repository for Humulus, located in Corvallis, Oregon and operated by the USDA-ARS, houses a collection of approximately 700 accessions of Humulus germplasm that have been sourced from various locations. The accessions are maintained as a field collection comprising 62 core representative accessions that have undergone heat treatment to eliminate viruses. These accessions are stored in pots under a screen and as tissue cultures. The practise of cryopreservation is currently regarded as a feasible alternative for the extended preservation of clonally propagated germplasm. The process of formulating a novel cryopreservation or cold storage protocol for each accession or genus of extensive multi-crop collections can be a protracted and laborious undertaking, as noted by Reed (2001) and Salgotra and Chauhan (2023).

The application of in vitro cold storage and cryopreservation methods, which have been utilized for other genera of temperate crops, can be effectively employed for analogous crops with minimal adjustments. The objective of this investigation was to assess the feasibility of preserving a vast assemblage of temperate hops germplasm using methodologies that were originally devised for genera that are not phylogenetically related. The present investigation aims to delineate the reaction of varied Humulus germplasm genotypes towards in vitro cold storage and cryopreservation in liquid nitrogen, utilizing methodologies analogous to those employed for blackberries, pears, and strawberries (Chang and Reed 1999; 2000; 2001; Reed 2001; Salgotra and Chauhan 2023; Engelmann 1999). Our study aimed to investigate the reaction of clonally propagated Humulus germplasm to conventional medium- and long-term storage methods and to establish a base collection by storing cryopreserved accessions.

7.4.2.2 Gene Banks

Today, hundreds of hop cultivars are grown and evaluated for their potential as ingredients in beer. Many modern European cultivars were developed from ancestral, genetically identical variants. Therefore, there is no variation between European cultivars in terms of genetics, morphology, or ecology (Murakami et al. 2006; Van Holle et al. 2019; Yan et al. 2019). Analysis of the genetic makeup of ancient European landraces reveals striking similarities between them (Patzak et al. 2010; Van Holle et al. 2019). These include the Fuggler, Hallertau, Saaz, Spalt, and Tettnanger. Over centuries of homestead cultivation in Europe's villages and cottage gardens, domesticated hops have interbred and backcrossed with wild hops. In the past, farmers would grow and trade new types among themselves. Although the quality and scent of some older landraces like Saaz are highly prized, their yield may be lower than that of some newer cultivars. Pests and illnesses can do greater damage to them, and they are more susceptible to climate change (Mozny et al. 2009). Because of climate change, there will be greater demand for breeding as a result of hotter temperatures and more frequent dry spells. The research conducted by Eriksen et al. (2020) suggests that certain cultivars may thrive in tropical environments. Studying wild Italian samples of *H. lupulus* (Rodolf et al. 2018) revealed a significant level of variation in Europe, similar to that found in Asia and the Americas (Patzak et al. 2010;

Van Holle et al. 2017). While the European variant, *H. lupulus*, is known for its high alpha-acid content and resilience to verticillium wilt (a fungal disease), the American wild hop subspecies, *H. neomexicanus* and *H. lupuloides*, are more genetically varied. Additionally, it appears that two distinct *H. lupuloides* populations have developed due to isolation by the Appalachian Mountains in North America (McCallum et al. 2019; Murakami et al. 2006), European settlers brought hop cultivars to the New World, which changed the genetic makeup of native American varieties. They are still genetically separated from the European cluster (Patzak et al. 2010; Rodolf et al. 2018). There is a need to introduce the American gene pool to Europe since American cultivars have superior resilience to pests and illnesses (Murakami et al. 2006; Patzak et al. 2010).

Traditional brewing techniques and traditional flavor preferences have led the selection process for hop varieties used in traditional beer production (Murakami et al. 2006). However, new grown varieties are in demand, particularly as a result of the proliferation of craft breweries with diverse tastes (McCallum et al. 2019; Rodolf et al. 2018).

Rules in the certification of hop plants and hop products and the development of novel DNA-based analytical technologies have resulted from the necessity of confirming the authenticity and regional origin of hop batches acquired by brewers and other users (Ocvirk et al. 2019). Saazer and Shinshu Wase, two cultivars, and *H. lupulus* var. cordifolius (Miq.) Maxim. ex Franch. & Sav., a Japanese wild hop, were used to construct the *H. lupulus* draft genome in 2015 (Natsume et al. 2015). The application of this kind of genomic data to hop breeding and research is promising. Shinshu Wase's de novo RNA sequencing investigation uncovered the developmental control of genes involved in beer flavor and aroma metabolic pathways. Further analysis revealed an increase in fragrance and flavor biosynthesis genes in Shinshu Wase compared to wild hop.

Following this, researchers at Oregon State University created HopBase (http://hopbase.cgrb.oregonstate.edu), a genetic database dedicated to the hop plant. Hop Base's goals include tracing the origin of muddled hop samples, supporting the development of cultivars and types with novel genetic diversity, and identifying hop strains that are both prolific and resistant to pathogenic microorganisms. The database will presumably also aid in determining the plant's biological history and its distribution, as well as locating the hop's progenitor (Vergara et al. 2016; Dessalle and Tattersall 2019). Hop's vast, repetitive, and heterozygous genome has complicated efforts to deduce the metabolic processes responsible for desirable features. The Cascade hop cultivar's genome, estimated to be 2.7 Gb in size, was recently assembled (Padgitt-Cobb et al. 2019). This not only improves our knowledge of the hop genome but could be used to investigate other complicated, big genomes as well. The precise editing of the targeted hop genome using the CRISPR/Cas9 system has recently been demonstrated, offering a viable path for hop breeding, according to the work of Awasthi et al. (2021).

7.4.2.3 Cytogenetics

Both cytogenetic studies and genome assembly analyses (Danilova et al. 2003; Natsume et al. 2015) indicate a highly repetitive and large hop genome, approximately 2.8 Gb in size. The inability to assemble across repetitive areas is a problem for short-read genome assembly when dealing with large volumes of repetitive DNA. Therefore, the repetitive areas are too big for mate-pair technology and must be assembled using high-priced long-read sequencing techniques. Even though the assemblies produced by the large and thorough efforts employing short-read sequencing techniques are not yet complete, they are available (Natsume et al. 2015; Soliman et al. 2017, 2023).

7.5 Traditional Breeding

Typically, basic phenotypic selection techniques were used by breeding programs to determine which offspring were superior before releasing them as new types. Open-pollinated progeny bred for better alpha levels and high yields led to the emergence of many early varieties in the first part of the twentieth century, such as 'Brewers Gold' and 'Bullion. Open-pollinated seedlings from a wild North American hop were the outcome of early attempts at breeding at Wye College to establish varieties with resistance to Verticillium wilt (Havill 2023; Salmon 1949). It wasn't until the 1930s and afterward that Zattler at the Huell in Germany and Keyworth in Wye began repeated breeding efforts. Zattler worked on developing downy mildew-resistant lines, whereas Keyworth prioritized wilt-resistant varieties. U.S. hop breeders set out to create high-yielding fragrance strains that thrived in the region's cool, wet climate. Neve (1991) provides a comprehensive and good overview of hop breeding for pest resistance.

Hop varieties can be developed through hybridization in several ways, with open pollination of known female cultivars being one of the simplest. In early hop breeding operations, the preferred female cultivar was generally grown alongside conventional male hops for pollination. The process of phenotypic selection involved collecting seeds after pollination and seed set, then planting them in the field. The best offspring were culled, cloned, and cultivated on a commercial scale. If a choice seemed good, it was grown in commercial plots and gathered for small-scale brewing experiments. Brewing trials were conducted to determine whether or not the experimental line should be issued as a cultivar. Brewers probably care a lot about the beer's flavor, and since that flavor is significantly affected by the hops used in the brewing process, this makes sense. Consequently, brewers have been adamant about sticking to hop cultivars that reliably deliver that flavor and hesitant to experiment with new varieties. Since the contributions from males could not be regulated, open pollination breeding techniques only allowed for partial control of the offspring's genetics. Early breeders realized that the transfer of unwanted features from unimproved male lines was slowing or preventing the widespread adoption of new cultivars in the brewing industry. Controlled crosses with pollen from particular males, some of which were seedlings from open pollination on known female cultivars, were soon being used by breeders in all three locations. This strategy, which is very similar to animal breeding practices, marked the beginning of what is now known as "pedigree breeding."

Hops developed from seeds can occasionally be better than those obtained by cuttings, as highlighted by Lance in his treatise on hops 1, in which he cited Mylius's (1750) novel sexual theory for plants. Mendel's research on the sexual transmission and segregation of traits in the garden pea was rediscovered around 1900, marking the beginning of scientific hop breeding. The approach used the systematic, controlled crossing of plants to produce offspring with genetic differences. The first modern hop breeding program began in Britain at Wye College in 1906, when Professor Salmon raised and planted seedlings arising from crosses made under the direction of Howard a few years earlier. These German and American programs were short-lived.

While records show that hops were cultivated in Europe as early as the ninth century and in Britain as early as 1524, the history of the plant and how it was farmed remain shrouded in mystery. However, even the earliest documents provide a detailed description of how hop plants can be propagated vegetatively by dividing a rootstock or removing underground stems. So, the hop farmers learned how to multiply the good plants. By 1669, researchers had identified several hops species, each with its unique characteristics. These would likely be varieties of plants that thrived in the local soils and climate. Hop varieties that become the norm in a given region often take their common name from either the name of the region or a defining characteristic of the hops themselves. So, Percival lists several different kinds of hops, including the Farnham Hop, the Canterbury Whitebine, and the Golden Tips. The Saaz region of the Czech Republic, the Hallertau region of Germany, and the Tettnang region of Austria are all well-known for their locally-produced varietals.

Natural clonal alterations or the accidental emergence of a seedling are two potential sources of diversity in addition to local adaptability. In cases where these plants were chosen, the credit was usually given to the cultivator who first saw their promise. Mr. Golding's hop and Mr. Fuggle's hop both came about in this way, eventually becoming shortened to Goldings and Fuggles, respectively. Cluster hops, the classic American variety, are said to be an accidental seedling of European hops brought to the New World by the Massachusetts Company in 1629 and Native American hops.

Many unique cultivars emerged as a result of growers' selection over several centuries. By the turn of the century, scientists had identified over 20 distinct species in England and over 60 across the rest of Europe. Using contemporary techniques for testing essential oils, it has been possible to determine which of the earlier kinds are genetically the same and which are separate. The Golding variety was selected around 1790, and its offspring include the Rodmersham, Mercers, Mathon, East Kent, Canterbury, Cobbs, Bramling, Eastwell, and Early Bird. Golding clones range widely in maturity, crop production, and disease resistance. Fuggle, on the other hand, was proven to be a separate variety despite the urban legend that it was picked out of a random seed in 1861. Oil research, however, reveals that Fuggle and Savinjski Golding are genetically identical cousins.

While most pre-1900 cultivars have been rejected, some of the high-quality types that were developed at this time are still widely grown and highly valued.

The complex chemical profile, perennial nature, and specialized production infrastructure required for aroma hop enhancement make this a difficult task. The lengthy juvenility stage in hop establishment (often three growing seasons in Western Oregon; Neve 1991) further makes these problems worse. Although the connection between chemical profile and subsequent beer qualities is not entirely understood (Sharpe and Laws 1981), the acids and essential oils produced in hop contribute to a spectrum of flavor and aroma in beer. For example, hop essential oil has a very intricate composition, and little is known about the inheritance and coinheritance of essential oil components. Developing a new cultivar of aroma hop using the 13 time-tested techniques is a laborious and time-consuming procedure that can take up to 10 or more years. According to Haunold (1981), mutation breeding, pedigree breeding, phenotypic recurrent selection, and mass selection are the most important methods for creating new hop cultivars.

Natural mutations or the application of chemical and physical mutagens to create ploidy or random mutations are also valid methods of mutation breeding (Haunold 1981; Trojak-Goluch and Skomra 2013).

Pedigree breeding is a method of selective breeding in which desirable features from multiple parents are recombined. The offspring of a hybrid often varies greatly from either parent. Breeding is a process in which good qualities are preserved while undesirable ones are bred out (Sleper and Poehlman 2006). The hop industry does not heavily employ either mutant breeding or pedigree breeding (Henning 2006). Recurrent phenotypic selection is the repeated process of improving a breeding population. To do this, the most successful male and female plants in a population are chosen and then randomly interbred to create offspring. This is done again and over again until the desired results are achieved (Bernardo 2002; Sleper and Poehlman 2006). More opportunities for recombination between genes with each passing generation means that novel allelic combinations might emerge, expanding the range of possible genotypes and phenotypes. Phenotypic recurrent selection is used to boost the number of beneficial alleles in a breeding population, which should, in theory, make it easier to single out superior lines. It helps enhance features controlled by many genes, and it helps keep genetic variety high in a population of 14 thanks to the frequent intermingling of heterozygous lines (Allard 1999; Sleper and Poehlman 2006). Phenotypically superior plants, numbering in the hundreds or thousands, are chosen in a mass selection process. The plants are allowed to mate with one another, and then the resulting seed is collected and stored without first conducting a progeny test. This is the most basic and traditional technique used by breeders to improve crops. Because pollination cannot be controlled, the number of generations needed to produce a viable population may increase (Allard 1999; Sleper and Poehlman 2006) when selected plants are pollinated by both superior and inferior pollen parents.

7.6 Role of Biotechnology

Numerous primary and secondary sources (Aron and Shellhammer 2010; Miranda et al. 1999; Stevens et al. 1997), as well as a high marker density genetic map (Henning et al. 2016a, b), RNA sequencing datasets (Clark et al. 2013; Natsume et al. 2015), and a draft genome assembly are all available. Another female hop variant has been deep-sequenced, compiled, and annotated by our team. In addition to identifying male-specific DNA and pseudoautosomal portions of the sex chromosomes, the assembly was used to guide the assembly of the first male hop genome (USDA 21422M) (Hill et al. 2016). There is no public annotation available for any of these sources, and no effort has been taken to compile the available data into a single location. Genome annotation and bioinformatics face significant difficulties in unifying and standardizing data (Attrill et al. 2016; Berardini et al. 2015; Howe et al. 2016).

7.6.1 *Molecular Breeding*

In the family Cannabaceae, *Humulus lupulus* L., the commercial hop, is a dioecious perennial. Based on differences in leaf shape and distribution, this species can be broken down into five distinct botanical subtypes (Small 1978a, b, 1981). (Small 1978a, b, 1981) These include the European hop (*H. lupulus* var. lupulus E. Small), the Asian hop (*H. lupulus* var. cordifolius (Miq.) Maxim.), and three endemic hops from North America (*H. lupulus* var. lupuloides E. Small, H. lupulus var. neomexicanus A. Although various subspecies of hop plants have been added to the breeding gene pool, most cultivated hop plants are descendants of the Euro-pean hop. Almost all hop varieties are diploid ($2n = 2x = 20$), with two sex chromosomes and 18 autosomal chromosomes. There are triploid cultivars and tetraploid breeding lines (Bassil et al. 2008; Haunold 1971).

The mature female inflorescences of the hop plant are the most valuable component of the plant. These are the cones. Manufacturing occurs in several different countries and continents, including the United States, Europe, China, Australia, New Zealand, and South Africa (Jakse et al. 2011; Henning et al. The lupulin lands found in female cones are responsible for the production of aromatic and bittering compounds (Jakse et al. 2008; Mafakheri et al. 2020) that are integral to beer's flavor. When male plants are not used in the breeding process, the seed set is reduced and lupulin makes up a larger percentage of hop cone weight (Jakse et al. 2008; Henning et al. 2015a, b). The National Agricultural Statistics Service (NASS) estimates that the value of the 51 thousand tonnes of cones gathered in the United States in 2019 is $637 million.

In 2016, the value of production was $65 million, roughly tripling from 2015. The growth of the craft beer business has been a major factor in the uptick in output. The number of taprooms in the United States has increased from zero in 2010 to 3471 in 2020 (Brewers Association 2021). This is an increase of 298% in the past decade. According to the 2016 research by the Barth-Haas Group, there are now 146 hop cultivars in use around the world, an increase of 17 from 2009. Farmers and brewers are responding to this need by planting hop gardens with specific varieties chosen for their ability to add distinctive scents and flavors to beer (Teghtmeyer 2018). The number of small, independent brewers in the United States has increased by at least 50% throughout all 50 states in the past decade (Brewers Association 2021).

Craft brewers are constantly on the lookout for new hop cultivars with interesting aromatic qualities that can be used to produce flavors beyond the standard citrus, floral, and herbal offerings. New varieties are being actively developed by breeding programs at the United States Department of Agriculture (USDA), Agricultural Research Services (ARS), private companies, universities, and most recently, Oregon State University (OSU) (Pigg 2017; Rodger 2021).

Hop cultivation originated in Europe in the 1100s, when monasteries started brewing beer and decided to add hops for flavor, a decision that had the unintended consequence of making the beer more sanitary in areas where the water supply was unfit for human use (Alberts 2020). Wild hops from the Saaz region of the Czech Republic, the Backa region of Serbia, and the Hallertau and Tettnang regions of Germany have all become famous for their distinctive attributes and flavors. After the turn of the century, hop breeding efforts began using long-term studies based on controlled crossings to improve plant quality (Hornsey 1999, 2003). Geographical isolation or striking physical characteristics were used to classify hop cultivars grown in greenhouses (Small 1978a, b, 1981). Hop plants were bred via root cuttings, but the importance of keeping pure lines of a desired variety alive that is, cloning was not understood until the twentieth century. Thus, fluctuations in rainfall and insect pressure led to widely varying plant performance and production. Once every decade or so, farmers might see a full harvest, which would cause them to overplant. This resulted in significant price swings between bumper and bust hop harvests. Mycologist E. S. Salmon at Wye College (South Eastern Agricultural College), Kent, England, initiated a hop development program in 1904 to combat disease through the use of advanced plant breeding techniques. Hop was first analyzed on a big scale by Salmon and the East Malling Research Station in 1917 (Hornsey 1999). To enhance disease resistance and resin content and to enhance scent profiles (Bassil et al. 2008; Darby 2005a, b), Salmon employed wild American material (*H. lupulus* var. lupuloides) from Manitoba, Canada. Salmon's 'Brewer's Gold,' 'Northern Brewer,' and 'Bullion' cultivars are all highly sought-after results of the Wye breeding program. These three varieties have been utilized as parents in the majority of high-alpha or bitter breeding projects to date (Darby 2005a, b).

Breeding programs at OSU and UNL, as well as the more established USDA-ARSFSCR in Corvallis, Oregon and Prosser, Washington, are working to create locally adapted cultivars with improved yields and brewing qualities, as well as resistance to diseases like powdery mildew and downy mildew. Wild germplasm is still used in modern breeding projects to create novel cultivars with unique disease resistance and aroma/flavor profiles (Mafakheri et al. 2020; Peredo et al. 2010). Each of these initiatives keeps a repository of European cultivars, wild accessions, and experimental breeding material in its germplasm collections. Hop breeding programs can't move forward without first assessing the available germplasm for diversity.

The NCGR at Corvallis, Oregon, oversees a world hop collection with 647 accessions from 21 countries to assist these breeding programs and preserve historically important cultivars for future generations. The collection is incredibly varied, with items ranging from cultivated European varieties to wild North American accessions. Since morphology can change due to environmental influences like growing conditions (Korbecka-Glinka et al. 2016; Mafakheri et al. 2020; McAdam et al. 2014), the NCGR uses molecular markers to ensure that each clonal accession is correctly identified (Bassil et al. 2008).

Complete maintenance at the USDA and other germplasm collections requires the deployment of a DNA-based economic instrument to confirm cultivar identity. Microsatellites, also known as simple sequence repeats (SSRs), and single nucleotide polymorphisms (SNPs) are two types of codominant DNA-based markers used in germplasm identification. Polymorphism, reproducibility, cost-effectiveness, the possibility of automation, and the ease of transfer between laboratories are only a few of the many benefits of simple sequence repeats (SSRs; 1-6 bp long tandem repeats (Bassil et al. 2008; Korbecka-Glinka et al. 2016; Li et al. 2002). In addition, less DNA is needed for these markers than for others (Robinson et al. 2004). One of the most effective DNA-based strategies for utilizing genetic studies in many crop species is employing SSR markers, as has been demonstrated repeatedly (Palumbo and Barcaccia 2018).

Among all DNA markers, SNPs are by far the most common (Henning et al. 2015a, b). High throughput genotyping, cultivar identification, and Quantitative trait loci (QTLs) analyses have all been accomplished with their help because of their automation friendliness (Henning et al. 2015a, b; et al. 2016; Howard et al. 2011; Matthews et al. 2013; Yamauchi et al. 2014). The management and breeding of hops require a variety of fingerprinting technologies that are both flexible and sensitive enough to discriminate between siblings and clarify links among cultivars and between the cultivated and wild gene pools (Abd El-Moneim et al. 2021; El-Mansy et al. 2021; ALshamrani et al. 2022; Safhi et al. 2022). Recently, hop has gotten a boost from the genetic revolution. Genomic (Brady et al. 1996; Hadonou et al. 2004; Jakse et al. 2002; Stajner et al. 2005) and genic (Bassil et al. 2008; Koelling et al. 2012; Jakse et al. 2011; Stajner et al. 2008) SSR markers have been isolated more than a thousand times. SSR markers in hop have been useful for a wide variety of purposes, including genotype identification, parent age analysis, diversity assessments, linkage mapping, marker-assisted selection, and more. (Brady et al. 1996; Cerenak et al. 2004; Jakse et al. 2011; Mafakheri et al. 2020; Murakami et al. 2006)

HI-AGA7, a simple sequence repeat marker, has been utilized to identify males and females (Jakse et al. 2008; McAdam et al. 2013). High polymorphism and allelic diversity have made the HI-AGA7 marker useful for genotyping (Jakse et al. 2008; Korbecka-Glinka et al. 2016; Mafakheri et al. 2020; Rodolfi et al. 2018). Hop breeding and genetics have benefited greatly from high throughput genotyping approaches based on SNPs. Methods like diversity array technology (DArT) (Henning et al. 2011; Rodolfi et al. 2018), genotyping-by-sequencing (GBS) (Henning et al. 2015a, b; 2016a, b; Hill et al. 2016; Zhang et al. 2017; VanHolle et al. 2019), and competitive allele-specific PCR (KASP) (Jiang et al. 2018). Linkage mapping, QTL analysis, and genome-wide association studies have all shown the utility of SNP genotyping (Henning et al. 2011, 2015a, b, 2017a, b; Zhang et al. 2017), even though many SNP genotyping methods can be costly to run on small sample sizes. Powdery mildew resistance (Henning et al. 2015a, b, 2017b), sex determination (Hill et al. 2016; Zhang et al. 2016), short stature (Henning et al. 2017a), and drought stress tolerance (Zhang et al. 2016) were all studied using SNP markers in the hop. The use of single-nucleotide polymorphisms (SNPs) for cultivar identification has garnered a lot of attention recently (Henning et al. 2015a, b; Jiang et al. 2018; Matthews et al. 2013; Yamauchi et al. 2014; VanHolle et al. 2019). Yamauchi (Yamauchi et al. 2014) characterized 21 hop cultivars by using transcriptome sequences obtained via Sanger sequencing to discover four SNP-containing areas. Jiang et al. (2018) employed 12 KASP markers to differentiate between 16 hop cultivars, whereas Matthews et al. (2013) and Van Holle et al. (2019) used GBS to identify between 178 and 56 commercial cultivars, respectively. Cultivar identification using previously discovered SNPs requires expensive and time-consuming sequencing using Sanger and Illumina technology. There is a need in the rap game industry for an SNP-based fingerprinting tool that is simple to use and assess on a budget.

7.6.2 Molecular Marker-Assisted Breeding

When it comes to rapidly increasing complicated features like a plant's resistance to downy and powdery mildews, marker-assisted selection procedures (Kearsey and Luo 2003) hold a lot of promise. Over the past 5 years, researchers have uncovered more about the molecular mechanisms that govern hop characteristics and the mechanisms behind genetic variety (Henning 2006). Disease resistance and other traits have been the subject of some QTL research as well (Cerenak et al. 2006, 2009; Henning et al. 2011; Koie et al. 2005; Seigner et al. 2004). Powdery mildew is another significant hop pathogen, and QTLs linked to resistance to it have been identified (Henning et al. 2011; Seigner et al. 2004). Molecular markers (DArT markers) were also utilized by McAdam et al. (2013) to identify QTLs associated with multiple hop characteristics. There is a dearth of literature on hop-related genome-wide association studies.

As a result of recent advancements in marker identification and development, breeders may be able to choose quantitatively inherited traits by the use of molecular markers that span the entire genome. The genes or gene families that regulate the expression of characteristics in other crop species, in animals, and in individuals with hereditary illnesses have been identified by genome-wide association studies (GWAS) and genotyping-by-sequencing (GBS). This strategy was utilized by Huang et al. (2010) to identify 14 economically valuable features in rice that are governed by genetic components. One hundred and fifty thousand molecular markers were discovered using next-generation sequencing (NGS). They identified 80 markers from this large set that regulate the expression of these characteristics. The manifestation of a trait was marginally, but noticeably, influenced by other factors (Abd

El-Moneim et al. 2021; ALshamrani et al. 2022; EL-Mansy et al. 2021; Safhi et al. 2022). Using this knowledge, scientists devised a method for locating the extremely uncommon offspring of greenhouse-grown plants that possessed all or nearly all of the desirable features. Benefits of GWAS have been established (Buckler et al. 2009; Sherman et al. 2009; Waugh et al. New research demonstrates how these techniques have allowed for a significant improvement in precision breeding (Huang et al. 2010; Poland and Rife 2012).

To determine if an SNP is associated with a particular phenotype, GWAS is typically performed on natural populations. But typical GWAS in natural populations yields many false negatives and false positives, as discovered by Visscher et al. (2012). These misclassifications are due to the incorrect structure of the population, which current statistical tools cannot do much to correct. One option to reduce the frequency of such errors is to conduct simple QTL studies on half-sib or full-sib families instead of utilizing linkage maps to examine the association between a single marker and a QTL. Mitchell-Olds (2010) argues that there are advantages and disadvantages to utilizing GWAS or QTL to link SNPs to phenotypes. While GWAS can uncover a significantly higher number of SNPs that are handed down from parents, QTL research utilizing bi-parental mapping populations can only look at SNPs that are passed down from parents. It is essential to create a high-density genetic map for use in QTL research. Finally, once the QTL analysis is complete, it is common practice to retest the markers (Mitchell-Olds 2010). Post-test validation of markers discovered by GWAS is also recommended due to the high prevalence of Type I and Type II errors in GWAS. The use of a mapping population with high phenotypic segregation for marker association with GBS is not frowned upon in light of these drawbacks, provided adequate post-test validation is planned and executed. This is especially true when money is tight, there are many distinct qualities in the mapping population, and straightforward marker-assisted selection methods are desired. There is a lack of understanding of the genetic basis for disease resistance, cone quality attributes, and agronomic performance due to the current lack of information about the hop genome. Next-generation sequencing has made it possible to analyze genomes rapidly and cheaply. These methods allow for a comprehensive genome-level comparison of two or more disease-resistant hop cultivars or accessions. Genomic regions influencing disease resistance might be identified, for instance, by comparing resistant and susceptible accessions to one another. The hop industry has not yet implemented the findings of genotyping by sequencing research.

7.6.3 Functional Genomics

Understanding the biological and genetic basis for each of the morphological features of the ideotype can greatly improve breeding for short-stature hop varieties. Darby (2001) offered a straightforward rationale for why hops tend to be short-statured. At the time, genetic characterization and marker systems were still in their infancy, making it impossible to perform a fine-grained dissection of the feature. The author categorized people into two groups, "'short'" (less than 4.5 m) and "'tall'" (more than 4.5 m), to make the data easier to understand. This paper uses information gathered from earlier attempts to breed hop varieties that are both small and conventionally tall (greater than 4.5 m in height). Darby (2001) concluded that the 5:3 segregation of progeny in this mapping population (Tall versus Short) was indicative of the presence of a single dominant gene controlling height and an '"epistatic major gene'" modulating the expression of absolute height. There has been a lack of literature detailing the genetics of small height in the hop. To boost selection efficiency and speed up selection procedures, molecular markers selected for small stature could be used in the greenhouse. QTL mapping is a method for locating markers that are linked to target traits (for a review, see Miles and Wayne 2008). The small stature of the hop population was identified using a genetic map of the population. QTL mapping analysis can make use of several different statistical tests. First, there is the "Single Marker Analysis," which entails running marker regression for each marker and the characteristic. Several false positives are possible with this test since other markers associated with the characters can have an impact. Interval mapping (IM) is an approach established by Lander and Botstein (1989) to help mitigate the effects of confounding variables and false positives. Composite interval mapping (CIM), a method introduced by Jansen (1993), attempts to solve some of the issues with IM. Previous QTL studies in the hop were referenced as Henning et al. (2015a, 2015b) elaborated on the numerous QTL analyses. Genome-wide association studies (GWAS) are another method for locating molecular markers associated with features of interest. Genome-wide association studies (GWAS) entail analyzing a large number of genotypes that collectively stand for a breeding stock's germplasm pool. Potential molecular markers that do not map to a specific linkage group can be discovered using the same methods on a bi-parental mapping population. Unusual segregation of the marker in the mapping population is the most prevalent reason for markers not being mapped to a certain linkage group. In other cases, like with the hop, the draft genome is either insufficient or incomplete, or the marker is associated with a repeating stretch of DNA that is located on several linkage groups. Association studies were used by Henning et al. (2015a, 2015b) to examine downy mildew resistance in a hop bi-parental mapping population. Finding molecular markers associated with qualities of interest that were not included on genetic maps of the same population is one of the many successes of applying GWAS methods to a bi-parental mapping population (Henning et al. 2015a, b).

7.7 Bioinformatics

Taking advantage of the fact that transcripts, like mate-pair reads, provide order information about the genome, transcript-guided assembly (TGA) is a method for improving genome assemblies. To utilize this knowledge, DNA reads matching the genomic sequence covering their associated genes are retrieved using the assembled mRNA sequences. This results in an assembly of the genomic sequence including and surrounding the genes, along with any promoters, introns, and another flanking sequence. Repetitive introns do not disrupt the assembly's ability to properly organize these regions. Our method is similar to a recently published (Aluome et al. 2016), but it also involves contig ordering and gap closing.

Xie et al. (2014) describe the initial step of transcriptome-guided assembly as a de novo transcriptome assembly using SOAP de novo trans (http://soap.genomics.org.cn/SOAPdenovo-Trans.html). The Shinsuwase paper, which contained the RNA-seq reads, was utilized. All Shinsuwase-related RNA-seq libraries (DRA002630) were obtained from the DNA Databank of Japan. A total of 1,102,071 scaffolds were assembled, with an N50 of 431 indicating a large number of very modest-sized scaffolds. To eliminate scaffolds that only contained small gene fragments, the contigs were filtered to be at least 1000 bp in length. Next, BLAST was used to search the NR database for potential contaminants in the remaining contigs (Pruitt et al. 2007). Significant matches that did not correspond to plant sequences were removed from the analysis. The final count was 43,926, with an N50 of 2765. These transcript sequences were then aligned to the draft genome assembly. Transcript-guided assembly (TGA) was carried out on misaligned transcripts. These transcripts were then subjected to a BLASTN alignment with whole genome data. The results were combined with Velvet using default settings (Zerbino and Birney 2008) and a K value of 51. Exonerate (Slater and Birney 2005) was used to match the resulting contigs back to the initial transcript. This "order" was then applied to the constructed contigs, which were sorted and scaffolded with Ns between them. Gap Closer 1.0 was utilized to fill in the gaps. Each transcript under went this process five times. The NG50 was 41,006 bp, and the final assembly size was 1,766,890,029 bp.

One possible drawback of the TGA method is that it may lead to incorrect genome assembly due to misassembled transcripts (Steijger et al. 2013). To solve this problem, we used a severe filtering method to exclude contigs that did not significantly correspond with plant genes in NR after being generated from transcript models based on length. The fact that TGA resulted in better public hop ESTs than the Shinsuwase assembly does provide some credibility to the technique. To create novel repetitions, a repeat library was assembled using k-mers with a high copy number (Li and Waterman 2003). Next, we used Jellyfish (http://www.cbcb.umd.edu/software/jellyfish/) and k 14 31 (Kurtz et al. 2008) to locate high-copy k-mers in the 173-bp library. Repeated k-mers were defined as those with over 120 copies, or 6 times the predicted coverage. Velvet was then used to combine these k-mers into a primary set of repeat sequences. We cut off the 64 bp sequences. Nussbaumer et al. (2013) used BLAST to compare leftover sequences to the MIPS repeat element database (mips-REdat) and NR for identification. Aligned contigs were sorted into chloroplast, mitochondrial, and ribosomal RNA groups. Plant-functionally annotated, non-redundant sequences were also culled from the library and flagged for further study. The final repetition set included 9615 repeats with an N50 of 212. Repeats were labeled with TE-class models that have already been trained (Abrusan et al. 2009). Hierarchical classification is used by TE-class, which successfully categorized 98% of repetitions, 86.2% of retrotransposons, and 14.6% of DNA transposons. As with other angiosperms, this is a pattern. This library was used to generate the final repetition library for masking, along with maps-REdat.

Annotating the genome was a multi-stage process. Before proceeding with additional annotation, the genome was masked using RepeatMasker and the previously described repeat database. StringTie version 1.2.0 was used to assemble the transcriptome from the RNA-seq reads described before, and HISAT version 0.1.6 was used to align the reads to the genome assembly (Kim et al. 2015; Pertea et al. 2015). This led to an unexpectedly large amount of RNA-seq, which produced over 1,000,000 transcripts. Single-exon genes with low read coverage are more likely to be false positives. Outlier detection with a sci-kit-learn-trained one-class SVM was used to filter the genes (Pedregosa et al. 2011). This initial group of "genes" was derived from the population's outliers. The StringTie transcripts were used to run MAKER-P, a pipeline for the automatic annotation of plant genomes (http://www.yandell-lab.org/software/maker-p.html), on the masked genome (Campbell et al. 2014). To locate genes in the supplied Arabidopsis models, we employed Augustus and SNAP (Korf 2004; Stanke et al. 2006). BLASTP was then used to link the retrieved peptide sequences from the remaining genes to the protein sequences from the mitochondria and chloroplasts of TAIR10 Arabidopsis (Johnson et al. 2008). We could not proceed with further investigation until the E-value was less than or equal to 1E-4. In addition, we extracted from the main annotation all genes tagged with the terms "gag", "pol" "Retrotransposon," and "Retroelement."

The remaining genes were BLASTP-searched against the TAIR 10 database of known hop genes and Uniprot (Rhee et al. 2003; Consortium 2008) for functional annotations. This annotation was chosen because it has the lowest E-value. This resulted in a total of 22,201 annotated genes for the Shinsuwase annotation and 16,161 for the Teamaker annotation. Assembly methods explain the discrepancy between the number of genes and the number of annotated genes. Because of its reputation for conservatism, ALLPATHS-LG may not include highly heterozygous or damaged genes in its output. A similarly aggressive assembler might double-scaffold these genes. Regardless, there is still a long way to go before the hop draft genome can be considered finished.

7.8 Tissue Culture Application

7.8.1 Micropropagation Approaches

Hop cultivars are often vegetatively reproduced by burying young vines or taking softwood nodal cuttings in the spring or summer (Neve 1991). Tissue culture methods offer an alternate strategy for propagating plants. Micropropagation refers to the process of growing plant clones in a laboratory. Asexual reproduction allows for the spread of identical copies of an organism, which is the essence of micropropagation. Aseptic clonal propagation technologies, in contrast to traditional approaches, allow for the rapid multiplication of a single plant into a large number of copies in a small area in a short amount of time (Chawla 2002). Starting from node cuttings, apical tip explants, meristems, and even shoot or leaf tissue (Batista et al. 1996; Peredo et al. 2006), in vitro micropropagation of hops is possible.

Roy et al. (2001) examined the impact of plant growth regulators and different media formulations on the efficiency of shoot multiplication using nodal explants of the hop variety H138. On Murashige and Skoog (MS) media supplemented with indoleacetic acid (IAA, 0.57 M) and 6-benzyl amino purine (BA 2.22 M), they were able to successfully establish cultures (96.6% of explants). Media supplemented with the cytokinin N-phenyl-N'-1,2,3-thidiazol-5-ylurea (TDZ) produced more shoots and nodes per explant than those supplemented with BA, 6-(, -dimethylallyl amino) purine (2iP), or kinetin (KIN). Using low sugar, starch-Gelrite media, Smykalova et al. (2001) created an effective in vitro micropropagation system for Czech and various foreign hop meri clones. Cost-effective micropropagation of non-vitrified hop cultures was achieved using this medium.

It is crucial to maintain the genetic integrity of recovered plants in clonal replication of plants, especially commercial cultivars of hops. Regenerated plants may have a different phenotype from the mother plant in some situations because of genetic and epigenetic modifications created by the in vitro culture environment (Kaeppler and Phillips 1993). Molecular techniques are frequently described in the literature as a method for assessing the genetic stability of plants developed from tissue culture. This is because molecular markers can characterize the phenomenon of somaclonal variation with much greater precision and less effort than karyological and phenotypic analyses. Peredo et al. (2006) characterized the genetic and epigenetic variation in hop plants regenerated from sequential subcultures of organ genic calli by using the amplified fragment length polymorphism (AFLP) and methylation-sensitive amplified polymorphism (MSAP) procedures. Since none of the AFLP loci were polymorphic, they were unable to discover any large-scale genomic rearrangements in the callus-derived plants; nonetheless, epigenetic modifications caused by a demethylation process were detected using the MSAP method. When the calli were kept in tissue culture for longer, the epigenetic modifications were found to be more pronounced. When using the meristem culture method to create virus-free stock plants, micropropagation is a common method for mass propagation.

7.9 Genetic Engineering and Gene Editing

7.9.1 Transgenic Cultivars

Horlemann et al. (2003) released the first comprehensive study of hop genetic modification. Scientists used Agrobacterium to insert the reporter gene 'GUS' into Tettnanger plants, allowing for the regeneration of non-meristematic tissue before successfully transforming hop. Horlemann et al. (2003) gave a great account of how they overcame these challenges, with special attention paid to two key factors: the environment in which donor tissue was collected, and the hormonal treatments utilized to stimulate organogenesis. In vitro-grown shoot, internodes were found to be more regenerative than those from field-grown plants. The internodes of the shoots on greenhouse-grown plants never matured past the callus stage. They also discovered that the best hormonal milieu for regeneration was achieved by adding 1.43 M indoleacetic acid (IAA) and 9.08 M TDZ (Thidiazuron) to a 2% glucose-supplemented Murashige and Skoog (1962) basal medium. The overall frequency of regenerated plantlets developing roots was found to be 40% using this method. Out of a total of 1440 explants, The Breeding of Hop 117 kanamycin selected 83 for further study; of these 83, 42 proved positive for GUS expression before being planted in soil. The transgene was successfully detected by a polymerase chain reaction and GUS expression in a total of twenty-one greenhouse-grown plants. Plants expressing the integrated chitinase transgene showed reduced growth of powdery mildew, but definitive quantitative data was not yet available. This was accomplished by using the cultivars 'Saazer' and 'Hallertauer mittel frueh' to transform hop with the pBin 19 transformation vector (Bevan 1984), which contained the GUS reporter gene (Henning and Moore 1999). The expression of the chitinase gene is now being characterized, and additional research is being conducted to confirm the stable transformation of these cultivars.

It has taken nearly three decades for genetic engineering, or the transfer of genetic information from one species to another, to become conceivable. Hop breeding with the help of genetic engineering has clear benefits (Awasthi et al. 2021; Mishra et al. Improvements in hops' agronomic and qualitative attributes can be achieved without altering the genome in any other way, in contrast to traditional breeding efforts that rely on cross-pollination. Since it was discovered that certain hop compounds have exceptional bioactive qualities demonstrating high estrogenic and anticancer activity (Bolton et al. 2019), hop has been considered a promising candidate for molecular farming.

Hop is one of the natural hosts of this soil bacteria, Agrobacterium tumefaciens, making the genetic transformation with Agrobacterium tumefaciens the most promising method (De Cleene and De Ley 1976). Although this method has been used to attempt the genetic transformation of the hop, just a handful of papers have been published, and the results have either been unsuccessful (Becker 2000) or cannot be replicated (Oriniaková and Matouek 1996). Horlemann et al. (2003) were the first to report that hop tissues could be transformed by *A. tumefaciens* and that transgenic plants of cv. Tettnanger could be regenerated from these tissues. *A. tumefaciens* mediated system cv. Osvald clone 72 (Okada et al. 2003; Kof and Luthar 2005 (cv. Aurora)] has been used in only two previous articles to successfully produce transgenic hop plants carrying gus reporter and/or nptII selection genes. Recently, Schwekendiek et al. (2007) reported using the technique devised by Horlemann et al. (2003) to alter hop (cv. Tettnanger) with a grapevine stilbene synthase gene. In comparison to Agrobacterium-mediated transformation, biolistic transformation has shown to be an effective and flexible method. Recently, Batista et al. (2008) built an effective particle bombardment-based hops transformation system. They developed a reliable protocol for transgenic plant regeneration from hop var. Eroica seeds.

7.10 Breeding with Mutants

To speed up the normal mutation process and produce novel genetic characteristics, mutation breeding exposes the tissue to a mutagen. The breeder then chooses the plants with the best mutations for increased yield or better quality. A breeder who discovers a mutant with desirable qualities may choose to either release it as a new cultivar or use it as a parent to introduce those traits into a more conventional line. Induced mutagenesis is not regarded as a GMO in the United States, despite the widespread media coverage of GMOs and transgenic species. USDA (2013) classifies mutagenesis as part of classical/traditional plant breeding procedures that are compatible with organic systems.

The craft beer business can benefit greatly from the use of mass selection and hybridization to create new varieties. Both strategies leverage hop heterozygosity through sexual recombination to generate novel smells and tastes. However, mutagenesis has the potential to be more effective for improving current aroma cultivars (developing a replacement cultivar) due to the 15 ways in which it can tweak a few agronomic parameters of an acceptable cultivar without drastically altering a brewing profile that is already widely recognized in the industry.

7.10.1 Mutation Breeding

Physical and chemical agents are the two most common types of mutagens. Changes in DNA sequence, and thus the appearance and features of the treated organism, are the primary effects of chemical and physical mutagens. This occurs when a therapy with a mutagen modifies DNA sequences or causes chromosomes to break. Chemical mutagens, which can be used to alter many different types of plant material, often contain alkylating chemicals like ethyl methane sulfonate (EMS). The conventional method in seed-propagated crops is the mutagenic treatment of seeds with chemical mutagens because of the ease with which they may be handled, stored, and transported (Kodym and Afza 2003).

The earliest colchicine-induced triploid seedlings, Willamette and Columbia, are examples of hops generated through mutation breeding employing chemical mutagens. These two varieties were bred and distributed as improved, seedless alternatives to Fuggle (Haunold et al. 1976a, b). By exposing Fuggle to colchicine, a tetraploid form of Fuggle was created, and this tetraploid form was then open-pollinated with a diploid male hop of unknown origin, resulting in the first triploid ($3x = 30$) cultivars, Willamette and Columbia. Willamette is widely produced in Oregon because of its value as a finishing hop, although Columbia is not. Additionally, the USDA-ARS has released Santiam, a 16 triploid developed by crossing diploid Tettnanger with a tetraploid male (Henning and Haunold 1999), Liberty, a half-sister triploid to Mt. Hood (Haunold et al. 1992), and Crystal, a half-sister triploid to Mt. Hood (Haunold et al. 1995).

While the New Zealand breeding program has expanded on the use of polyploids and continues to release triploid cultivars, the USDA breeding program has done so to a lesser extent. When further enhancement of outstanding cultivars is needed and it is essential to generally maintain the traits of the existing cultivar, physical mutagens can be a beneficial tool in vegetatively propagated crops like hop (Shu et al. 2012b; Suprasanna et al. 2014, 2015).

Radiation sources like gamma rays, X-rays, ultraviolet light, and particle radiation like fast and thermal neutrons, beta, and alpha particles are commonly used in physical mutagens (Kodym and Afza 2003; Penna et al. 2023). Ionizing radiation has been used to cause mutations in organisms since the early 20th century. Once improvements were shown to be possible in maize, barley, and wheat in 1928 and 1930 (Stadler 1928), plant breeders began using radiation-induced mutations as a tool for plant breeding.

X-rays were previously utilized in most attempts to cause mutations in plants, but with the advent of nuclear research facilities, gamma radiation has become the preferred method. Since gamma rays are ionizing radiation, they disrupt plant development at multiple levels, including cellular, biochemical, physiological, and morphological (Wi et al. 2005). Plant size, bloom time, plant architecture, yield, fruit quality, and harvest date are all factors that have improved thanks to gamma radiation in vegetatively propagated crops (Shu et al. 2012a). The LD50 of hop nodal explants exposed to gamma radiation in tissue culture has been reported to be less than 5 Gy in previous studies. There is a general dearth of writing on the subject of using gamma radiation to enhance hop. Increased efficiency of mutagenic treatments for variation induction, manageability of large populations, and rapid cloning of selected variants are all benefits of using plant tissue culture in aroma hop improvement via induced mutations (Predieri 2001).

7.11 Conclusions and Prospects

In light of the discoveries made over the previous decade, it is fair to say that hop breeding has undergone something closer to a revolution than an evolution. It was only a few decades ago that many hop breeding efforts prioritized the systematic selection of clonal duplicates. A decade ago, the majority of breeders only used straightforward pedigree methods. Genetic fingerprinting is being used to find heterotic male and female pairs for hybrid work, genes involved with important economic and morphological characteristics are being isolated and identified, and molecular markers are being used to select specific traits. Indeed, hop genetics is entering a new and promising phase.

While we were able to cover a wide range of recent and classic research in this chapter, we were unable to do so for works published before the early 1990s. Neve's (1991) Hops is a monograph that gives detailed treatment on breeding for plant diseases and pest resistance and hence is a significant source of information on topics not addressed here. Numerous plant pests, including viruses, are also extensively covered. The millennium overview by Moir (2000) is another great place to go for details about the development, cultivation, and breeding of hops. Although not solely focused on breeding, this page discusses the origins, early years, and methods of hop production and selection. For further background information, check to see Burgess's (1964) Hops: Botany, Cultivation, and Utilization. The history of hop breeding and genetics before the early 1990s is covered very well by the information offered in these three sources.

References

Abd El-Moneim D, ELsarag EIS, Aloufi S et al (2021) Quinoa (*Chenopodium quinoa* Willd.): genetic diversity according to ISSR and SCoT markers, relative gene expression, and morpho-physiological variation under salinity stress. Plants 10:2802

Abrusan G, Grundmann N, DeMester L et al (2009) Teclass—a tool for automated classification of unknown eukaryotic transposable elements. Bioinformatics 25:1329–1330

Alberts L (2020) Zatec, cradle of Saazhopsandl and mark of commercial hop cultivation. Brew Hist 181:43–50

Almaguer C, Schönberger C, Gastl M et al (2014) *Humulus lupulus*—a story that begs to be told. Areview. J Inst Brewing 120(4):289–314

Alonso-Esteban JI, Pinela J, Barros L et al (2019) Phenolic composition and antioxidant, antimicrobial and cytotoxic properties of hop (*Humulus lupulus* L.) seeds. Ind Crop Prod 134:154–159

Aluome C, Aubert G, Alves Carvalho S et al (2016) De novo construction of a Gene-space for diploid plant genome rich in repetitive sequences by an iterative process of extraction and assembly of NGS reads (iPEA protocol) with limited computing resources. BMC Res Notes 9:81

Amoriello T, Fiorentinio S, Vecchiarelli V, Pagano M (2020) Evaluation of spent grain biochar on hop (*Humulus lupulus* L.) growth by multivariate image analysis. Appl Sci 10:533. https://doi.org/10.3390/app10020533

Aron PM, Shellhammer TH (2010) A discussion of polyphenols in beer physical and flavour stability. J Inst Brew 116:369–380

Astray G, Gullón P, Gullón B et al (2020) *Humulus lupulus* L. as a natural source of functional biomolecules. Appl Sci 10:5074. https://doi.org/10.3390/app1015507.

Attrill H, Falls K, Goodman JL et al (2016) The FlyBase consortium; FlyBase: establishing a gene group resource for drosophila melanogaster. Nucleic Acids Res 44(D1):786–792. https://doi.org/10.1093/nar/gkv1046

Awasthi P, Kocábek T, Mishra AK et al (2021) Establishment of CRISPR/Cas9 mediated targeted mutagenesis in Hop (*Humulus lupulus*). Plant Physiol Biochem 160:1–7

Barth (2005) The Barth Report 2004/2005. Nuremberg, Germany

Barth HJ, Klinke C, Schmidt C (1994) The hop atlas, the history and geography of the cultivated plant. Joh Barth & Sohn, Nuremberg

Bassil NV, Gilmore B, Oliphant JM et al (2008) Genic SSRs for European and North American hop (*Humulus lupulus* L.). Genet Resour Crop Evol 55(7):959. https://doi.org/10.1007/s10722-007-9303-9

Batista D, Fonseca S, Serrazina S et al (2008) Efficient and stable transformation of hop (*Humulus lupulus* L.) var. Eroica by particle bombardment. Plant Cell Rep 27:1185–1196

Batista D, Sousa MJ, Pais MS (1996) Plant regeneration from stem and petiole-derived callus of *Humulus lupulus* L. (hop) clone Branganca and var. Brewers's Gold. In Vitro Cell Dev Biol Plant 32:37–41

Becker J (2000) Untersuchungen zur *in vitro* regeneration und genetischen transformation von hopfen (*Humulus lupulus* L.). Dissertation, University of Kaiserslautern, Kaiserslautern, Germany

Bedini S, Flamini G, Girardi J et al (2015) Not just for beer: evaluation of spent hops (*Humulus lupulus* L.) as a source of eco–friendly repellents for insect pests of stored foods. J Pest Sci 88:583–592

Berardini TZ, Reiser L, Li D et al (2015) The arabidopsis information resource: making and mining the gold standard annotated reference plant genome. Genesis 53:474–485

Bernardo RN (2002) Breeding for quantitative traits in plants. Stemma Press, Woodbury, Minn

Bolton JL, Dunlap TL, Hajirahkimkhan A et al (2019) The multiple biological targets of hops and bioactive compounds. Chem Res Toxicol 32:222–233

Bortoluzzi C, Menten JFM, Romano GG et al (2014) Effect of hops β–acids (*Humulus lupulus*) on performance and intestinal health of broiler chickens. J Appl Poult Res 23:437–443

Boutain JR (2014) On the origin of hops: genetic variability, phylogenetic relationships and ecological plasticity of *Humulus* (Cannabaceae), Doctoral dissertation, University of Hawaii at Manoa

Brady JL, Scott NS, Thomas MR et al (1996) DNA typing of hops (*Humulus lupulus*) through application of RAPD and microsatellite marker sequences converted to sequence tagged sites (STS). Euphytica 3:277–284. https://doi.org/10.1007/BF00033088

Brewers Association (2021) National beer sales & production data. www.brewersassociation.org/statistics-and-data/national-beer-stats

Brits G (2001) Advances in breeding strategies of the South African breweries hop farms in George, South Africa. In: Seigner E (ed) Proceedings of the scientific commission of the international Hop growers convention, Canterbury, UK, p 122

Buckler ES, Holland JB, Bradbury PJ et al (2009) The genetic architecture of maize flowering time. Science 325:714–718

Burgess AH (1964) Hops. World Crops Series, London, Leonard Hill

Campbell MS, Holt C, Moore B et al (2014) Genome annotation and curation using MAKER and MAKER-P. Curr Prot Bioinform 48:1–39. https://doi.org/10.1002/0471250953.bi0411s48

Carter PR, Oelke EA, Kaminski AR et al (1990) Hop. Alternative field crops manual. University of Wisconsin Cooperative Extension and University of Minnesota Center for Alternative Plant and Animal Products. www.hort.purdue.edu/newcrop/afcm/hop.html

Chilliard Y, Glasser F, Faulconnier Y et al (2009) Ruminant physiology: digestion, metabolism, and effects of nutrition on reproduction and welfare. Wageningen Academic Publishers

Cerenak A, Jakse J, Javornik B et al (2004) Identification and differentiation of hop varieties using simple sequence repeat markers. J Am Soc Brew Chem 62(1):1–7. https://doi.org/10.1094/ASBCJ-62-0001

Cerenak A, Satovic Z, Jakse J et al (2009) Identification of QTLs for alpha acid content and yield in hop (*Humulus lupulus* L.). Euphytica 170:141–154

Cerenak A, Satovic Z, Javornik B et al (2006) Genetic mapping of hop (*Humulus lupulus* L.) applied to the detection of QTLs for alpha-acid content. Genome 49(5):485–494. https://doi.org/10.1139/g06-007

Chang Y, Reed BM (1999) Extended cold acclimation and recovery medium alteration improve regrowth of Rubus shoot tips following cryopreservation. CryoLetters 20:371–376

Chang Y, Reed BM (2000) Extended alternating-temperature cold acclimation and culture duration improve pear shoot cryopreservation. Cryobiol 40:311–322

Chang Y, Reed BM (2001) Preculture conditions influence cold hardiness and regrowth of *Pyrus cordata* shoot tips after cryopreservation. HortSci 36(7):1329–1333

Chawla HS (2002) Introduction to plant biotechnology. Science Publishers, p 538
Clark SM, Vaitheeswaran V, Ambrose SJ et al (2013) Transcriptome analysis of bitter acid biosynthesis and precursor pathways in hop (*Humulus lupulus*). BMC Plant Biol 13:12
Consortium U (2008) The universal protein resource (UniProt). Nucleic Acids Res 36:190–195
Danilova TV, Danilov SS, Karlov GI (2003) Molecular genetic polymorphisms of cultivars of common hops (*Humulus lupulus* L.) using ISSR-PCR analysis. Genetika 39:1484–1489
Darby P (1999a) Economic yield potential of dwarf hop varieties. In: New procedures in hop growing, proceedings of international symposium, Hüll, Bavaria, Federal Ministry of Food, Agriculture and Forestry, Germany, 17–18 May 1999, pp 36–38
Darby P (1999b) Genetic improvement research leading to the breeding, development and commercialisation of varieties of dwarf hops. Preisverleihung der Rudolf Hermanns Stiftung, Geisenheim, Germany, pp 20–24
Darby P (2001) Single gene traits in hop breeding. In: Seigner E (ed) Proceedings of the scientific commission of the international hop growers convention, Canterbury, UK, pp 76–80
Darby P (2005) The assessment of resistance to diseases in the UK hop breeding programme. In: Seigner E (ed) Proceedings of the scientific commission of the international hop growers convention, George, South Africa, pp 7 11
Darby P (2005) The history of hop breeding and development. Brew Hist 121:94–112
Darby P, Farris MD (1991) The development of dwarf hops. In: Lander JB (ed) Annual Booklet for 1991, Association of Growers of New Varieties of Hops, pp 39–42
Darby P, Walker C (2002) Where next for the hop industry? Brewers Guardian 131:22–25
DeGrandi-Hoffman G, Ahumada F, Probasco G, Schantz L (2012) The effects of beta acids from hops (*Humulus lupulus*) on mortality of Varroa destructor (Acari: Varroidae). Exp Appl Acarol 58:407–421
De Cleene M, De Ley J (1976) The host range of crown gall. Bot Rev 42:389–466
Desalle R, Tattersall I (2019) A natural history of beer. Yale University Press, New Haven. https://doi.org/10.1093/nar/gkv1046
Donner P, Pokorný J, Ježek J et al (2020) Influence of weather conditions, irrigation and plant age on yield and alpha-acids content of Czech hop (*Humulus lupulus* L.) cultivars. Plant Soil Environ 66:41–46
EL-Mansy AB, Abd El-Moneim D, ALshamrani SM et al (2021) Genetic diversity analysis of tomato (*Solanum lycopersicum* L.) with morphological, cytological, and molecular markers under heat stress. Hort 7:65
Engelmann F (1999) Management of field and in vitro germplasm collections. In: Proceedings of a consultation meeting, 15–20 January 1996, CIAT Cali, Colombia. Rome (IT), IPGRI, p 165
Eriksen RL, Rutto LK, Dombrowski JE, Henning JA (2020) Photosynthetic activity of six hop (*Humulus lupulus* L.) cultivars under different temperature treatments. HortScience 55:403–409
Gallo C, Dallaglio K, Bassani B, Rossi T, Rossello A, Noonan DM, Albini A (2016) Hop derived flavonoid xanthohumol inhibits endothelial cell functions via AMPK activation. Oncotarget 7(37):59917
Gerhäuser C (2005) Beer constituents as potential cancer chemopreventive agents. Eur J Cancer 41(13):1941–1954
Glazier N (2007) Hop production-coming back to New York? Cornell University Cooperative Extension
Haas GJ, Barsoumian R (1994) Antimicrobial activity of hop resins. J Food Prot 57:59–61
Hadonou AM, Walden R, Darby P et al (2004) Isolation and characterization of polymorphic microsatellites for assessment of genetic variation of hops (*Humulus lupulus* L.). Mol Ecol Notes 4(2):280–282. https://doi.org/10.1111/j.1471-8286.2004.00641.x
Hamel PB, Chiltoskey MU (1975) Cherokee plants: their use. A 400-year history. Cherokee Publications
Haunold A (1971) Cytology, sex expression, and growth of a tetraploid × diploid cross in hop (*Humulus lupulus* L.). Crop Sci Madison 11(6):868–871. /https://doi.org/10.2135/cropsci1971.0011183X001100060031x

Haunold A, Nickerson GB, Gampert U et al (1993) Agronomic and quality characteristics of native North American hops. J Amer Soc Brew Chem 51(3):133–137

Havill JS (2023) Phenotypic and genotypic characterization of hop (*Humulus lupulus* L.) germplasm resources for powdery mildew resistance, Doctoral Dissertation, University of Minnesota, USA

Henning J, Gent D, Twomey M et al (2016) Genotyping-by sequencing of a bi-parental mapping population segregating for downy mildew resistance in hop (*Humulus lupulus* L.). Euphytica 208:545–559

Henning J, Hill S, Darby P et al (2017a) QTLexamination of a bi-parental mapping population segregating for short-stature in hop (*Humulus lupulus* L.). Euphytica 213(3):1–15. https://doi.org/10.1007/s10681-017-1848-x

Henning JA (2006) The breeding of hop. In: Bamworth CW (ed) Brewing: new technologies. Woodhead Publishing Limited, Cambridge

Henning JA, Coggins J, Peterson M et al (2015a) Simple SNP-based minimal marker genotyping for *Humulus lupulus* L. identification and variety validation. BMC Res Notes 8(1):542. https://doi.org/10.1186/s13104-015-1492-2PMID:26438052

Henning JA, Gent DH, Townsend MS et al (2017b) QTL analysis of resistance to powdery mildew in hop (*Humulus lupulus* L.). Euphytica 213(4):1–13. https://doi.org/10.1007/s10681-017-1849-9

Henning JA, Gent DH, Twomey MC et al (2015b) Precision QTL mapping of downy mildew resistance in hop (*Humulus lupulus* L.). Euphytica 202(3):487–98. https://doi.org/10.1007/s10681-015-1356-9

Henning JA, Gent DH, Twomey MC et al (2016) Genotyping-by-sequencing of a bi-parental mapping population segregating for downy mildew resistance in hop (*Humulus lupulus* L.). Euphytica 208(3):545–559. https://doi.org/10.1007/s10681-015-1600-3

Henning JA, Townsend MS, Gent DH et al (2011) QTL mapping of powdery mildew susceptibility in hop (*Humulus lupulus* L.). Euphytica 180(3):411–20. https://doi.org/10.1007/s10681-011-0403-4

Henning JA, Townsend MS, Mahaffee WF, Kenny S, Haunold A (2004) Registration of Newport hop. Crop Sci 44:1018–1019

Hill ST, Coggins J, Liston A et al (2016) Genomics of the hop pseudoautosomal regions. Euphytica 209(1):171–9. https://doi.org/10.1007/s10681-016-1655-9

Horlemann C, Schwekendiek A, Höhnle M, Weber G (2003) Regeneration and *Agrobacterium*-mediated transformation of hop (*Humulus lupulus* L.). Plant Cell Rep 22:210–217

Hornsey IS (1999) Brewing. Cambridge, UK, The Royal Society of Chemistry

Hornsey IS (2003) A history of beer and brewing. UK, The Royal Society of Chemistry, Cambridge

Howard EL, Whittock SP, Jakse J et al (2011) High-throughput genotyping of hop (*Humulus lupulus* L.) utilizing diversity arrays technology (DArT). Theor Appl Genet 122(7):1265–1280. https://doi.org/10.1007/s00122-011-1529-4

Howe KL, Bolt BJ, Cain S et al (2016) WormBase 2016: expanding to enable helminth genomic research. Nucleic Acids Res 44(D1):774–780. https://doi.org/10.1093/nar/gkv1217

Huang X, Wei X, Sang T et al (2010) Genome-wide association studies of 14 agronomic traits in rice landraces. Nat Genet 42:961–967. https://doi.org/10.1038/ng.695

Ibrahim AA, Abdel Razik ES, Abd-Ellatif S, Ramadan S (2023a) *Nigella sativa* and stressful conditions. In: Latef AAHA (ed) Medicinal plant responses to stressful conditions. CRC Press. https://doi.org/10.1201/9781003242963

Ibrahim AA, Salem KFM, Ramadan S (2023b) *Allium sativum* and stressful conditions. In: Latef AAHA (ed) Medicinal plant responses to stressful conditions. CRC Press. https://doi.org/10.1201/9781003242963

Jakse J, Bandelj D, Javornik B et al (2002) Eleven new microsatellites for hop (*Humulus lupulus* L.). Mol Ecol Notes 2(4):544–546. https://doi.org/10.1046/j.1471-8286.2002.00309.x

Jakse J, Satovic Z, Javornik B (2004) Microsatellite variability among wild and cultivated hops (*Humulus lupulus* L.). Genome 47(5):889–899

Jakse J, Stajner N, Kozjak P et al (2008) Trinucleotide microsatellite repeat is tightly linked to male sex in hop (*Humulus lupulus* L.). Mol Breed 21(2):139–148. https://doi.org/10.1007/s11032-007-9114-x

Jakse J, Stajner N, Luthar Z et al (2011) Development of transcript-associated microsatellite markers for diversity and linkage mapping studies in hop (*Humulus lupulus* L.). Mol Breed 28(2):227–239. https://doi.org/10.1007/s11032-010-9476-3

Jiang P, Song T, Jiang W et al (2018) SNP-based kompetitive allele-specific PCR (KASPTM) method for the qualification and quantification of hop varieties. J Am Soc Brew Chem 76(3):185–189. https://doi.org/10.1080/03610470.2018.1483702

Johnson M, Zaretskaya I, Raytselis Y et al (2008) NCBI BLAST: a better web interface. Nucleic Acids Res 36:5–9

Kaeppler SM, Phillips RL (1993) DNA methylation and tissue culture-induced variation in plants. In Vitro Cell Dev Biol Plant 29:125–130

Karabín M, Hudcová T, Jelínek L, Dostálek P (2016) Biologically active Compounds from hops and prospects for their use. Compr Rev Food Sci Food Saf 15:542–567

Karlov G, Danilova T, Horlemann C et al (2003) Molecular cytogenetics in hop (*Humulus lupulus* L.) and identification of sex chromosomes by DAPI-banding. Euphytica 132:185–190

Kearsey MJ, Luo ZW (2003) Mapping, characterization and deployment of quantitative trait loci. In: Newbury HJ (ed) Plant molecular breeding. CRC Press, Boca Raton, pp 1–29

Kenneth J, Jeffery M, Pires J et al (2002) Urticalean rosids: circumscription, rosid ancestry and phylogenetics based on *RBCL, TRNL-F,* and *NDHF* sequences. Am J Bot 89(9):1531–1546. https://doi.org/10.3732/ajb.89.9.1531

Keyworth WG (1942) Verticillium wilt of the hop (*Humulus lupulus*). Ann Appl Biol 29:346–357

Kim D, Langmead B, Salzberg SL (2015) HISAT: a fast spliced aligner with low memory requirements. Nat Method 12:357–360

Kneen R (2001) Small scale and organic hops production. Left Fields, British, Columbia

Kneen R (2003) Small scale and organic hops production. Kneen, Left fields, British, Columbia

Koelling J, Coles MC, Matthews PD et al (2012) Development of new microsatellite markers (SSRs) for *Humulus lupulus*. Mol Breed 30(1):479–84. https://doi.org/10.1007/s11032-011-9637-z

Koie K, Inaba A, Okada Y et al (2005) Construction of the genetic linkage map and QTL analysis on hop (*Humulus lupulus* L.). Acta Hort 668:59–66

Kolenc Z, Langerholc T, Hostnik G et al (2023) Antimicrobial properties of different hop (*Humulus lupulus*) genotypes. Plants 12(1):120. https://doi.org/10.3390/plants12010120

Korbecka-Glinka G, Skomra U, Olszak-Przybys H et al (2016) Cultivar identification in dry hop cones and pellets using microsatellite loci. Eur Food Res Technol 242(9):1599–605. https://doi.org/10.1007/s00217-016-2715-z

Korf I (2004) Gene finding in novel genomes. BMC Bioinformatics 5:59

Krajnović T, Drača D, Kaluđerović GN et al (2019) The hop-derived prenylflavonoid isoxanthohumol inhibits the formation of lung metastasis in B16-F10 murine melanoma model. Food Chem Toxicol 129:257–268. https://doi.org/10.1016/j.fct.2019.04.046

Kurtz S, Narechania A, Stein JC et al (2008) A new method to compute K-mer frequencies and its application to annotate large repetitive plant genomes. BMCGenomics 9:517

Li X, Waterman MS (2003) Estimating the repeat structure and length of DNA sequences using L-tuples. Genome Res 13:1916–1922

Li YC, Korol AB, Fahima T et al (2002) Microsatellites: genomic distribution, putative functions and mutational mechanisms: a review. Mol Ecol 11(12):2453–2465

Liberatore CM, Mattion G, Rodolf M et al (2018) Chemical and physical pre-treatments to improve in vitro seed germination of *Humulus lupulus* L., cv Columbus. Sci Horticult 235:86–94

Mafakheri M, Kordrostami M, Rahimi M et al (2020) Evaluating genetic diversity and structure of a wild hop (*Humulus lupulus* L.) germplasm using morphological and molecular characteristics. Euphytica 216(4):58. https://doi.org/10.1007/s10681-020-02592-z

Matthews PD, Coles MC, PitraNJ et al (2013) Next generation sequencing for a plant of great tradition: application of NGSto SNP detection and validation in hops (*Humulus lupulus* L.). Brewing Sci 66:185–191
McAdam EL, Freeman JS, Whittock SP et al (2013) Quantitative trait loci in hop (*Humulus lupulus* L.) reveal complex genetic architecture underlying variation in sex, yield and cone chemistry. BMC Genomics 14:360. https://doi.org/10.1186/1471-2164-14-360
McAdam EL, Vaillancourt RE, Koutoulis A et al (2014) Quantitative genetic parameters for yield, plant growth and cone chemical traits in hop (*Humulus lupulus* L.). BMC Genet 15(1):22. https://doi.org/10.1186/1471-2156-15-22
McCallum JL, Nabuurs MH, Gallant ST et al (2019) Phytochemical characterization of wild hops (*Humulus lupulus* ssp. lupuloides) germplasm resources from the maritime region of Canada. Front Plant Sci 10:1438
Mesfer ALshamrani S, Safhi FA, Alshaya DS et al (2022) Genetic diversity using biochemical, physiological, karyological and molecular markers of *Sesamum indicum* L. Front Genetics 13:1035977. https://doi.org/10.3389/fgene.2022.1035977
Miranda CL, Stevens JF, Helmrich A et al (1999) Antiproliferative and cytotoxic effects of prenylated flavonoids from hops (*Humulus lupulus*) in human cancer cell lines. Food Chem Toxicol 37:271–285
Mitchell-Olds T (2010) Complex trait analysis in plants. Genome Biol 11:113–115. https://doi.org/10.1186/gb-2010-11-4-113
Montell S (1994) Hops in the backyard: from planting to harvest and the hazards in between. Brew Techn 2(3):1–6
Mozny M, Tolasz R, Nekovar J et al (2009) The impact of climate change on the yield and quality of Saaz hops in Czech Republic. Agric Forest Meteorol 149:913–919
Murakami A, Darby P, Javornik B et al (2006) Microsatellite DNA analysis of wild hops *Humulus lupulus* L. Genet Resour Crop Evol 53(8):1553–62. https://doi.org/10.1007/s10722-005-7765-1
Murashige T, Skoog F (1962) A revised medium for rapid growth and bioassays with tobacco tissue cultures. Physiol Plant 15:473–497
Natsume S, Takagi H, Shiraishi A et al (2015) The draft genome of Hop (*Humulus lupulus*), an essence for brewing. Plant Cell Physiol 56:428–441
Negrão R, Costa R, Duarte D, Taveira Gomes T, Mendanha M et al (2010) Angiogenesis and inflammation signaling are targets of beer polyphenols on vascular cells. J Cell Biochem 111(5):1270–1279. https://doi.org/10.1002/jcb.22850
Neve RA (1991) Hops. Chapman and Hall, London
Neve RA (2012) Hops. Springer Science & Business Media
Nikolic D, Li Y, Chadwick LR, van Breemen RB (2006) In vitro studies of intestinal permeability and hepatic and intestinal metabolism of 8-prenylnaringenin, a potent phytoestrogen from hops (*Humulus lupulus* L.). Pharm Res 23:864–872
Nussbaumer T, Martis MM, Roessner SK et al (2013) MIPS PlantsDB: a database framework for comparative plant genome research. Nucleic Acids Res 41:1144–1151
Ovcirk M, Necemer M, Kosir J (2019) The determination of the geographic origins of hops (*Humulus lupulus* L.) by multi–elemental fingerprinting. Food Chem 277:32–37
Okada Y, Saeki K, Inaba A et al (2003) Construction of gene expression system in hop (*Humulus lupulus* L.) lupulin gland using valerophenone synthase promoter. J Plant Physiol 160:1101–1108
Ono T (1955) Studies in hop. I. Chromosomes of common hop and its relatives. Bull Brew Sci 2:1–65
Oriniaková P, Matoušek J (1996) Viroid infection of hop (*Humulus lupulus* L.) mediated by *Agrobacterium tumefaciens* and conditions for hop transformation. Rostl Výr 52:233–239
Padgitt–Cobb LK, Kingan SB, Wells J et al (2019) A phased, diploid assembly of the Cascade hop (*Humulus lupulus*) genome reveals patterns of selection and haplotype variation. bioRxiv. https://doi.org/10.1101/786145

Palumbo F, Barcaccia G (2018) Rediscovery of landraces as a resource for the future. In: Grillo O (ed) Critical aspects on the use of microsatellite markers for assessing genetic identity of crop plant varieties and authenticity of their food derivatives. UK, Intech Open, pp 129–160. https://doi.org/10.5772/intechopen.70756

Patzak J, Nesvadba V, Henychova´A et al (2010) Assessment of the genetic diversity of wild hops (*Humulus lupulus* L.) in Europe using chemical and molecular analyses. Biochem Syst Ecol 38(2):136–145. https://doi.org/10.1007/s10722-005-7765-1

Pedregosa F, Varoquaux G, Gramfort A et al (2011) Scikitlearn: machine learning in Python. J Mach Learn Res 12:2825–2830

Penna S, Shirani Bidabadi S Jain SM (2023) Mutation breeding to promote sustainable agriculture and food security in the era of climate change. In: Penna S, Jain SM (eds) Mutation breeding for sustainable food production and climate resilience. Springer, Singapore, pp 1–23. https://doi.org/10.1007/978-981-16-9720-3_1

Peredo EL, Arroyo-Garcia R, Martinez-Zapater JM, Revilla MA (2005) Evaluation of microsatellite detection using autoradiography and capillary electrophoresis in hops. J Am Brew Chem 63:57–62. https://doi.org/10.1094/ASBCJ-63-0057

Peredo EL, Revilla MÁ, Arroyo-García R (2006) Assessment of genetic and epigenetic variation in hop plants regenerated from sequential subcultures of organogenic calli. J Plant Physiol 163(10):1071–1079

Peredo EL, Revilla MA, Reed BM et al (2010) The influence of European and American wild germplasm in hop (*Humulus lupulus* L.) cultivars. Genet Resour Crop Evol 57(4):575–586. https://doi.org/10.1007/s10722-009-9495-2

Pertea M, Pertea GM, Antonescu CM et al (2015) StringTie enables improved reconstruction of a transcriptome from RNAseq reads. Nat Biotechnol 33:290–295

Pigg K (2017) Aroma hops breeding program develops, grows new strata hop. The Daily Barometer. Available from: https://dailybaro.orangemedianetwork.com/7625/daily-barometer-news/aroma-hops-breeding-program-develops-grows-new-strata-hop

Poland JA, Rife TW (2012) Genotyping-by-sequencing for plant breeding and genetics. Plant Genome 5(3):92–102

Possemiers S, Rabot S, Espín JC et al (2008) Eubacterium limosum activates isoxanthohumol from Hops (*Humulus lupulus* L.) into the potent phytoestrogen 8-prenylnaringenin in vitro and in rat intestine. J Nutr 138(7):1310–1316

Predieri S (2001) Mutation induction and tissue culture in improving fruits. Plant Cell Tissue Organ Cult 64(2):185–210

Pruitt KD, Tatusova T, Maglott DR (2007) NCBI reference sequences (RefSeq): a curated non-redundant sequence database of genomes, transcripts and proteins. Nucleic Acids Res 35(suppl 1):61–65

Reed BM (2001) Implementing cryogenic storage of clonally propagated plants. Cryo Lett 22:97–104

Rhee SY, Beavis W, Berardini TZ et al (2003) The Arabidopsis Information Resource (TAIR): a model organism database providing a centralized, curated gateway to arabidopsis biology, research materials and community. Nucleic Acids Res 31:224–228

Robinson AJ, Love CG, Batley J et al (2004) Simple sequence repeat marker loci discovery using SSR primer. Bioinformatics 20(9):1475–1476. https://doi.org/10.1093/bioinformatics/bth104

Rodger E (2021) Cornell Agritech launches shops breeding program. Cornell Chronicle

Rodolfi M, Silvanini A, Chiancone B et al (2018) Identification and genetic structure of wild Italian *Humulus lupulus* L. and comparison with European and American hop cultivars using nuclear microsatellite markers. Genet Resour Crop Evol 65(5):1405–1422. https://doi.org/10.1007/s10722-018-0622-9

Rossini F, Virga G, Loreti P et al (2021) Hops (*Humulus lupulus* L.) as a novel multipurpose crop for the Mediterranean region of Europe: challenges and opportunities of their cultivation. Agriculture 11(6):484. https://doi.org/10.3390/agriculture11060484

Roy AT, Leggett G, Koutoulis A (2001) Development of a shoot multiplication system for hop (*Humulus lupulus* L.). In Vitro Cell Dev Biol Plant 37:79–83

Safhi FA, ALshamrani SM, Jalal AS et al (2022) Genetic characterization of some Saudi Arabia's accessions from *Commiphora gileadensis* using physio-biochemical parameters, molecular markers, DNA barcoding analysis and relative gene expression. Genes 13:2099

Salgotra RK, Chauhan BS (2023) Genetic diversity, conservation, and utilization of plant genetic resources. Genes 14:174. https://doi.org/10.3390/genes14010174

Schaefer O, Hümpel M, Fritzemeier KH et al (2003) 8-Prenyl naringenin is a potent ERα selective phytoestrogen present in hops and beer. J Steroid Biochem Mol Biol 84(2–3):359–360

Schwekendiek A, Spring O, Heyerick A et al (2007) Constitutive expression of a grapevine stilbene synthase gene in transgenic hop (*Humulus lupulus* L.) yields resveratrol and its derivates in substantial quantities. J Agric Food Chem 55:7002–7009

Seefelder S, Ehrmaier H, Schweizer G et al (2000) Genetic diversity and phylogenetic relationships among accessions of hop *Humulus lupulus* as determined by amplified fragment length polymorphism fingerprinting compared with pedigree data. Plant Breed 119(3):257–63. https://doi.org/10.1046/j.1439-0523.2000.00500.x

Serwe A, Rudolph K, Anke T, Erkel G (2012) Inhibition of TGF-β signaling, vasculogenic mimicry and proinflammatory gene expression by isoxanthohumol. Invest New Drugs 30(3):898–915. https://doi.org/10.1007/s10637-011-9643-3

Seigner E, Lutz A, Radic-Miehle H et al (2004) Breeding for powdery mildew resistance in hop (*Humulus lupulus* L.): strategies at the hop research center, Huell, Germany. Acta Hort 668:19–30

Sherman EL, Nkrumah JD, Moore SS (2009) Whole genome single nucleotide polymorphism associations with feed intake and feed efficiency in beef cattle. J Anim Sci 88(1):16–22

Shephard HL, Parker JS, Darby P, Ainsworth CC (2000) Sexual development and sex chromosomes in hop. New Phytol 148(3):397–411

Shu, QY, Forster, BP, Nakagawa H (2012a) Principles and applications of plant mutation breeding. In: Plant mutation breeding and biotechnology. CABI, Wallingford, pp 301–325

Shu Q, Forster BP, Nakagawa H (2012b) Plant mutation breeding and biotechnology. CABI

Simieli MB, Gazola RPD, Pagliarini MK et al (2021) Development and production of hop in a high-temperature region. Res Soc Dev 10(13):1–14

Slater GS, Birney E (2005) Automated generation of heuristics for biological sequence comparison. BMC Bioinform 6:31

Small E (1978) A numerical and nomenclatural analysis of morpho-geographic taxa of *Humulus*. Syst Bot 3(1):37–76. https://doi.org/10.2307/2418532

Small E (1978) A numerical and nomenclatural analysis of morpho-geographic taxa of Humulus. In: Systematic botany, pp 37–76

Small E (1981) A numerical analysis of morpho-geographic groups of cultivars of *Humulus lupulus* based on samples of cones. Can J Bot 59(3):311–24. https://doi.org/10.1139/b81-044

Smykalova I, Ortova M, Lipavska H et al (2001) Efficient in vitro micropropagation and regeneration of *Humulus lupulus* on low sugar, starch-Gelrite media. Biol Plantarum 44:7–12

Suprasanna P, Mirajkar SJ, Patade VY, Jain SM (2014) Induced mutagenesis for improving plant abiotic stress tolerance. In: Tomlekova NB, Kozgar MI, Wani MR (eds) Mutagenesis: exploring novel genes and pathways. Wageningen Academic Publishers, Wageningen, pp 347–376

Suprasanna P, Mirajkar SJ, Bhagwat SG (2015) Induced mutations and crop improvement. In: Bahadur B, Venkat Rajam M, Sahijram L, Krishnamurthy KV (eds) Plant biology and biotechnology, plant diversity, organization, function and improvement, vol I. Springer, India, pp 593–617

Soliman MI, Ibrahim AA, El-Metwally MA, Eshak DA (2017) Cytotoxicity effects of biological control and antioxidants using *Vicia faba* chromosomal aberration assay. J Environ Sci Technol 10:230–237. https://doi.org/10.3923/jest.2017.230.237

Soliman MI, Mohammed NS, EL-Sherbeny G et al (2023) Antibacterial, antioxidant activities, GC-Mass characterization, and cyto/genotoxicity effect of green synthesis of silver nanoparticles using latex of *Cynanchum acutum* L. Plants 12:172. https://doi.org/10.3390/plants12010172

Stajner N, Jakse J, Kozjak P et al (2005) The isolation and characterization of microsatellites in hop (*Humulus lupulus* L.). Plant Sci 168(1):213–221. https://doi.org/10.1016/j.plantsci.2004.07.031

Stajner N, Satovic Z, Cerenak A, Javornik B et al (2008) Genetic structure and differentiation in hop (*Humu luslupulus* L.) as inferred from microsatellites. Euphytica 161(1–2):301–311. https://doi.org/10.1007/s10681-007-9429-z

Stanke M, Keller O, Gunduz I et al (2006) AUGUSTUS: ab initio prediction of alternative transcripts. Nucleic Acids Res 34:435–439

Steijger T, Abril JF, Engstrom PG et al (2013) Assessment of transcript reconstruction methods for RNA-seq. Nat Meth 10:1177–1184

Stevens JF, Ivancic M, Hsu VL et al (1997) Prenylflavonoids from *Humulus lupulus*. Phytochem 44:1575–1585

Stevens PF (2001) Angiosperm phylogeny website. Retrieved 2016 September 20 from 9http://www.mobot.org/MOBOT/research/APweb/

Sung B, Chung JW, Bae HR et al (2015) *Humulus japonicus* extract exhibits antioxidative and anti-aging effects via modulation of the AMPK-SIRT1 pathway. Exp Therap Med 9:1819–1826

Teghtmeyer S (2018) Hops. J Agric Food Inf 19(1):9. https://doi.org/10.1080/10496505.2018.1403248

USDA NOP (2013) Section G. Policy memos, agricultural marketing service. USDA, National Organic Program

Van Cleemput M, Cattoor K, De Bosscher K et al (2009) Hop (*Humulus lupulus*) derived bitter acids as multipotent bioactive compounds. J Nat Prod 72:1220–1230

Van Holle A, Muylle H, Van Landschoot A et al (2019) Single nucleotide polymorphisms and biochemical markers as complementary tools to characterize hops (*Humulus lupulus* L.) in brewing practice. J Agric Food Chem 67(13):3761–3771. https://doi.org/10.1021/acs.jafc.9b00816

Van Holle A, Van Landschoot A, Roldan–Ruiz I et al (2017) The brewing value of Amarillo hops (*Humulus lupulus* L.) grown in Northwestern USA: a preliminary study of terroir significance. J Inst Brewing 123(3):312–318

Vergara D, White KH, Keepers KG, Kane NC (2016) The complete chloroplast genomes of Cannabis sativa and *Humulus lupulus*. Mitochondrial DNA Short Commun Part A 27(5):3793–3794. https://doi.org/10.3109/19401736.2015.1079905

Verzele M, De Keukeleire D (1991) Chemistry and analysis of hop and beer bitter acids. Elsevier, Amsterdam

Visscher PM, Brown MA, McCarthy MI, Yang J (2012) Five years of GWAS discovery. Am J Hum Genet 90(1):7–24. https://doi.org/10.1016/j.ajhg.2011.11.029

Wendakoon C, Gagnon D, Koenig M, Dwarakanath S (2018) Hops (*Humulus lupulus*) strobile extract and its major components show strong antibacterial activity against methicillin-resistant *Staphylococcus aureus*. J Med Act Plant 7(1):1–4

Wi SG, Chung BY, Kim JH et al (2005) Ultrastructural changes of cell organelles in Arabidopsis stem after gamma irradiation. J Plant Biol 48(2):195–200. https://doi.org/10.1007/BF03030408

Xie Y, Wu G, Tang J et al (2014) SOAPdenovo-Trans: de novo transcriptome assembly with short RNA-Seq reads. Bioinformatics 30:1660–1666

Yamauchi H, Mukouzaka Y, Taniguchi T et al (2014) Newlydevel-opedSNP-based identification method of hop varieties. J Am Soc Brew Chem 72(4):239–45. https://doi.org/10.1094/ASBCJ-2014-1006-01

Yan DD, Wong YF, Shellie RA et al (2019) Assessment of the phytochemical profiles of novel hop (*Humulus lupulus* L.) cultivars: a potential route to beer crafting. Food Chem 275:15–23. https://doi.org/10.1016/j.foodchem.2018.09.082

Young J, Oakley WR, Fox G (2023) *Humulus lupulus* and microbes: exploring biotic causes for hop creep. Food Microbiol 114:104298

Zerbino DR, Birney E (2008) Velvet: algorithms for de novo short read assembly using de Bruijn graphs. Genome Res 18:821–829

Zhang D, Easterling KA, Pitra NJ et al (2017) Non-Mendelian single-nucleotide polymorphism inheritance and atypical meiotic configurations a prevalent in hop. Plant Genome 10(3). https://doi.org/10.3835/plantgenome2017.04.0032

Zhang D, Pitra NJ, Coles MC et al (2016) Non-mendelian inheritance of SNP markers reveals extensive chromosomal translocations in dioecious hops (*Humulus lupulus* L.). bioRxiv:069849. https://doi.org/10.1101/069849

Chapter 8
Biotechnological, Genetic, Molecular and Breeding Attempts and Strategies for the Analysis and Production of Pharmaceutically Important Tropane Alkaloids in *Atropa belladonna* L.: A Review

Usmangani Abdul Attar, Saurabha Bhimrao Zimare, Ashwini Venkatrao Mohite, and Prachi Saurabha Zimare

Abstract *Atropa belladonna* L. (Solanaceae), commonly known as 'Deadly nightshade' or 'Belladonna'. It is a valuable plant that produces anticholinergic tropane alkaloids (TAs). Hyoscyamine, scopolamine, and atropine are the extensively investigated TAs. Plants usually accumulate low levels of TAs to meet the growing market demand. For over years, significant advancements are being developed to comprehend the TA biosynthetic pathways on the basis of molecular biology and metabolic regulation. Traditional use of *A. belladonna* as hallucinogens, mydriasis, sedative, poison, and in the treatment of headache, asthma, black cough, abnormal urination, and many more disorders attracted attention of scientific community for further exploration of bioactive principles and bioactivities. Studies revealed that scopolamine and hyoscyamine had numerous pharmacological properties, and were extensively employed in mitigating spasm, treating motion sickness, quitting drug addictions, and Parkinson's disease, anesthetic, mydriatic, antiemetic, as sedative to inhibit bronchospasm especially in asthma and whooping cough, as anti-cholinergic and cough

U. A. Attar
Department of Botany, Devchand College, Arjunnagar, Kolhapur, Maharashtra, India

S. B. Zimare
Department of Botany, Sir Parashurambhau College (Autonomous), Pune, Maharashtra, India

A. V. Mohite
Department of Botany, Yashvantrao Chavan Institute of Science, Satara, Maharashtra, India

P. S. Zimare (✉)
Department of Botany, Comrade Godavari Shamrao Parulekar College of Arts, Science and Commerce, Talasari, Palghar 401606, Maharashtra, India
e-mail: prachik16@gmail.com

J. M. Al-Khayri et al. (eds.), *Biodiversity and Genetic Improvement of Medicinal and Aromatic Plants I*, Advances in Plant Breeding Strategies 9,
https://doi.org/10.1007/978-3-031-81846-2_8

suppressant, as an analgesic, in neuralgia, rheumatoid arthritis, psychiatric disturbances, and hyperhidrosis. There is a huge demand for TAs of *A. belladonna*, especially scopolamine, as it has fewer side-effects as compared to hyoscyamine. But, the supply of scopolamine is insufficient as it is produced in very less quantities in *A. belladonna.* Although the whole plant is utilized for medicinal purposes, the major alkaloids are synthesized in the roots. The roots take around 3–4 years for maturation to produce root biomass that can yield extractable amounts of compounds of interest. Wild populations are unable to meet the growing industrial demand for these pharmacologically valuable TAs. Numerous challenges are encountered such as naturally growing plants take longer time for accumulating metabolites, huge investments yield sparse production, and chemical synthesis of these compounds is uneconomical. There is a need for developing strategies in addition to existing traditional methods used. Due to the immense importance of commercial value of *A. belladonna* development of promising methods to enhance TA production such as biotechnological, genetic, and molecular approaches can have great relevance.

Keywords Deadly nightshade · Hyoscyamine · Scopolamine · Solanaceae · Tropane alkaloids

8.1 Introduction

Solanaceae is one of the largest cosmopolitans and most valuable families unveiling a diversity of utilizations comprising of most important crop plant species, numerous species possessing pharmaceutical importance owing to their secondary metabolites, several species of commercial importance, and toxic species as well. Since time immemorial, numerous members of Solanaceae were well-known for their poisonous properties and abundant information is available of its use in folk medicines by different traditional groups. These poisonous Solanaceae members are at the moment categorized under genera *Atropa, Brugmansia, Datura, Duboisia, Hyoscyamus* and *Scopolia* (Grynkiewicz and Gadzikowska 2008). A German chemist Carl F. Wilhelm Meissner who coined the term 'alkaloid' (alkaline-like) in 1819, was the first to realize that the active principles in these poisonous plants showed alkaline character, and could be isolated. Although it is very difficult to define alkaloid, many researchers have defined and classified alkaloids based on different studies and characteristics. The basic and wide description of alkaloid is the compound having at least one ring in the molecule, not necessarily heterocyclic. Tropane alkaloids (TAs) is a specific class of alkaloids and is precisely defined as an alkaloid possessing a tropane ring system (Kohnen-Johannsen and Kayser 2019). TAs are one of the earliest known medicines to mankind having hallucinogenic, an aesthetic, mydriatic, antiemetic, antispasmodic, and bronchodilator activities, and are prominent in members of Solanaceae. Individual isolation of these TAs began after 1830, initiating with atropine from *Atropa belladonna*, followed by hyoscyamine from *Hyoscyamus niger* and scopolamine (Grynkiewicz and Gadzikowska 2008). TAs have a simple chemical structure and are characterized by their core structure 8-azabicyclo [3.2.1] octane (nortropane) ring (DeSimone et al. 2008). Although TAs are synthesized by many plant families, Solanaceae is by far the largest family whose 29 genera have the potential to produce TAs, of which *Atropa, Datura, Hyocyamus, Duboisia, Scopolia, Anisodus*, and *Brugmansia* are predominant. Roots are the primary site for the accumulation of tropane alkaloids (Jirschitzka et al. 2013; Bedewitz et al. 2014).

One such noteworthy genus of family Solanaceae is *Atropa* which includes five species (*A. acuminata, A. komarovii, A. pallidiflora, A. baetica* and *A. belladonna*) (POWO, 2022) widely distributed in the Southern and Western Europe, Mediterranean region, and Asia (Lee 2007; Majid et al. 2012; Fattahi et al. 2021). *Atropa belladonna* L., commonly known as 'Deadly nightshade' or 'Belladonna' has been extensively investigated by researchers for its TAs such as atropine, hyoscyamine, and scopolamine (Griffin and Lin 2000). Folklores of the use of *A. belladonna* extracts as hallucinogens, mydriasis, sedative, poison, and in the treatment of diseases such as headache, asthma, black cough, abnormal urination, and many more disorders (Hedayati et al. 2021) attracted attention of scientific community for further exploration of bioactive principles and bioactivities. Among the TAs, scopolamine is one of the most significant alkaloids having enormous properties of pharmaceutical importance, along with less side effects as compared to hyoscyamine (Wang et al. 2011). However, studies have shown that *A. belladonna* plants mainly produce about tenfold higher hyoscyamine than scopolamine (Yun et al.1992). This gave an impetus in the development of novel varieties *A. belladonna* with high yield of scopolamine. Due to the immense importance of commercial value of *A. belladonna*active research is being carried out employing biotechnological, genetical, and molecular approaches to improve the yield of TAs, which have been reviewed under this chapter.

8.1.1 Taxonomy and Distribution

Highest species diversity of Solanaceae is found in America, and center of diversity of this taxonomic group is in South America (Pigatto et al. (2015). The taxonomic profile of *A. belladonna* has been given in Table 8.1. *A. belladonna* is distributed from Western Europe to the Himalayas (Lee 2007) and is native to Albania, Austria, Belgium, Bulgaria, France, Germany, Greece, Hungary, Italy, Netherlands, Poland, Portugal, Romania, Spain, Switzerland, Ukraine, Great Britain, Czechoslovakia, Trans Caucasus, North Caucasus, Algeria, Morocco, Sardegna, Sicily, Iran, Turkey, Lebanon, Syria, Yugoslavia, and has been introduced in Denmark, Sweden, Ireland, Illinois, Missouri, South Australia, New Zealand South, Baltic States, Southeast China (POWO 2022) (Fig. 8.1). Fattahi et al. (2021) has stated that *A. belladonna* is an endangered medicinal plant native to Europe and Asia. According to Majid et al. (2012) roots of *A. belladonna* were exported from Himalayas during World War—I, and presently it is cultivated in England, Central Europe and USA for medicinal uses.

Table 8.1 Taxonomical classification of *A. belladonna*

Kingdom	Plantae
Division	Angiospermae
Class	Magnolipsida
Order	Solanales
Family	Solanaceae
Tribe	Hyoscyameae
Genus	*Atropa*
Species	*belladonna* L

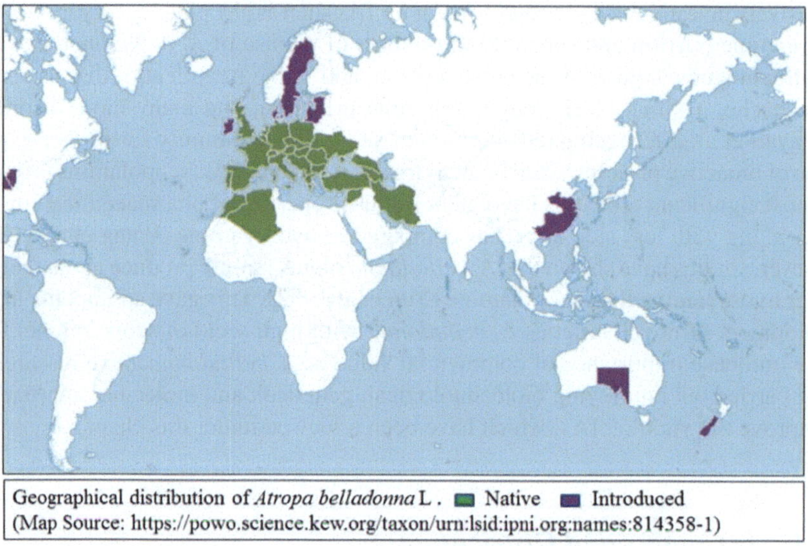

Fig. 8.1 Geographical distribution of *A. belladonna* (POWO 2022)

8.1.2 Economic Importance and Health Benefits

Whole plant or plant parts of *A. belladonna* have been used for medicinal purposes in traditional systems of medicine in various cultures in the form of herbal preparations like herbal teas, extracts, and creams. *A. belladonna* extract was used as eye-drops to dilate pupils (mydriasis), ointments were used for intoxication effects followed by hallucinations, herbal teas exhibited aphrodisiac property. It also showed analgesic, toxic, an aesthetic, and cholinergic activities, and was used for toothache, for bronchorelaxation in asthma and as pain-killers (Kohnen-Johannsen and Kayser 2019). Due to these and many more medicinal applications of *A. belladonna*, researchers developed interest in exploring the therapeutic effects and the phytochemical constituents responsible for the activities. Grynkiewicz and Gadzikowska (2008) reviewed the advances in the chemistry and therapeutics of the TAs in Solanaceae members and their derivatives taking into account their future potential in commercial pharmaceutics application. The alkaloids of *A. belladonna* exhibited multipurpose medicinal properties for instance, mydriatic, antiemetic, antispasmodic, as sedative to inhibit bronchospasm especially in asthma and whooping cough, as anti-cholinergic and cough suppressant, as an analgesic, in neuralgia, rheumatoid arthritis, colic, and in the treatment of hemorrhoids, psychiatric disturbances, and hyperhidrosis (Grynkiewicz and Gadzikowska 2008; Tian et al. 2018). Studied revealed the pharmacological effects of scopolamine which had minimal side-effects and was extensively employed in mitigating spasm, treating motion sickness, as an anesthesia prior operation, quitting drug addictions, pesticide poisoning, and Parkinson's disease (Wang et al. 2011). Hyoscyamine acts as an anti-cholinergic drug which affects on the parasympathetic nervous system, thus used as sedatives, anti-spasmodics, as pre-operative medication, and for treating arrhythmias, cardiac and gastrointestinal disorders, Parkinson's disease, and organo-phosphate poisoning; while anisodamine in curing gastrointestinal colic and vascular spasms (Zeng et al. 2021; Fattahi et al. 2021). The medicinal potential of *A. belladonna* is mainly because of the alkaloids. Various analytical methods like TLC, GC, HPLC-MS, Capillary electrophoresis coupled with electrospray ionization and mass spectrometry (CE-ESI-TOF-MS) have been employed in the identification of alkaloids from different plant parts. The identified alkaloids were apoatropine, hyoscyamine, scopolamine, tropine, norhyoscyamine, 6β-hydroxy-hyoscyamine and belladonnine (Fig. 8.2). Cuscohygrine (Baralle and Gros 1969) and flavonoids namely, kaempferol and quercetin (Cicek et al. 2020) are also reported from *A. belladonna* (Fig. 8.2). *A. belladonna* is widely used in numerous herbal products like Belladonna extract powder, *A. belladonna* root tablets, Belladonna plaster, Belladonna Homaccord—Drops, Belladonna perfumes, etc. (Atropa belladonna products 2022). In 2004, the international sale of pharmaceuticals containing scopolamine as key constituent was around 1.5 billion dollars. In China, *A. belladonna* is the only plant source of scopolamine commercially used. Even though the whole plant is utilized for medicinal purposes, the scopolamine content is comparatively low in different plant parts, and is relatively higher in tender roots (Wang et al. 2011). The major alkaloids in *A. belladonna* are synthesized in the roots, and the roots take around 3–4 years for maturation to produce root biomass that can yield extractable amounts of compounds of interest. After a period of nine months, *A. belladonna* yields 0.1% dry weight (DW) scopolamine (Shahzad et al. 2013). This implies that there is limited production behind a huge production cost.

8.1.3 Domestication

Native drug plants are becoming threatened and endangered in their natural habits due to clearing of forests, increased activities of indiscriminate collection, and expansion of land for cultivation. Sustainable utilization, management, and protection is essential. Furthermore, sustainable utilization of such medicinally important plants

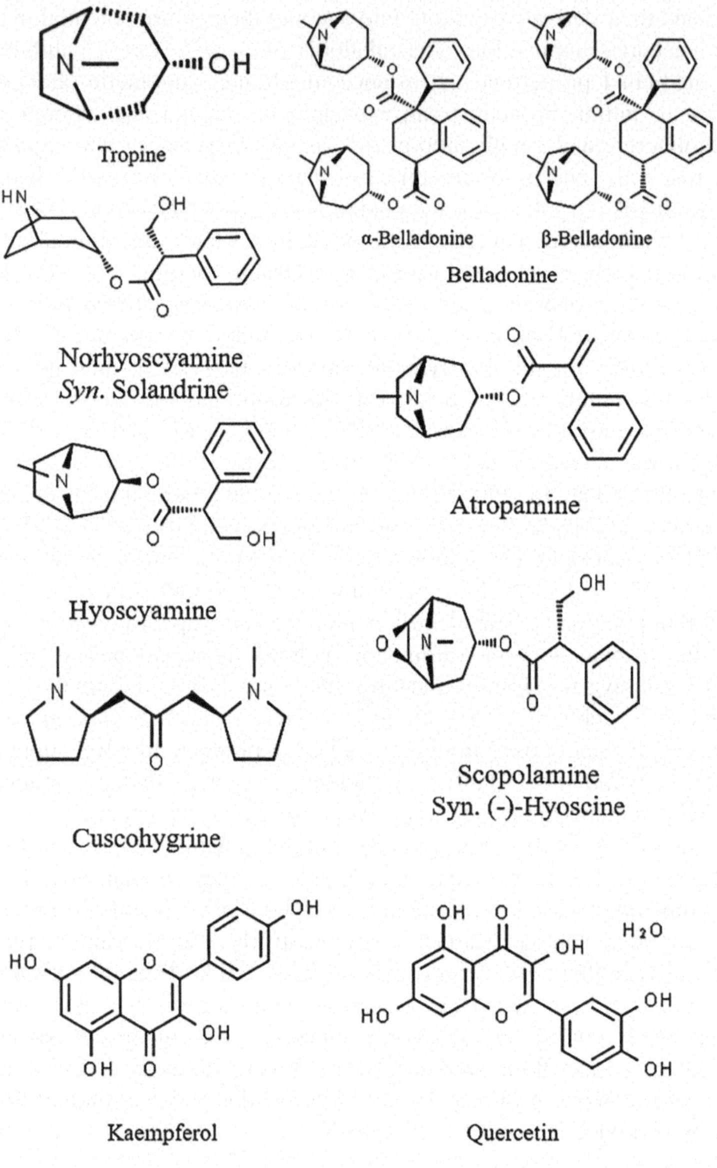

Fig. 8.2 Alkaloids and flavonoids in *A. belladonna*

can generate source of income for the native inhabitants. The leaves and roots of *A. belladonna* are a rich source of phytochemicals. Since ancient time, it has been in demand for the leaves and roots. *A. belladonna* has always been in priority for its domestication using wild individuals because of it economically and pharmaceutically important tropane alkaloids, especially atropine. Thus, several approaches have been made to cultivate and domesticate the belladonna plants in open field and green house. From the investigations several key observations regarding ecology, geography, edapho-climatic conditions required and their impact on the growth and yield of belladonna crops were recorded. These studies have documented significant success in the cultivation and domestication of *A. belladonna* (True 1905; Miller 1914; Stockberger 1915, 1919). The agricultural experimentation station located at Berkeley, California cultivated belladonna and demonstrated that it can be successfully and profitably grown for commercial production (Schneider 1916).

8.2 Current Cultivation Practices and Challenges

8.2.1 Current Cultivation Practices

A. belladonna grows abundantly on waste and arid lands, quarries, shady places, on wooded hills and on chalk or limestone. It is found growing as a weed species naturally as open vegetation (Griffin and Lin 2000). *A. belladonna* is a temperate crop, giving maximum herbage, and alkaloid yield in temperate climate. Seeds obtained from berries, mostly from September to November are used for propagation. For better germination, seeds are treated with 80% sulphuric acid before sowing. In nursery, seeds are directly sown and maintained under shade. Nursery is raised after mid-May upto the end of September to October in autumn. The land should be well ploughed before cultivation. *A. belladonna* thrives well in fertile soil rich in humus, heavy clay soil which causes water-logging should be avoided. Beds of size 3×1 m are raised surrounded with irrigation channels. Well decomposed FYM is applied to the soil. Fungicides pretreated seeds mixed with fine soil should be broadcasted in nursery beds, and covered with a layer of FYM and straw followed by watering. Seed germination takes around 3 weeks. When the seedlings are of 15–20 cm height after 8–12 weeks, these are ready for field planting. Ideally, March–April or October–November are favourable for transplantation. Seedlings should be treated with fungicide before transplanting. Transplantation of seedlings is safe on raised beds having wide strips of 1 m with ridges to prevent water-logging, distance maintained at 60×70 cm. Harvesting must be carried out once it starts flowering, since this is the prime time of accumulation of higher alkaloid content. Leaves should be dried immediately after harvesting. The roots can be harvested after 3 years. After harvesting they should be washed, cut into small pieces, and then shade- or sun-dried.

8.2.2 Current Challenges

After approximately 12 months, *A. belladonna* grows up to 4 feet height. During flowering season, the plant is cut few inches above the ground and leaves and flowering inflorescences are segregated. Second harvest is performed in August–September from these cuttings. Around 2–3 harvests are carried out in the second and third year, and roots are harvested in the fourth year. New plantation is initiated after four years. The collected plant material takes a couple of day for drying. Absolutely dried drug decomposes and releases ammonia. Thus, it is stored in air-tight containers kept away from light and moisture. Belladonna is prone to be attacked by the fungus *Phytophthora* causing wilt and root rot. Infected plants need to be dug out from soil and burnt to avoid further infection and loss. It is also attacked by flea beetle, which eats mostly old and large leaves. Cultivation and harvesting period are longer, and it takes about four years to procure roots. The cultivation of *A. belladonna* is done to obtain tropane alkaloids, but it is challenging to know the accurate season for the alkaloid production. This conventional method of propagation is not sufficient to fulfil the increasing demand, and is time consuming. Farmers should be introduced to novel technologies for biomass production. Apart from the recommendations of researchers for encouraging domestication of such medicinally important plants, it is also necessary to introduce the modern biotechnological and genetic approaches to uplift the production.

8.2.3 Genetic Improvement Objectives

A. belladonna is mostly explored for the TAs, precisely for scopolamine and hyoscyamine, which are synthesized in the roots. Although whole plant is utilized for medicinal purposes, roots are of more industrial importance. The roots take longer duration to accumulate extractable amounts of secondary metabolites of interest, which cannot suffice the growing demand of pharmaceutical industries. Environmental factors and biotic and abiotic factors play a crucial role in the secondary metabolite accumulation in plants. These challenges can be surpassed using novel biotechnological, genetic and molecular approaches such as genetic engineering, plant tissue culture, molecular breeding, gene editing, micropropagation, and hybridization etc.

8.3 Germplasm Biodiversity and Conservation

A broad array of traits of plant is influenced by the overall amount of all allelic sources comprises the plant genetic resources, and the genetic material passed down through generations is the germplasm. Germplasm conservation not only aids to protect information about wild or extinct plant species, but it as certain secure management and preservation of the commercially valuable plant germplasm. Accelerated progress employing various technologies like cryopreservation, in vitro culture technology, and molecular markers have opened up new avenues for the for studies in plant diversity and genetic resource management.

8.3.1 Germplasm Diversity

Hyoscyamine is the prime compound of *A. belladonna*, while the derivatives of hyoscyamine, namely anisodamine and scopolamine are the minor compounds. Due to similarity in their structures, it becomes difficult, time consuming, laborious, and uneconomical for isolation of each compound. There is a need to develop technology to separate these compounds at lower cost. Development of new varieties that accumulate higher TAs can also be an option to fulfil the growing demands of pharmaceutical industries. Zeng et al. (2021) produced *A. belladonna* plant lines with reduced *AbH6H* function using *CRISPR/Cas9* system. This novel germplasm line showing homozygous mutations in *AbH6H* produced enhanced hyoscyamine without the synthesis of anisodamine and scopolamine, that can be applied practically to produce hyoscyamine at a competitive price. Effect of gamma rays on the callus of *A. belladonna* was studied along with the molecular level of protein and genetic analysis using SCoT markers. Lower irradiation of gamma rays favoured callus growth and improved secondary metabolites. SDS-PAGE and SCoT-PCR technique can be used to assess variation in *A. belladonna* (Elshorbagy et al. 2021).

8.3.2 Genetic Resources Conservation Approaches

The population explosion throughout the world is accountable for the unsustainable use of natural resources leading in the deterioration of biodiversity. The ultimate goal of conservation is to compensate for the loss of biodiversity by preserving the environment, the species, and the genetic diversity for the future. Conservation of the genetic variability of the germplasm can be achieved by the implementation of in situ and ex situ conservation strategies that can be attained by means of various techniques.

8.3.2.1 Ex Situ Conservation

Seed Bank

The Majella National Park has implemented a unique integrated in situ-ex situ strategy for the preservation of the national heritage of Majella. The Majella Seed Bank plays an important role in the ex situ conservation. *A. belladonna* is one of the taxa which has been preserved in this seed bank (Di Cecco et al. 2020).

8.3.2.2 Cryopreservation

Cryopreservation is a valuable technique used in the long-term conservation of biological materials such as plant cell, tissues and organs, and vegetatively maintained crops. The germplasm can be preserved for infinite period with affordable price and little space. Apart from the conservation of genetic resources, plant tissues with specific features can be securely stored. Various techniques like pre-growth, desiccation, encapsulation, vitrification, and droplet-freezing can be implemented for cryopreservation (Radha et al. 2012). The survival of the frozen embryos is influenced by a number of factors. Depending on the size and developmental stage, young and comparatively small embryos usually demonstrate better survival rate as compared to the developing embryos, that require desiccation prior to freezing. Higher viability was recorded in the pollen embryos of *Atropa* at the globular and heart-shaped stages as against the latter embryo stages, which held up with the −196 °C freezing (Bajaj 1977). Number a cryoprotectants are used in the studies on embryos such as glycol, glycerol, DMSO, sucrose, and sorbitol. The level of success depends on the nature and concentration of these cryoprotectants, as well as their use as single or in combination. a mixture of DMSO, sucrose, and glycerol was successfully used in the pollen embryos of *Atropa* (Bajaj 1978).

Touno et al. (2006) cryopreserved hairy roots (clone M8) of *A. belladonna* using the vitrification technique. No significant difference was observed in the viability of the cryopreserved hairy roots for different periods, and the re-growth of vitrified hairy roots was much the same to that of the untreated hairy roots.

8.3.2.3 Molecular Studies

Since there were no studies relating to the selection and assessment of reference genes in *A. belladonna*, Li et al. (2014) conducted a study using eight potential reference genes, namely 18S ribosomal RNA (18S), GAPDH, actin (ACT), beta-tubulin (TUB), phosphoglycerate kinase (PGK), elongation factor 1α (EF-1α), cyclophilin CYC, and β-fructosidase (FRU) to determine reference genes procured from *A. belladonna* plant parts under different abiotic stress conditions for gene expression analysis using quantitative real-time PCR normalization. Genes were cloned due to unavailability of the sequences of the eight candidate reference genes in the public NCBI GenBank database. The amplified sequences were provided and deposited in GenBank (Table 8.2).

8.3.3 Cytogenetics

Krumbiegel and Schieder (1979) performed a fusion experiment between diploid ($2n = 24$) and tetraploid ($4n = 48$) *Datura innoxia* and wild-type *A. belladonna* ($2n$

Table 8.2 Candidate reference genes and target genes used in the study

Gene name	Accession number	Function	Primer sequence (5′ to 3′) (forward/reverse)	Amplicon length (bp)	Melting temperature (°C)	E (%)	R2
18S	KC881109	40S ribosome subunit RNA component	ATGATAACTCGACGGATCGC CTTGGATGTGGTAGCCGTTT	169	56	99.6	0.999
ACT	JX154681	Cytoskeleton contraction	TTGTGTTGGACTCTGGTGATGG CCGTTCAGCAGTGGTGGTG	167	60	98.3	0.999
CYC	JX154678	Protein folding, RNA splicing	CCCAGGGTTCATGTGTCAAGG ATTAGTTCCAGGTCCAGCATTAGC	154	59	97.1	0.997
FRU	JX154680	Hydrolysis sucrose	AGACTCAAACGGGCTGGACAC GCCTCAACCCAATACCACAATCC	170	60.4	102.5	0.998
EF-1α	JX154677	Protein translation	TGAAGAATGGTGATGCTGGTATGG CTTCTGAGCAGCCTTGGTGAC	191	59.5	91.1	0.998
GAPDH	JX154679	Glucose metabolic process	GCTGCTTCATTCAACATCATTCC ACAGTAAGGTCCACAACAGAAAC	140	56.5	96.6	0.999
PGK	JX154676	Glucose metabolic process	TCGCTCTTGGAGAAGGTTGAC CTTGTCGGCAATCACTACATCAG	195	59.5	97.4	0.999
TUB	JX154675	Structural constituent of cytoskeleton	CTTGCTGTTTGAGATTCCCTGGTC CGCACATCATATTCTTGGCATCCC	195	59	99.7	0.996
PMT	AB018570	Putrescine N-methyltransferase	CCTACTTACCCTACTGGTGTTATC GCGAAAGATGGCAAAATAAAAGC	178	57	95.3	0.998

E amplification efficiency, R^2 correlation coefficients
Source Li et al. (2014)

= 72) for hybridization as their chromosomes varied in sizes, *D. innoxia* chromosomes being twice in size than the nearly punctiform *A. belladonna.* Gleba et al. (1988) studied the genetic constitution of inter-tribal hybrids generated from the protoplast fusion obtained from a nitrate reductase-deficient mutant of *Nicotiana plumbaginifolia* ($n = 10$) and wild type *A. belladonna* ($n = 36$), following lethal doses of gamma-rays irradiation, which produced asymmetric nuclear hybrids. Fusion of mesophyll protoplasts of *A. belladonna* (kanamycin-resistant) and mesophyll protoplasts of *Nicotiana tabacum* (phosphinothricin-resistant) produced highly asymmetric somatic hybrids, whose cytological analysis showed a whole genome of *N. tabacum* ($2n = 48$) large chromosomes and one small extra chromosome (not more than 2.8%) of *A. belladonna* genome ($2n = 72$) (Babiychuk et al. 1992).Volkov et al. (2017) applied a combination of karyological and molecular analysis for investigating chromosomal structure, molecular organization and evolution of ribosomal DNA (rDNA) in *A. belladonna* one of the oldest known allohexaploid among flowering plants. Due of their abundance and specific molecular organization, 45S and 5S rDNA are extensively used in taxonomic and evolutionary studies.

8.4 Molecular Breeding

8.4.1 Functional Genomics and Bioinformatics

Littorine is a TA and a main biosynthetic intermediate in the biosynthetic pathways of hyoscyamine and scopolamine. Nevertheless, formation of littorine via phenyllactate and tropine precursors remains unknown. Explication of the biosynthesis of littorine was done by implementing functional genomics approach and with two novel genes encoding phenyllactate UDP-glycosyltransferase (*UGT1*) and littorine synthase (*LS*). In *A. belladonna* these genes are highly/specifically expressed in the secondary roots. Suppression of any of these genes interrupted littorine biosynthesis and the TA derivatives namely, hyoscyamine and scopolamine. Identification of these novel genes offered lacking connection in littorine biosynthesis (Qiu et al. 2020). Sadre et al. (2022) investigated the potential functions of two *A. belladonna* cytochrome P450 genes*AbP450-5021* (CYP71 family) and *AbP450-116623* (CYP82 family) using reverse genetics along with discovery metabolomics and multivariate statistical analysis. This was further succeeded by *in planta* transient assays for the exploration of alkaloid diversity. The involvement of *AbP450-5021* and *AbP450-116623* in the complex and dynamic alkaloid sub-network in *A. belladonna* root which channelizes the pseudotropine as precursor for the production of modified tropane alkaloids was revealed by the discovery metabolomic-guided approach. The catalytic transformations of pseudotropine-derived intermediates of *AbP450-5021* and *AbP450-116623* had a double role of N-demethylase/ring-hydroxylase and ring-hydroxylase, respectively. *A. belladonna* genome analysis underlies the evidences of conserved biosynthetic pathways of TAs such as conserved gene cluster with gene duplication, branching genes, which will be helpful for further advancement of biotechnological approach for TAs production (Zhang et al. 2023).

8.5 Tissue Culture Applications

Hyoscyamine and scopolamine are the most significant TAs of *A. belladonna* with wide-ranging applications in medicine. Due to its indiscriminate exploitation from the natural habitat, *A. belladonna* is now enlisted in the endangered medicinal plant species category (Anonymous 1948; Shahzad et al. 2013). The industrial demand is unable to be sufficed by the wild populations. The demand for hyoscyamine and scopolamine is on a rise, and numerous challenges are encountered such as naturally growing plants take longer time for accumulating metabolites, huge investments yield sparse production, chemical synthesis of these compounds is uneconomical. There is a need for developing strategies in addition to existing traditional methods used. Plant tissue culture applications can be the alternative for enhancing the secondary metabolites in a short duration for commercial purposes.

8.5.1 Micropropagation Approaches

A detailed review of literature on the diverse aspects of *invitro* culture studies and *invitro* alkaloids production in *A. belladonna* has been reviewed by Bajaj and Simola (1991) from 1969 to 1989 (Tables 8.3 and 8.4). Doubly transformed hairy roots were initiated by re-infection in transgenic *Atropa belladonna* (*Agrobacterium rhizogenes* ATCC 15,834) leaf segments with another stain of *A. rhizogenes*, MAFF 03–01,724, which produced shoots and viviparous leaves. Tropane alkaloid analysis showed that the doubly transformed hairy roots had intermediate quantity as compared to the singly transformed roots (Jaziri et al. (1994). Hilton and Wilson (1995) examined the ability of *A. belladonna* transformed root cultures in the uptake of carbon source, some anions and cations in a Gamborg B5 medium batch culture for 28 days. Total consumption of NH_4^+ and PO_4^{3-} was observed in all cultures, while for ions the utilization rate differed, but the complete consumption was not observed. Ten hairy roots clones were isolated and selected from the in vitro *A. belladonna* cultures infected by *Agrobacterium rhizogenes* (R1601) for alkaloid production. The #K_4, #K_5 and #K_8 clones with best hairy root growth were selected and cultivated in the Gamborg B5 medium for biomass production. These clones showed spontaneous reorganization of shoot and leaf formation. Highest hyoscyamine and scopolamine content was recorded in the #K_5 clone, and overall alkaloid content was higher in the clones as compared to the reorganized plants (Hank et al. 2004). Mera and Takayama (2012) established hairy root cultures from adventitious buds of *A. belladonna* with *Agrobacterium rhizogenes* (MAFF 301724) and evaluated the effect of culture media components and shear conditions on plant regeneration and mass propagation. Multiple shoots with roots were obtained on MS medium, which was superior to other studied media. Al-Ashaal et al. (2013) established the *invitro* (callus and plant differentiation) cultures from leaves of *A. belladonna* on MS media to enhance tropane alkaloid production, and studied its bioactivities. The HPLC analysis revealed higher amount of tropane alkaloids in differentiated plants as compared to the original plants. Shoot tip and nodal explants of were inoculated on growth media containing different concentrations of kinetin and BAP to define the ideal explant source and optimal PGRs concentrations for micropropagation. The results revealed that the use of shoot tip was better for organogenesis at 1 mg/ l of BAP which produces 3.25 shoot explant^{-1} with an average length of 3.03 cm, having 7.0 leaves explant^{-1} (Ahmed et al. 2021). The positive effects on growth of callus and induce production of TAs were recorded on MS medium augmented with 2,4-D, Kinetin and salicylic acid and He–Ne laser treatments (Nabil et al. 2022).

8.5.2 *Other in Vitro Applications that Pertain to the Species*

8.5.2.1 Nanoparticle (NP)-Mediated Production of Secondary Metabolites

Nanoparticles is the delivery system that is designed with an objective to control the particle size, surface properties and to release pharmacologically active agents to achieve the site-specific action of the compound at the optimum rate and dose treatment (Mohanraj and Chen 2006). Tian et al. (2018) scrutinized the effect of manganese oxide nanoparticles (Mn_2O_3 NPs) on the morphological, physiological, biochemical attributes and secondary metabolism of *A. belladonna* in vitro shoot tip cultures, and its effect on elicitation of pharmaceutically important secondary metabolites.

8.6 Genetic Engineering and Gene Editing

Extensive research on genetic engineering of TAs has been reviewed and highlighted that natural yield of hyoscyamine and scopolamine in *A. belladonna* is far too low for commercialization and the conversion of hyoscyamine to scopolamine remains a challenge and major objectives of the future studies. In addition, it reviewed the successful metabolic engineering attempts to know the rate limiting steps and respective genes of biosynthetic pathways of TAs. The metabolic engineering of plants pertains to the second-generation category of plant genetic engineering, which is an effective method of genetically transform the plant metabolites of industrial importance. Therefore, with the help of procured metabolic engineering knowledge, it might be possible to enhance the production of desired TAs in *A. belladonna*

Table 8.3 In vitro culture studies of *A. belladonna*

Culture system	Observations/results	References
Callus/suspension culture	Established callus and suspension cultures; differentiation of roots and shoots, regeneration of plants	Bajaj (1977, 1978, 1995), Bajaj and Simola (1991)
	Formation of embryo-like structures from root callus	
	Callus, buds, plants	
	Effects of auxins on several enzymes in suspension culture	
	Organogenesis and embryogenesis in root, callus, and suspension cultures	
	Ultrastructural studies and changes in enzymes activity during differentiation of roots	
	Development of chloroplasts in stem callus	
	Plants from haploid and diploid callus	
	Effect of methionine on callus growth and ethylene production	
	Effect of nitrogen sources and amino acids on callus growth and nitrogen metabolism	
	Phenylalanine transport in suspension cultures	
	Feeding experiments with precursors of tropane alkaloids	
Anther culture/Haploid production	Haploid plants and the study of stability in cell cultures	
	Plants showed various level of ploidy ranging from haploids to hexapodids	
Pollen mother cells	Repeated cell division	
Pollen tetrads	Additional microspore formation	
Protoplast culture/Somatic hybridization	Embryos from cell suspension-derived protoplasts	
	Plants from haploid mesophyll protoplasts showing genetic variability	
	Somatic hybridization with *Datura innoxia*	

(continued)

Table 8.3 (continued)

Culture system	Observations/results	References
	Somatic hybridization with *Nicotiana plumbaginifolia*	
Cryopreservation	Cell suspension survived	
	Pollen embryos formed plants	
	Mesophyll protoplasts formed embryos, plants	
Transformation	Excised roots infected with *Agrobacterium rhizogenes* Hairy root cultures with alkaloid production	

through genetic modification and exact regulation of the TAs biosynthetic pathway (Oksman-Caldentey and Arroo 2000).

8.6.1 Pathway for Tropane Alkaloids in A. belladonna

The accumulation of tropane alkaloids (hyoscyamine and scopolamine) is a multi-step pathway and has been partly elucidated in *A. belladonna* and other members of Solanaceae family (Bedewitz et al 2014; Qiu et al 2018) (Fig. 8.3). However, the enzymes involved in the production of hyoscyamine and scopolamine were reviewed by Zhang et al (2005). The pathway originates from two amino acids, ornithine and arginine (Patterson and O'Hagan 2002) which work as initial precursors for tropane alkaloid accumulation. Ornithine get converted to the putrescine by ornithine decarboxylase (*OrnDC*), while arginine produces putrescine via arginine decarboxylase (*ArgDC*) with an intermediate product, agmatine. Putrescine is an organic compound works as precursor for the polyamines and tropane alkaloids production (Hashimoto and Yamada 1986, 1987; Hibi et al. 1992; Zhang et al. 2005).

Putrescine *N*-methyltransferase (*PMT*; EC 2.1.1.53) an enzyme engaged in the conversion of putrescine to the *N*-Methyl-putrescine (Suzuki et al. 1999). PMT is also involved in removal of putrescine from the polyamine pool and the tropane ring (tropane alkaloids) and pyrrolidine ring (nicotine) is obtained from putrescine by *N*-methylputrescine synthesis (Richter et al. 2005; Zhang et al 2005). The first important step in the production of tropane alkaloids in *A. belladonna* and *H. niger* is the *N*-methylation of putrescine which is carried by PMT (Matsuda et al. 1991; Hibi et al. 1992). Further, *N*-methyl putrescine is converted to tropinone which is a branching point for the production of tropine and pseudotropine. Tropinone reductase I and II reduces tropinone to tropine and pseudotropine, respectively (Koelen and Gross 1982; Drager et al. 1988). Tropine is considered as the precursor for the accumulation of different tropane alkaloids such as hyoscyamine, scopolamine, and calystegines (Richter et al. 2005; Zhang et al 2005). Phenylalanine is an aromatic

Table 8.4 In vitro production of alkaloids in *A. belladonna*

Culture system	Observations/results	References
Root culture, callus	Synthesis, of atropine (dl-hyoscyamine) in root callus, no alkaloids were detected in stem and leaf callus	Bajaj and Simola (1991)
Root cultures of seedling and callus origin, shoots of callus origin	Atropine, scopolamine, tropine and cuscohygrine detected	
Root-bearing aggregates, organ-free aggregates from suspension cultures with root formation	Seven Dragendorff-positive compounds including atropine and scopolamine	
Callus	Tropine, atropine, and scopolamine detected	
Callus cultures (stem origin), cell clumps, green embryoids	Quantitative estimation of total tropane alkaloids	
Callus with roots, redifferentiated roots, plantlets	Quantitative estimation of total tropane alkaloids	
Root cultures	13 alkaloids from root cultures, their pattern was almost identical with that of roots from the intact plants	
Infected roots with *Agrobacterium rhizogenes*	Amount of atropine and scopolamine was higher in the hairy roots	
Shoot cultures	Tropane alkaloids	
Root cultures infected with *A. rhizogenes*	Atropine, scopolamine, cuscohygrine detected	
	Atropine and scopolamine detected	
Callus culture	Studied the effect of various factors on alkaloids production in calli, synthesis of atropine in stem callus	
Cell suspension	Feeding experiments with precursors of tropane alkaloids (hygrine, tropinone, and tropic acid)	

amino acid that converts into phenyllacetic acid which is along with tropine gives rise to the hyoscyamine. Hyoscyamine-6-hydroxylase (*H6H*; EC 1.14.11.11) is a key enzyme required for the conversion of hyoscyamine to the 6*β*-hydroxyhyoscyamine and scopolamine (Yun et al. 1992; Richter et al. 2005), while pseudotropine leads to the production of calystegines A3 (Matsuda et al. 1991; Hashimoto and Yamada 1994; Zhang et al. 2005).

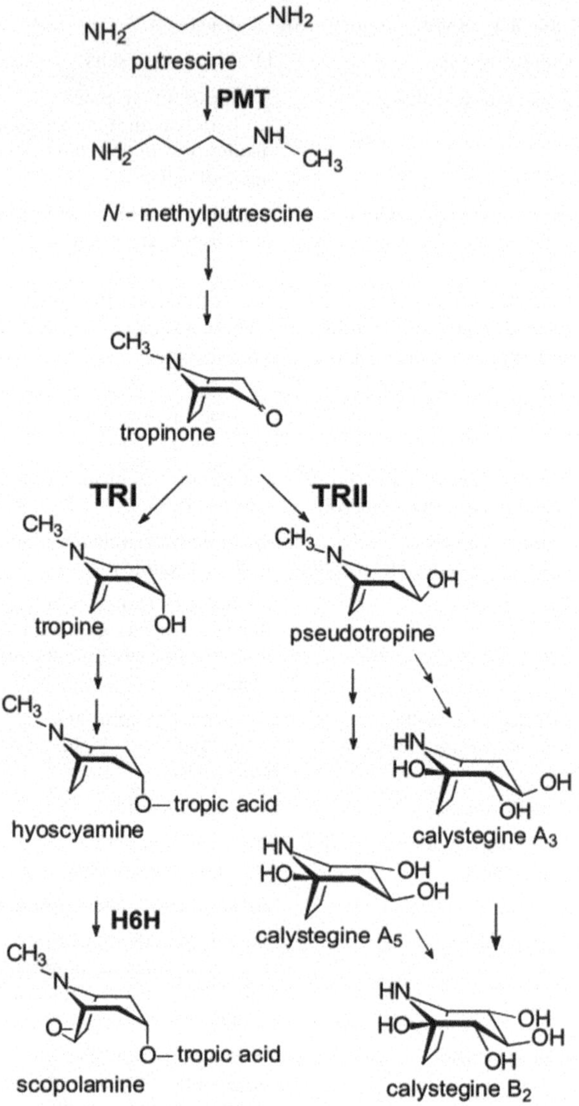

Fig. 8.3 Biosynthetic pathway for the accumulation of tropane alkaloids in *A. belladonna.* Source: Richter et al. 2005

8.6.2 Genetic Engineering for TAs in A. belladonna

Is it known that scopolamine is formed from hyoscyamine via 6*β*-hydroxyhyoscyamine. *H6H* catalyzes the hydroxylation of hyoscyamine to 6*β*-hydroxyhyoscyamine, along with the epoxidation of 6*β*-hydroxyhyoscyamine to scopolamine (Jirschitzka et al. 2013). Overexpression of *H6H* in transgenic *A. belladonna* effectively converted hyoscyamine to scopolamine (Yun et al. 1992). *H6H* gene was introduced in *A. belladonna* and expressed constitutively in hairy roots. Five-fold increased concentration of scopolamine along with increased (2- to fourfold) transformed roots was recorded in transgenic *A. belladonna* hairy roots as compared to wild-type hairy roots due to overexpression of *H6H* gene (Hashimoto et al. 1993). This successful metabolic engineering raised expectations for biotechnological approaches for enhancement of secondary metabolites for industrial production. *PMT* is the first enzyme committed for the biosynthesis of tropane alkaloids. Overexpression of *PMT* in transgenic *A. belladonna* as well hairy roots, however did not enhance the accumulation of TAs (Sato et al. 2001). Further, Rothe et al. (2003) regenerated transgenic *A. belladonna* as described by Sato et al. (2001). The effect of elevated sucrose level and addition of auxins on the production of alkaloids was studied on the T_2 generation of transgenic *A. belladonna* plants and root cultures. The *PMT* overexpressed roots did not respond to the high sucrose level, while the auxin treatment inhibited formation of TAs in 35S-PMT roots. This work ascertained that increased *PMT* expression alone was not enough to enhance TA production in *A. belladonna* plants and root cultures. Simultaneous overexpression of multiple enzymes in the pathway should be carried out for the enhanced production of TAs. Overexpression of multiple key enzyme genes or manipulation of regulatory genes in the target bioengineering pathway is an assuring approach to modify the biosynthetic pathways for accumulation of secondary metabolites (Sato et al. 2001). Two rate-limiting enzyme genes *PMT* and *H6H*were introduced into *A. belladonna.* Significant enhancement (1.3–2.5 folds) in scopolamine accumulation as compared to the wild type, was produced due to the simultaneous over expression of the *PMT* and *H6H* genes in the transgenic hairy roots of *A. belladonna* (Liu et al. 2010). This work laid the conceptual basis for the genetic manipulation in scopolamine biosynthetic pathway, along with the cultivation of transgenic *A. belladonna* plants producing higher scopolamine. By means of co-transformation, two key enzymatic genes involved in biosynthetic pathway of scopolamine that encoded for upstream and downstream key enzymes *PMT* and *H6H*, respectively in transgenic *A. belladonna* were simultaneously overexpressed. Gene expression analysis revealed that co-transformation of *PMT* and *H6H* was a promising approach to develop biosynthesis potential in *A. belladonna* for increased yields of scopolamine (Wang et al. 2011). In a study carried out by Xia et al. (2016), *Nicotiana tabacum PMT* (*NtPMT*) and *Hyoscyamus niger H6H* (*HnH6H*) were overexpressed in T1 progeny of transgenic *A. belladonna* to investigate biosynthesis of TAs and distribution in the aerial (ripe fruits, leaves, and stem) and underground (primary and secondary roots) parts under field conditions. The four endogenous TA biosynthetic genes (*AbPMT*, *AbTRI*, *AbCYP80F1*, and *AbH6H*) had the highest expression levels in secondary roots, implying that TAs were mostly produced in secondary roots. Transgenic *A. belladonna* T1 progeny exhibited remarkable chemotype that was rich in scopolamine content, and which had the potential to improve the commercial and pharmaceutical value of *A. belladonna.* Qiu et al (2018) identified a novel root-expressed gene that encoded a phenylpyruvic acid reductase (*AbPPAR*) essential for TA biosynthesis in *A. belladonna. AbPPAR* was particularly expressed in the root pericycles and endodermis. *AbPPAR* catalyzed the reduction of a precursor of TAs phenyl pyruvic acid to phenyl lactic acid, while the suppression of *AbPPAR* disturbed TA biosynthesis via reduction of phenyl lactic acid levels. A missing enzyme was recognized in the TA biosynthetic pathway providing an innovative perception in the biosynthesis of TAs. Ornithine and arginine synthesize TAs through a key precursor putrescine (Hashimoto et al. 1989). Albeit which of these contributes substantially in the biosynthesis of TAs is a controversial subject. Ornithine decarboxylase (*ODC*) has an essential role in numerous biological processes, nonetheless its function in plant secondary metabolism, particularly in TA biosynthesis remains ambiguous. *ODC* played a vital role than arginine decarboxylase (*ADC*) in TA biosynthesis was confirmed using enzyme-specific inhibitors difluoromethyl ornithine (DFMO) and difluoromethyl arginine (DFMA). The characterized *AbODC AbPMT*, *AbAT4*, *AbPPAR AbPYKS*, *AbCYP82M3*, *AbTRI*, *AbUGT1*, *AbLS*, *AbCYP80F1* and *AbH6H*, were highly or else specifically expressed in secondary roots. The enzymatic kinetics of the recombinant *AbODC* revealed that *AbODC* had the ability to catalyze the ornithine decarboxylation to produce putrescine. Elevated transcript levels of *AbODC* was observed in the transgenic root cultures in gene expression study. *AbODC* overexpression stimulated putrescine, N-methylputrescine, hyoscyamine and anisodamine production, except scopolamine. All these investigations indicated the key role of *AbODC* in TA biosynthesis, confirming its prospective in the development of high TA yielding cultivars of *A. belladonna* (Zhao et al. 2020). *AbCaM1*, a calmodulin gene has calcium-binding activity, and is highly expressed in *A. belladonna* secondary roots. Homozygous transgenic *A. belladonna* was produced by co-overexpression of the novel *AbCaM1* calmodulin gene and a reported EPSPS gene (*G2-EPSPS*) by means of glyphosate as selection pressure. The generated transgenic homozygous lines yielded higher TAs as compared to the wild-type plants. Up-regulation of seven key genes of TA biosynthesis due to *AbCaM1* was observed in the gene expression analysis. Commercially recommended glyphosate dose for field was effectively endured by the transgenic homozygous lines. These novel *A. belladonna* varieties showed the potential to produce enhanced TAs along with glyphosate resistance, worthy for industrial production at reduced costs (Zhang et al. 2022). Gong et al. (2022) reviewed the advances in medicinal tropane alkaloids engineering through over expression of single gene or multiple genes like *AbODC*, *AbHDH* in *A. belladonna.* The metabolic engineering pyrrolidine alkaloids revealed their importance and relief complexity in tropane alkaloids biosynthesis in *A. belladonna* (Parks et al. 2023).

8.7 Mutation Breeding

Mutation breeding has played a pivotal role in the improvement of crop plants. Seeds of *A. belladonna* were exposed to 35 Kr of gamma irradiation to induce mutation. These effects were observed through the M_1 and M_2 generations. The macromutations affected the morphological characters of the selected 15 mutants. Higher alkaloid content was recorded in the mutants as compared to untreated plants

(Kaul et al. 1974). The effect of gamma radiation (50, 80, 110, 150 Gy) and gibberellin (20, 40, 60, 80, 100, 120 ppm) on *A. belladonna* seed germination and alkaloid production was studied from M_1 and M_2 generations. There was an increase in seed germination with the increase of gamma radiations exposure until 150 Gy. Highest germination percentage and alkaloids content were recorded from seeds treated with 110 Gy radiation and 100 ppm gibberellin. Moreover, three alkaloid rich mutants were observed from M_2 generation (Abdel-Hady et al. 2008). Hana (2021) performed an experiment to germinate seeds of *A. belladonna* using different techniques which included X-rays irradiation. Low doses (40–60 kV) did not show any effect, while increased irradiation for 6 s. decreased seed germination and root growth. Exposure to X-rays for 1 s. at 110 kV stunted the vegetative growth. Chemically induced mutations were found to increase seed germination, alkaloids content, qualitative and quantitative traits in *A. belladonna. A. belladonna* seeds were treated with chemical mutagens namely, methyl nitrosoguanidine (NG) and ethylenimine (EI) and different quantitative parameters and alkaloid content were taken in to account from M_1, M_2, and M_3 generations. The significantly increased plant height, tiller number, leaf length, and leaf width were recorded from the plants raised from the seeds treated with NG (1 and 2 mM) and EI (0.05%). All the treatments (except NG, 1 mM) showed considerable increase in alkaloid content as compared to the control (Kaul and Choudhary 1975).

8.8 Hybridization

8.8.1 Somatic Hybridization

Extensive work on somatic hybridization has gained importance in the production of the secondary metabolites since it is possible to practically utilize the obtained hybrid cell lines or plants. Somatic hybridization has the ability to develop cells having new genetic constitution. Not only species having diverse degrees of genetic variation can be combined into functional hybrids, but also those species that are from distinct tribes can also be hybridized. Protoplast fusion is a scientific tool employed in plant breeding experiments for the production of cybrids which is alternative to the sexual crossing (Kushnir et al 1987). This technique enables researchers to produce cybrids of different taxa by discounting any kinds of genetic barrier/ constrains which were not possible in the conventional breeding. Thirteen somatic hybrids were selected after the fusion of protoplasts from chlorophyll-deficient mutant of *Datura innoxia* Mill. ($2n = 24$ and $4n = 48$) with green wild-type protoplasts of *A. belladonna* ($2n = 72$). Morphological and cytological investigations verified the hybrid nature. The almost punctiform chromosomes of *A. belladonna* could be differentiated from the large chromosomes of *D. innoxia* (Krumbiegel and Schieder 1979). In their further investigation, chromosome elimination was observed demonstrating strong somatic incompatibility in these somatic hybrids (Krumbiegel and Schieder 1981). Intertribal asymmetric hybrids were obtained between nitrate reductase-deficient mutant of *Nicotiana plumbaginifolia* and a wild type *A. belladonna* after screening huge number of gamma-fusion products, caryological examinations of clones, and biochemical and genetic analysis of progeny plants Gleba et al. (1988). Functional somatic hybrids were obtained after the fusion of mesophyll protoplasts of *Nicotiana tabacum* (plastome chlorophyll-deficient, streptomycin-resistant) and wild type *A. belladonna.* Cytogenetic and biochemical studies revealed that most of the obtained cybrids were diploid, morphologically and phenotypically identical to *N. tabacum.* First fertile plants were reported from *Nicotiana, Atropa* cybrids possessing *Nicotiana* genome and *Atropa* plastome from different genera and tribe (Kushnir et al. 1987). On the other hand, non-functional (chlorophyll-deficient) cybrids were obtained possessing *A. belladonna* nuclear genome and *N. tabacum* plastome (Kushnir et al. 1991). Highly asymmetric somatic hybrids were constructed after the fusion of mesophyll protoplasts of *A. belladonna* (kanamycin-resistant) and mesophyll protoplasts of *Nicotiana tabacum* (phosphinothricin-resistant). These asymmetric somatic hybrids contained a whole genome of *N. tabacum* with a small part (one chromosome, i.e. not more than 2.8%) of *A. belladonna* genome. Although both species were congruent somatically, no pre-treatment was given to the donor genome (Babiychuk et al. 1992). Protoplasts of *A. belladonna* ($2n = 72$) var *acuminata* derived from cell suspension culture was fused with the protoplast of *Hyoscyamus muticus* ($2n = 28$) (achlorophyllous mutant cell line) to obtain intergeneric somatic hybrid plants. The somatic hybrid plants expressed intermediate leaf characters or considerable characteristics reflecting occurrence of both genomes. This was likely because *Atropa* and *Hyoscyamus* belonged to the same tribe, complimenting at the intergeneric level, as in contrast with *Atropa* × *Nicotiana* and *Atropa* × *Datura* in previous works belonging to different tribes. Chromosome elimination of *Atropa* chromosomes was observed (Ahuja et al. 1993).

8.8.2 An Overview of Current Status

A. belladonna is an important medicinal plant due to the presence of the pharmaceutically valuable tropane alkaloids. There is an increasing demand for hyoscyamine, scopolamine, and other TAs due to its various therapeutic applications. Among the numerous Solanaceae members *A. belladonna* is the most important source of hyoscyamine and scopolamine. Vast investigation has been carried out on the pharmacological aspects of *A. belladonna* TAs, leading to the increasing demand in pharmaceutical industries. Alternative production systems like engineered plants, climate-independent production techniques such as cell cultures or heterologous microbial production have been explored in *A. belladonna.* Metabolic engineering including gene expression analysis, and functional identification of genes of interest has been used for the genetic modification of the scopolamine biosynthetic pathway in *A. belladonna.* Biosynthetic pathways to hyoscyamine and scopolamine have been elucidated. Novel varieties having potential to produce enhanced levels of TAs (hyoscyamine and scopolamine) are being developed. Engineering the biosynthesis of TAs has been achieved in hairy root cultures of *A. belladonna.* Overexpression of biosynthesis genes to promote metabolite production has been widely applied.

8.8.3 Recommendations of Future Research and Utilization

Although alternative approaches have been developed like engineered plants, climate-independent production systems such as plant cell cultures or heterologous microbial production, these are not economically reasonable for industrial application. Till date, the biosynthetic pathways concerning various classes of TA biosynthesis are inadequately elucidated. Not many enzymes have been characterized as yet, several corresponding genes are unidentified, rate-limiting steps have not yet been detected and the correlations of metabolic flux still remains unknown. Recently accessible "omics" approaches can contribute it the interpretation of these pathways.

Appendix I List of institutes engaged in Research on *A. belladonna*

Institute Name	Specialization Research Activities	Contact Information and Website
ServizioBiodiversità e Ricercascientifica, Majella Seed Bank, Parco Nazionaledella Majella	Conservation of plant biodiversity using in situ-ex situ conservation strategies	Ente Parco Nazionale Della Majella, Via Badia, 28, 67039 Sulmona AQ, Italy https://www.parcomajella.it/La-Banca-del-Germoplasma.htm
Michigan State University	Biochemistry, Molecular Biology, Horticulture	Michigan State University, East Lansing, MI 48824, USA https://msu.edu/
Centre for Medicinal Plants Research (CMPR)	Phytochemistry, Pharmacognosy	CMPR, AVS Square, Changuvetty, Kottakkal Malappuram, Kerala, India https://cmpr-avs.in/
Indian Council of Agricultural Research -National Bureau of Plant Genetic Resources (ICAR-NBPGR)	Genetic Diversity, Molecular Study	ICAR-NBPGR, Pusa Campus, New Delhi, Delhi, India http://www.nbpgr.ernet.in/

(continued)

(continued)

Institute Name	Specialization Research Activities	Contact Information and Website
Department of Horticultural Sciences, TarbiatModares University	Horticultural Sciences	Department of Horticultural Sciences, Agriculture Faculty, TarbiatModares University, Tehran, Iran https://www.modares.ac.ir/en/
School of Life Sciences, Southwest University	Plant Resource Conservation and Germplasm Innovation	School of Life Sciences, Southwest University, Chongqing 400715, China http://www.swu.edu.cn/

References

Abdel-Hady MS, Okasha EM, Soliman SSA, Talaat M (2008) Effect of gamma radiation and gibberellic acid on germination and alkaloid production in *Atropa belladonna*. Aust J Basic

Ahmed ZS, Salim AM, Alsoufi ASM (2021) In vitro micropropagation of *Atropa belladonna* L. using plant growth regulators. Plant Cell Biotech Mol Bio 22(3–4):169–174

Ahuja PS, Rahman LU, Bhargava SC, Banerjee S (1993) Regeneration of intergeneric somatic hybrid plants between *Atropa belladonna* L. and *Hyoscyamus muticus* L. Plant Sci 92:91–98

Al-Ashaal HA, Aboutabl ME, Maklad YA, El-Beih AA (2013) Tropane alkaloids of *Atropa belladonna* L.: *Invitro* production and pharmacological profile. Egypt Pharm J 12:130–135.https://doi.org/10.4103/1687-4315.124012

Anonymous (1948) The wealth of India: a dictionary of Indian raw materials and industrial products (vol I, pp 135–136). New Delhi: Council of Scientific and Industrial Research (CSIR). App Sci 2(3):401–405

Atropa Belladonna Products (2022) https://www.google.com/search?q=atropa+belladonna+products+2022&tbm=isch&ved=2ahUKEwjcjtnqmfD4AhWGjtgFHWMnDs4Q2-cCegQIABAA&oq=atropa+belladonna+products+2022&gs_lcp=CgNpbWcQAzoECCMQJ1DuA1jXrgJgg7MCaABwAHgAgAGmBIgB8guSAQkwLjMuMi41LTGYAQCgAQGqAQtnd3Mtd2l6LWltZ8ABAQ&sclient=img&ei=ScLLYpzGH4ad4t4P48648Aw&bih=647&biw=1362&rlz=1C1UEAD_enIN990IN990

Babiychuk E, Kushnir S, Gleba YY (1992) Spontaneous extensive chromosome elimination in somatic hybrids between somatically congruent species *Nicotiana tabacum* L. and *Atropa belladonna* L. Theor Appl Genet 84(1–2):87–91. https://doi.org/10.1007/BF00223985

Bajaj YP, Simola LK (1991) *Atropa belladonna* L.: *In Vitro* culture, regeneration of plants, cryopreservation, and the production of tropane alkaloids. In: Medicinal and aromatic plants

Bajaj YPS (1977) Survival of *Atropa* and *Nicotiana* pollen embryos frozen at -196°C. Curr Sci

Bajaj YPS (1978) Effect of super-low temperature on excised anthers and pollen-embryos of *Atropa*, *Nicotiana*, and *Petunia*. Phytomorphology 28:171–178

Bajaj YPS (1995) Cryopreservation of somatic embryos. In: Bajaj YPS (eds) Somatic embryogenesis and synthetic seed I. Biotechnology in agriculture and forestry, vol 30

Baralle FE, Gros EG (1969) Biosynthesis of cuscohygrine in *Atropa belladonna* from sodium acetate-2-14C. Phytochemistry 8(5):853–855

Bedewitz MA, Gongora-Castillo E, Uebler JB, Gonzales-Vigil E, Wiegert-Rininger KE et al (2014) A root-expressed L-phenylalanine:4-hydroxyphenylpyruvate aminotransferase is required for tropane alkaloid biosynthesis in *Atropa belladonna*. Plant Cell 26:3745–3762. https://doi.org/10.1105/tpc.114.130534

Cicek SS, Malek S, Zidorn C (2020) Flavonoids from *Atropa belladonna* (Solanaceae) leaves revisited. Biochem Syst Ecol 88:103990

DeSimone R, Margarucci L, De Feo V (2008) Tropane alkaloids: an overview. Pharmacologyonline 1:70–89

Di Cecco V, Di Santo M, Di Musciano M, Manzi A, Di Cecco M et al (2020) The Majella National Park: a case study for the conservation of plant biodiversity in the Italian Apennines. Ital Bot 10:1–24. https://doi.org/10.3897/italianbotanist.10.52952

Drager B, Hashimoto T, Yamada Y (1988) Purification and characterization of pseudotropine forming tropinone reductase from *Hyoscyamus niger* root cultures. Agr Biol Chem 52:2663–2667. https://doi.org/10.1080/00021369.1988.10869106

Elshorbagy MI, Ibrahim SM, Abo El-Seoud KA, Abd El-Maksoud AI (2018) Tropane alkaloids production from callus culture of *Atropa belladonna* L. as affected by elicitors and precursor feeding. Int Res J Pharm 9(7):116–125. https://doi.org/10.7897/2230-8407.097135

Fattahi F, Shojaeiyan A, Palazon J, Moyano E, Torras Claveria L (2021) Methyl β cyclodextrin and coronatine as new elicitors of tropane alkaloid biosynthesis in *Atropa acuminata* and *Atropa belladonna* hairy root cultures. Physiol Plant 172(4):2098–2111. https://doi.org/10.1111/ppl.13444

Gleba YY, Hinnisdaels S, Sidorov VA, Kaleda VA, Parokonny AS et al (1988) Intergeneric asymmetric hybrids between *Nicotiana plumbaginifolia* and *Atropa belladonna* obtained by "gamma-fusion." Theor Appl Genetics 76:760–766. https://doi.org/10.1007/BF00303523

Gong H, He P, Lan X, Zeng L, Liao Z (2022) Biotechnological approaches on engineering medicinal tropane alkaloid production in plants. Front Plant Sci 13:924413. https://doi.org/10.3389/fpls.2022.924413

Griffin WJ, Lin GD (2000) Chemotaxonomy and geographical distribution of tropane alkaloids. Phytochemistry 53:623–637. https://doi.org/10.1016/s0031-9422(99)00475-6

Grynkiewicz G, Gadzikowska M (2008) Tropane alkaloids as medicinally useful natural products and their synthetic derivatives as new drugs. Pharmacol Rep 60:439–463

Hana AA (2021) Study the effects of mechanical, chemical, growth regulators and irradiation on the germination of *Atropa belladonna in-vitro* conditions. J Phar Bio Sci 16(5):10–19

Hank H, Szoke É, Tóth K et al (2004) Investigation of tropane alkaloids in genetically transformed *Atropa belladonna* L. cultures. Chromatographia 60:S55–S59. https://doi.org/10.1365/s10337-004-0240-x

Hashimoto T, Matsuda J, Yamada Y (1993) Two-step epoxidation of hyoscyamine to scopolamine is catalyzed by bifunctional hyoscyamine 6 beta-hydroxylase. FEBS Lett 329:35–39.https://doi.org/10.1016/0014-5793(93)80187-y

Hashimoto T, Yamada Y (1986) Hyoscyamine 6β-hydroxylase, a 2-oxoglutarate-dependent dioxygenase, in alkaloid-producing root cultures. Plant Physiol 81:619–625. https://doi.org/10.1104/pp.81.2.619

Hashimoto T, Yukimune Y, Yamada Y (1989) Putrescine and putrescine N-methyltransferase in the biosynthesis of tropane alkaloids in cultured roots of *Hyoscyamus albus*. Planta 178:123–130. https://doi.org/10.1007/BF00392535

Hedayati A, Hemmaty S, Nourozi E, Amirsadeghi A (2021) Effect of yeast extract on h6h gene expression and tropane alkaloids production in *Atropa belladonna* L. hairy roots. Russ J Plant Physiol 68:102–109. https://doi.org/10.1134/S1021443721010088

Hibi N, Fujita T, Hatano M, Hashimoto T, Yamada Y (1992) Putrescine N-methyltransferase in cultured roots of *Hyoscyamus albus*. Plant Physiol 100:826–835. https://doi.org/10.1104/pp.100.2.826

Hilton MG, Wilson PD (1995) Growth and the uptake of sucrose and mineral ions by transformed root cultures of *Datura stramonium, Datura candida* × *aurea*, *Datura wrightii*, *Hyoscyamus muticus* and *Atropa belladonna*. Planta Med 61(04):345–350. https://doi.org/10.1055/s-2006-958097. III 1991, pp 1–23. Springer, Heidelberg

Jaziri M, Yoshimatsu K, Homes J, Shimomura K (1994) Traits of transgenic *Atropa belladonna* doubly transformed with different *Agrobacterium rhizogenes* strains. Plant Cell Tiss Organ Cult 38:257–262. https://doi.org/10.1007/BF00033885

Jirschitzka J, Dolke F, D'Auria JC (2013) Increasing the pace of new discoveries in tropane alkaloid biosynthesis. Adv Bot Res 68:39–72. https://doi.org/10.1016/B978-0-12-408061-4.00002-X

Kaul BL, Chaudhary BK (1975) The effect of chemical mutagens on quantitative characters in *Atropa belladonna*. Planta Med 27(4):337–342. https://doi.org/10.1055/s-0028-1097811

Kaul BL, Choudhary DK, Kachroo P (eds) (1974) Radiation induced polygenic mutations in *Atropa belladona*. Hindustan Publishing Corporation, India

Koelen KJ, Gross GG (1982) Partial purification and properties of tropine dehydrogenase from root cultures of *Datura stramonium*. Planta Med 44:227–230. https://doi.org/10.1055/s-2007-971454

Kohnen-Johannsen KL, Kayser O (2019) Tropane alkaloids: chemistry, pharmacology, biosynthesis and production. Molecules 24(4):796. https://doi.org/10.3390/molecules24040796

Krumbiegel G, Schieder O (1979) Selection of somatic hybrids after fusion of protoplasts from *Datura innoxia* Mill. and *Atropa belladonna* L. Planta 145(4):371–375. https://doi.org/10.1007/BF00388363

Krumbiegel G, Schieder O (1981) Comparison of somatic and sexual incompatibility between Daturainnoxia and *Atropa belladonna*. Planta 153:466–470. https://doi.org/10.1007/BF00394988

Kushnir S, Babiychuk E, Bannikova M, Momot V, Komarnitsky I, Cherep N, Gleba Y (1991) Nucleo-cytoplasmic incompatibility in cybrid plants possessing an *Atropa* genome and a Nicotiana plastome. Mol Gen Genet 225(2):225–230. https://doi.org/10.1007/BF00269852

Kushnir SG, Shlumukov LR, Pogrebnyak NJ, Berger S, Gleba Y (1987) Functional cybrid plants possessing a *Nicotiana* genome and an *Atropa* plastome. Mol Gen Genet 209:159–163. https://doi.org/10.1007/BF00329852

Lee MR (2007) Solaneceae IV: *Atropa belladonna*, deadly nightshade. J R Coll Physicians Edinb 37:77–84

Li J, Chen M, Qiu F, Qin B, Liu W et al (2014) Reference gene selection for gene expression studies using quantitative real-time PCR normalization in *Atropa belladonna*. Plant MolBiol Rep 32:1002–1014. https://doi.org/10.1007/s11105-014-0701-9

Liu X, Yang C, Chen M, Li M, Liao Z, Tang K (2010) Promoting scopolamine accumulation in transgenic plants of *Atropa belladonna* generated from hairy roots with over expression of pmt and h6h gene. J Med Plants Res 4(17):1708–1713. https://doi.org/10.5897/JMPR10.269

Majid R, Latief A (2012) Evaluation of anti-inflammatory potential of *Atropa acuminata* in carrageenan induced inflammation in rats. J Med Plant Res 6(43):5586–5592. https://doi.org/10.5897/JMPR12-160

Matsuda J, Okabe S, Hashimoto T, Yamada Y (1991) Molecular cloning of hyoscyamine 6β-hydroxylase, a 2-oxoglutaratedependent dioxygenase, from cultured roots of *Hyoscyamus niger*. J Biol Chem 266:9460–9464

Mera N, Takayama S (2012) Effect of medium components and shear conditions on the formation and growth of adventitious bud derived from hairy roots of *Atropa belladonna* L. Environ Control Boil 50(4):393–406. https://doi.org/10.2525/ecb.50.393

Miller FA (1914) The propagation of medicinal plants. Bull Torrey Botanical Club 41(2):105–129

Mohanraj VJ, Chen Y (2006) Nanoparticles—a review. Trop J Pharm Res 5(1):561–573. https://doi.org/10.4314/tjpr.v5i1.14634

Nabil A, EL-Shaer FA, Abd El-Hady MS, Nassar M (2022) Molecular genetic studies on *Atropa belladonna* L. under the effect of laser radiation and salicylic acid on Atropine, Scopolamine and Hyoscyamine by tissue culture techniques. Afr J Biol Sci 18(2):1–17. https://doi.org/10.21608/AJBS.2022.126803.1027

Oksman-Caldentey KM, Arroo R (2000) Regulation of tropane alkaloid metabolismin plants and plant cell cultures. In: Verpoorte R, Alfermann AW (eds) Metabolic engineering of plant secondary metabolism. Kluwer Academic Publishers, Dortrecht, pp 253–281

Parks HM, Cinelli MA, Bedewitz MA, Grabar JM, Hurney SM, Walker KD, Daniel Jones A, Barry CS (2023) Redirecting tropane alkaloid metabolism reveals pyrrolidine alkaloid diversity in *Atropa belladonna*. New Phytol 237:1810–1825. https://doi.org/10.1111/nph.18651

Patterson S, O'Hagan D (2002) Biosynthetic studies on the tropane alkaloid hyoscyamine in *Datura stramonium*: Hyoscyamine is stable to *in vivo* oxidation and is not derived from littorinevia a vicinal interchange process. Phytochemistry 61:323–329. https://doi.org/10.1016/S0031-9422(02)00200-5

Pigatto AG, Blanco CC, Mentz LA, Soares GL (2015) Tropane alkaloids and calystegines as chemotaxonomic markers in the Solanaceae. An Acad Bras Cienc 87:2139–2149. https://doi.org/10.1590/0001-3765201520140231

POWO (2022) Plants of the World Online, Facilitated by the Royal Botanic Gardens, Kew. Published on the Internet. http://www.plantsoftheworldonline.org/Retrieved. 04 July 2022. https://powo.science.kew.org/taxon/urn:lsid:ipni.org:names:687109-1

Qiu F, Yang C, Yuan L, Xiang D, Lan X et al (2018) A phenylpyruvic acid reductase is required for biosynthesis of tropane alkaloids. Org Lett 20(24):7807–7810. https://doi.org/10.1021/acs.orglett.8b03236

Qiu F, Zeng J, Wang J, Huang JP, Zhou W (2020) Functional genomics analysis reveals two novel genes required for littorine biosynthesis. New Phytol 225(5):1906–1914. https://doi.org/10.1111/nph.16317

Radha R, Decruse WS, Krishna P (2012) Plant cryopreservation. In: Current frontiers in cryopreservation. IntechOpen. https://doi.org/10.5772/35919

Richter U, Rothe G, Fabian AK, Rahfeld B, Dräger B (2005) Overexpression of tropinone reductases alters alkaloid composition in *Atropa belladonna* root cultures. J Exp Bot 56(412):645–652. https://doi.org/10.1093/jxb/eri067

Rothe G, Hachiya A, Yamada Y, Hashimoto T, Dräger B (2003) Alkaloids in plants and root cultures of *Atropa belladonna* overexpressing putrescine N-methyltransferase. J Exp Bot 54(390):2065–2070. https://doi.org/10.1093/jxb/erg227

Sadre R, Anthony TM, Graba JM et al (2022) Metabolomics-guided discovery of cytochrome P450s involved in pseudotropine-dependent biosynthesis of modified tropane alkaloids. Nat Commun 13:3832. https://doi.org/10.1038/s41467-022-31653-1

Sato F, Hashimoto T, Hachiya A, Tamura KI, Choi KB (2001) Metabolic engineering of plant alkaloid biosynthesis. Proc Natl Acad Sci USA 98(1):367–372. https://doi.org/10.1073/pnas.98.1.367

Schneider A (1916) Cultivation of belladonna in California. Bulletin No. 275, University of California Press, Berkeley

Shahzad A, Shaheen A, Kozgar MI, Sahai A, Sharma S (2013) Phytoactive compounds from *in vitro* derived tissues. In: Shahid M, Shahzad A, Malik A, Sahai A (eds) Recent trends in biotechnology and therapeutic applications of medicinal plants. Springer, Dordrecht. https://doi.org/10.1007/978-94-007-6603-7_1. Springer, Berlin, Heidelberg. https://doi.org/10.1007/978-3-662-03091-2_15

Stockberger WW (1915) Drug plants under cultivation. Farmers Bulletin, n° 663. Washington, D.C., US Department of Agriculture

Stockberger WW (1919) Commercial drug growing in the United States in 1918. J Am Pharma Assoc 8(1):809

Suzuki K, Yamada Y, Hashimoto T (1999) Expression of Atropa belladonna putrescine N-methyltransferase gene in root pericycle. Plant Cell Physiol 40:289–297. https://doi.org/10.1093/oxfordjournals.pcp.a029540

Tian H, Ghorbanpour M, Kariman K (2018) Manganese oxide nanoparticle-induced changes in growth, redox reactions and elicitation of antioxidant metabolites in deadly nightshade (*Atropa belladonna* L.). Ind Crops Prod 126:403–414. https://doi.org/10.1016/j.indcrop.2018.10.042

Touno K, Yoshimatsu K, Shimomura K (2006) Characteristics of *Atropa belladonna* hairy roots cryopreserved by vitrification method. Cryo Lett 27(2):65–72

True RH (1905) Progress in drug-plant cultivation

Volkov RA, Panchuk II, Borisjuk NV, Baranska MH, Maluszynska J, Hemleben V (2017) Evolutional dynamics of 45S and 5S ribosomal DNA in ancient allohexaploid *Atropa belladonna*. BMC Plant Biol 17(21):1–15. https://doi.org/10.1186/s12870-017-0978-6

Wang X, Chen M, Yang C, Liu X, Zhang L, Lan X, Tang K, Liao Z (2011) Enhancing the scopolamine production in transgenic plants of *Atropa belladonna* by overexpressing pmt and h6h genes. Physiol Plant 143(4):309–315. https://doi.org/10.1111/j.13993054.2011.01506.x

Xia KA, Xiaoqiang LA, Qiaozhuo ZA, Wei QA, Jianjun GB et al (2016) Promoting scopolamine biosynthesis in transgenic *Atropa belladonna* plants with pmt and h6h overexpression under field conditions. Plant Physiol Biochem 106:46–54. https://doi.org/10.1016/j.plaphy.2016.04.034

Yun DJ, Hashimoto T, Yamada Y (1992) Metabolic engineering of medicinal plants: transgenic *Atropa belladonna* with an improved alkaloid composition. Proc Natl Acad Sci USA 89:11799–11803. https://doi.org/10.1073/pnas.89.24.11799

Zeng L, Zhang Q, Jiang C, Zheng Y, Zuo Y et al (2021) Development of *Atropa belladonna* L. plants with high-yield hyoscyamine and without its derivatives using the CRISPR/Cas9 system. Int J MolSci 22(4):1731. https://doi.org/10.3390/ijms22041731

Zhang F, Qiu F, Zeng J, Xu Z, Tang Y, Zhao T, Gou Y, Su F, Wang S, Sun X, Xue Z, Wang W, Yang C, Zeng L, Lan X, Chen M, Zhou J, Liao Z (2023) Revealing evolution of tropane alkaloid biosynthesis by analyzing two genomes in the Solanaceae family. Nat Commun 14(1):1446. https://doi.org/10.1038/s41467-023-37133-4

Zhang L, Kai GY, Lu BB, Zhang HM, Tang KX et al (2005) Metabolic engineering of tropane alkaloid biosynthesis in plants. J Integr Plant Biol 47(2):136–143. https://doi.org/10.1111/j.1744-7909.2005.00024.x

Zhang Q, Liang M, Zeng J, Yang C, Qin J (2022) Engineering tropane alkaloid production and glyphosate resistance by overexpressing AbCaM1 and G2-EPSPS in *Atropa belladonna*. Metab Eng 72:237–246. https://doi.org/10.1016/j.ymben.2022.03.014

Zhao T, Li S, Wang J, Zhou Q, Yang C (2020) Engineering tropane alkaloid production based on metabolic characterization of ornithine decarboxylase in *Atropa belladonna*. ACS Synth Biol 9(2):437–448. https://doi.org/10.1021/acssynbio.9b00461

Chapter 9
Advances in Molecular Breeding for *Ginkgo biloba*

Maha M. Elshamy, Mohamed A. Abdein, Haifa A. S. Alhaithloul, Suliman M. S. Alghanem, and Mai M. Wahba

Abstract The robust *Ginkgo biloba* plant, a member of the Ginkgopsida family and a native to China, is referred to as a "living fossil" by the botanical community. Ginkgo has become widely cultivated around the world because of its high decorative and therapeutic benefits. Its high level of environmental adaptation, robust resistance to air pollution, and essentially total resistance to diseases and pests made it popular. It is a single member of the ginkgo family is the long-lived dioecious gymnosperm known as ginkgo. Because of its strong vitality, early origin, and great adaptability, ginkgo is highly prized for scientific studies on the development and genesis of this species. Ginkgo has significant economic importance added to being a significant ornamental and timber tree. Significant antioxidant, insecticidal, antibacterial, and other biological properties are present in ginkgo fruit. The ginkgo leaves are listed in the Chinese Pharmacopoeia as it is rich in flavonoids and terpenoids. The leaf extract has excellent therapeutic potential, scavenging free radicals, lowering oxidative stress, minimizing nerve damage, and lowering platelet aggregation, Inflammation-reducing, cancer-fighting, anti-aging activities and in cardiovascular diseases treatment. Therefore, it is crucial to do genetic investigations on the development, growth, and endurance of ginkgo. Due to the great survivability of this taxa and the replication of the genes

M. M. Elshamy · M. M. Wahba
Botany Department, Faculty of Science, Mansoura University, Mansoura, Egypt
e-mail: mahammy@mans.edu.eg

M. M. Wahba
e-mail: maiwahba@mans.edu.eg

M. A. Abdein (✉)
Seeds Development Department, El-Nada Misr Scientific Research and Development Projects, Turrell, Mansoura, Egypt
e-mail: abdeingene@yahoo.com

H. A. S. Alhaithloul
Biology Department, College of Science, Jouf University, Sakaka, Saudi Arabia
e-mail: haifasakit@ju.edu.sa

S. M. S. Alghanem
Department of Biology, College of Science, Qassim University, Buraydah, Saudi Arabia
e-mail: su.alghanem@qu.edu.sa

J. M. Al-Khayri et al. (eds.), *Biodiversity and Genetic Improvement of Medicinal and Aromatic Plants I*, Advances in Plant Breeding Strategies 9,
https://doi.org/10.1007/978-3-031-81846-2_9

causing stress and resistance, it is perfect for use in urban greenery designs. An excellent framework for analyzing the development and growth of ginkgo at the entire stage of genome will be made available by obtaining the whole sequencing of the ginkgo genome (chromosomes (2n = 24). In this chapter we discussed its background and historical context, botanical description, economic and health significance, cultivation requirements, photochemistry, biodiversity and germplasm conservation, cytogenetics, objectives and steps of breeding plans, and conventional breeding techniques. We also present the recent genetic engineering and marker-assisted breeding studies for plant breeding.

Keywords *Ginkgo* · Phytochemistry · Biodiversity · Conservation · Cytogenetics · Breeding techniques

9.1 Introduction

A medium-sized tree of the Ginkgo family and Ginkgo genus, *Ginkgo* (*Ginkgo biloba*) originated in China through Japan and Korea, the eighteenth century saw the diffusion of Chinese culture to Europe (McElwain 2013). It is the only surviving species of the gymnosperm *G. biloba*, which is why it is known as the "living fossil" of the plant world. *G. biloba* has a certain value in food, traditional medicine, and ornamental, and it is now widely cultivated worldwide. *G. biloba* has strong soil adaptability, strong resistance to harsh environments, and is not easily infested by pests and diseases. According to Singh et al. (2008), ginkgo trees have a long lifespan and are extremely resilient to pests, viral and bacterial diseases, and pollution of air. Due to the elevated survivability of this species and the genes replication causing stress and resistance, it is perfect for use in urban greenery designs. The leaves are also beautiful in form and golden in color during the deciduous period and are used as urban street trees or landscape trees. It is now regularly planted on campuses of university, in gardens and parks, or beside roadways and asphalt road all over the world. (Lin et al. 2022). The ginkgo is a significant commercial tree species that is widely utilized in daily life and production. Ginkgo biloba extract (GBE), which is a key source of economic value for ginkgo and contains flavonoids, is frequently utilized in clinical therapy (Zhao et al. 2019a, b). Ginkgo flavonoid content is influenced by a variety of elements, including fertilization (Wu et al. 2020a, b), salt, temperature, light (Xu et al. 2020), etc. To comprehend the regulatory network of ginkgo flavonoids, numerous transcription factors (TFs) and regulatory genes have been found (Ying et al. 2020).

In China, the *G. biloba* products are also used in traditional medicine. *G. biloba* leaves can be extracted with Ginkgetina and ginkgolide, which inhibit platelet aggregation and are anti-thrombotic (Liu et al. 2020). They are utilized, among other things, in the formulations employed to remedy cardiovascular and cerebral vascular illnesses. It was proven to be effective in treating cardiovascular disorders in the 1960s. *G. biloba* leaf extract is among the most widely used herbal remedies and dietary supplements in both Europe and the US. Outstanding therapeutic potential is shared by the decrease of platelet aggregation, reduction of oxidative stress, minimization of neuronal damage, and free radical scavenging. anti-inflammatory, anti-cancer Clinical studies support its positive effects in treating conditions of the central nervous system, including, among others, Alzheimer's disease, and cognitive impairments. Fighting, and anti-aging activities are also present (Toghueo 2020).

Furthermore, people in many different countries find inspiration from these trees for their art and religion (Chen et al. 2021). Around East Asia, antique examples are frequently discovered around temples, historic settlements, or close to streams. The ginkgo tree, often known as the national tree of China, is revered as a cultural representation of serenity and hope in that country (Lin et al. 2022). In the study of Zhao et al. (2019a, b), they identified the evolutionary history of this species by resequencing 545 ginkgo genomes. They utilized genomic information from the nuclear and chloroplast to investigate outstanding issues regarding the evolution of ginkgo. They specifically pinpoint four ancient genetic subgroups of Chinese ginkgo populations as well as three glacial refugia. Their demographic analyses show that population cycles have occurred repeatedly in response to Pleistocene climatic change besides decline glacial hybridization of remnant populations in northern China.

9.2 Taxonomy

Ginkgoaleans remained to be a minor to moderately important partner in mid- to high-latitude palaeofloras from the mid-Mesozoic to the mid-Cenozoic (Vakhrameev 1991). Long-standing disagreements existed regarding the taxonomic status of the ginkgo (syn. *Salisburia adiantifolia*). *Ginkgo macrophylla*, *Salisburia biloba* (L.) Hoffmanns., and small *Salisburia ginkgo*, K. Koch *Salisburia macrophylla*, *Rich Pterophyllus salisburiensis*, Reyn. J. Nelson, *Gingko biloba f. aurea G. biloba f. fastigiata* Beissn. (A. Henry) *G. biloba f. laciniata* Rehder (Carrière) *G. biloba f. microsperma*, Beissn. *G. biloba f. parvifolia* in Sugim. *G. biloba f. pendula* (Sugim) Van Geert *G. biloba f. variegata* (Beissn.) Carrière *G. biloba* variety aurea (Beissn.) *G. biloba* variety epiphylla, A. Henry *G. biloba* var. *fastigiata* Makino *G. biloba* var. *laciniata (Carrière)*, A. Henry *biloba var. pendula* (Van Geert) Carrière, *G. biloba* var. *variegata* (*Carrière*) *S. adiantifolia var. laciniata*, often known as *Carrière*. Soon after the *Ginkgo* was introduced to Europe, Carl Linnaeus (1771) (as Gertz and Kiefer 2004 mentioned) provided a scientific description of the tree. In contrast to pines, Renzaglia and Garbary (2001) noted the discharge of swimming sperm from pollen tubes after fertilization. Numerous methods have been used to establish the evolutionary relationship between Ginkgo and other gymnosperms. In terms of pollen shape, reproductive habits, and embryology, ginkgo was compared more favorably to cycads. (Norstog et al. 2004) (Fig. 9.1).

Fig. 9.1 *Ginkgo biloba* Linnaeus, Mant. Pl. 2: 313. 1771

Public Name	Gingko, maidenhair
Botanical Name	*Gingko biloba*
Clade	*Ginkgoalean*
Order	Ginkgoales
Family	Ginkgoaceae
Life form	Tree
Matured Dimensions	30–40 ft. wide, 50–80 ft. tall
Exposition to Sun	Full, partial
Kind of Soil	well-drained, Sandy
pH of Soil	Acidic, neutral, alkaline
Flowering Period	Spring
Color of Flowers	Green
Hardiness Zones	4–9 (USDA)
Native Rigon	China

9.3 Origin and Dispensation

G. biloba is cultivated throughout much of Europe, America, as well as the temperate zones of India, New Zealand, Argentina and native to China, Korea, and Japan (Fig. 9.2.). There is just one wild *G. biloba* tree left in the world, and it is in China's Zhejiang Province at a height of 1506 m above sea level (Singh et al. 2008). Due to the declining number of trees in their natural populations, *G. biloba* has been classified as an endangered species by the International Union for Conservation of Nature and Natural Resources (Belwal et al. 2018). Ginkgophytes appear to have reached their peak in terms of systematic diversity and phytogeographical distribution during the late Mesozoic and early Tertiary when they had a circumpolar distribution in the Northern Hemisphere and had extended into many areas of the Southern Hemisphere. The Miocene to Quaternary transition of the Tertiary reduction appears to have been followed by an Oligocene fall (Royer 2003). Natural woods in the Tianmu Mountains (China) contain only untamed trees. Approximately 28–350 million years ago, during the Upper Carboniferous period of the Paleozoic. About 140 million years ago, between the Late Jurassic and the Early Cretaceous, *G. biloba* plants had a time of great flourishing. Nearly all over the world, there are more than 20 genera and 150 species. Currently, *G. biloba* has gained international recognition as a "living fossil" and has once more broadened its geographic distribution (Royer 2003).

One of the oldest gymnosperms on the globe is *G. biloba*. During the Jurassic epoch, species belonging to this genus were discovered all over the globe. There are believed to have been 12 different Ginkgo species based on fossil data. The Tertiary glacial epochs resulted in the extinction of the bulk of species in this genus. However, due to the warm environment, *G. biloba* is thought to have persisted in southern China.

Numerous Ginkgo sites' sedimentological characteristics show that the tree grew largely in disturbed habitats near stream borders. Typically, these areas receive enough water. In these environments, open canopy woods are typical (Carr 1998). Two of the Ginkgo's more frequent flowering partners are the two species of Platanus divided by *P. guillelmae* and *P. raynoldsii* and *Cercidiphyllum genetrix*. *C. japonicum* Siebold and Zuccarini, C. genetrix's nearest living relative (NLR), is now present in

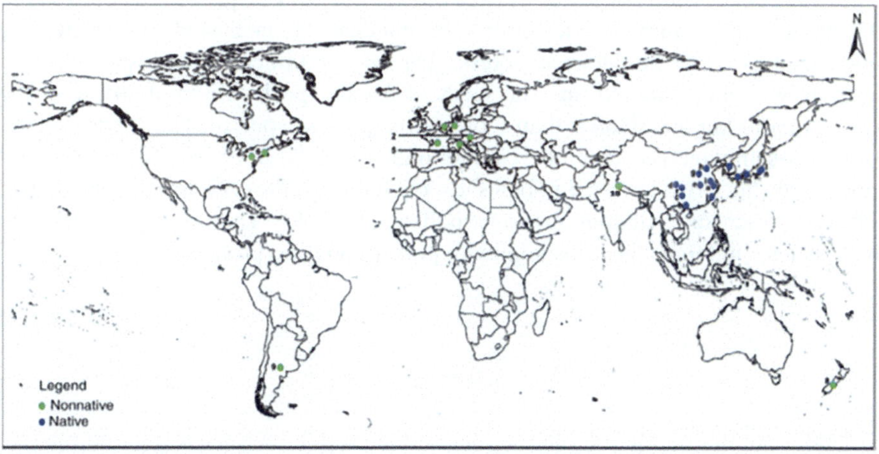

Fig. 9.2 Geographical distribution of *Ginkgo biloba* exhibiting both native and non-native occurrences (Belwal et al. 2018)

the wild in China and Japan. According to Seiwa and Kikuzawa (1996), it is an initiate species that grows quickly, frequently reproduces via sprouting. It is significant to note that *C. japonicum* may also be discovered in other highly degraded settings and on scree slopes at mid-elevations (1450–3500 m) in south-central Chinese mountains. Despite the rarity of surviving plant remains in alluvial fan facies (Isle of Mull, Scotland), ginkgo was discovered in fine-grained sediments inside an alluvial fan deposit (Tang and Ohsawa 1997).

9.4 Fossils Phenology of *Ginkgo biloba*

Ginkgophytes are among the strangest and most intriguing seed plants due to the extensive fossil record for a clade that is monotypic in the current flora and may not have natural populations in the wild (Li 1956). The systematic interest of ginkgophytes and their ostensibly important function in the development of seed plants are enhanced by a variety of apparently primitive characteristics. Ginkgo was one of the three main seed plant groupings according to Meyen (1987) current cladistic analysis Ginkgo should be ranked below all other live seed-tree species, or almost all of them. That it has undergone minimal evolutionary change since the earliest seeds plants (Rothwell and Serbet 1994). The bulk of the oldest leaf fossils from Permo-Carboniferous strata are still remarkably like the living *G. biloba* after extensive dissection. The Upper Cretaceous (Maastrichtian) assemblage found with significant dinosaur bones in Alberta, Canada, is a great example (Archangelsky and Cuneo 1990). These fossils are found in an assortment of locations and show a wide extant of organs like those seen in *Ginkgo.* A sandstone matrix preserves the most leaves. (Fig. 9.3a, b). The venation is dichotomous, just like in *G. biloba* in living form.

Two varieties of compacted seeds were found in Alberta's Upper Cretaceous strata. One has a rounded apex, slender sarcotesta, and chalaza that is slightly cordate. The other, seen in Fig. 9.3c, lacks sarcotesta, has two prominent sclerotestal ribs, and tapers to a slightly pointed tip. The seeds of fossile species resemble those of the live *G. biloba* but are smaller than those of the present species. The Maastrichtian Alberta assemblage contains one *Ginkgo* pollen cone (Fig. 9.3d). Terete attached sporophylls have two terminal pollen sacs. Even though the cone has degenerated, a dehisced pollen sac still contains some pollen (Fig. 9.3e). The fossil grain, like the live species (Fig. 9.3f), is ellipsoidal with pointy ends and prolate.

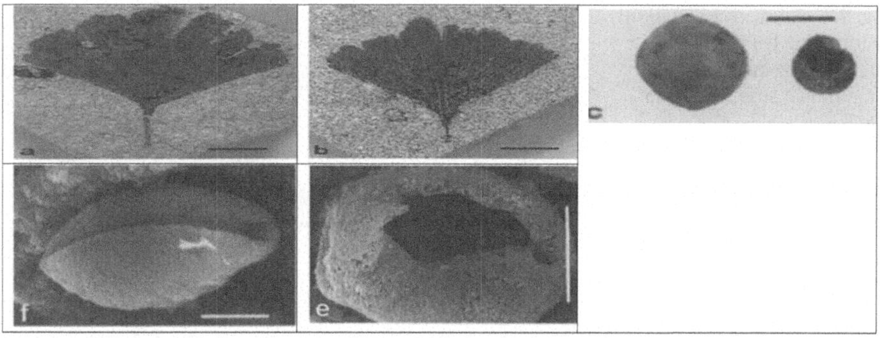

Fig. 9.3 **a–f** Ginkgo fossils from Central Alberta, Canada's Upper Cretaceous strata. **a**, **b** Pressed leaves showing the diversity of distal border dissections. **c** Ginkgo seeds that have the sarcotesta peeled out to show the seed coat. **d** Fossil pollen cone with one intact sporophyll and two terminal pollen sacs and sporophyll bases arranged in a helical pattern. **e** inside of a dehisced pollen sac from a cone, pollen grain. **d** with the distal sulcus facing out. **f** Using the pollen of living *G. biloba* to contrast it with prehistoric pollen. **e** Granular orientation like that of the fossil. According to Rothwell and Holt (1997)

9.5 *Ginkgo biloba* Morphology and Anatomy

9.5.1 Morphology

The broadleaf, deciduous ginkgo tree, often known as maidenhair, has a sculptural pyramid shape. This old tree is a conifer, losing its leaves in the winter. The typical tree is 40 m tall, with a trunk that may reach a diameter of 4 m, and it has longitudinally fissured bark that is light grey or greyish brown, especially on older trees; short branchlets that are blackish grey and have compact, irregularly elliptic leaf scars; long branchlets that are pale brownish yellow during the winter; and leaves that are primarily grey in the end with internodes that are 1–1.5 cm long. Petiole of leaves (3-)5-8(-10) cm, pale green blade (up to 13 8(-15) cm on young trees, but normally 5–8 cm broad), split into 2 lobes, each further divided; short branchlets with undulate distal and border notched tip. Pollen cones have a boat-like form, are ivory in colour, and range in size from 1.2 to 2.2 cm. The seeds are elliptic, narrowly obovoid, ovoid, or subglobose, measuring 2.5–3.5 1.6–2.2 cm; the sarcotesta is yellow or orange-yellow glaucous, white, and has two or three longitudinal ridges; and the endotesta is a light reddish-brown hue. When ripe, the sarcotesta has a rotten smell. In March to April, pollination takes place, and in September to October, seeds mature.

In Ginkgo's vegetative organs, Hara in 1997 found numerous rudimentary traits of seed plants. The fundamental structure of a ginkgo leaf is mostly made up of a cruciate-shaped dichotomy rather than a flabellate one, and this development lasts at least as far as the second bifurcation of the lamina, according to some primitive vegetative traits that have been seen. The tunica and corpus's cell structure is hidden by the shoot apex of the ginkgo tree.

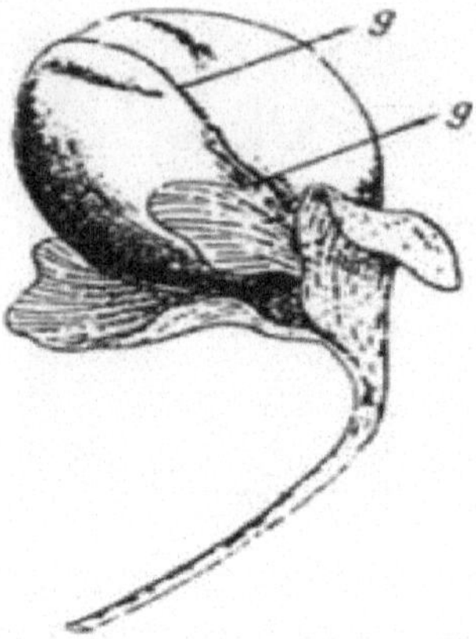

Fig. 9.4 A leaf of vegetation bearing ripe fruit. g, the pulpy part of the seed's longitudinal striation (Hara 1997)

Sik and Soh investigated how the vascular system formed in mature Ginkgo embryos and seedlings back in 1983. In contrast to the protoxylem, which develops sporadicy in the cotyledons and irregularly in the upper hypocotyl, the procambium and protophloem establish a continuous system that encompasses the epicotyl-hypocotyl root axis and cotyledons in the adult embryo. Arnott did note the discovery of vein unions, noting that 9.9% of them contained one or more anastomoses. 13.4% of the leaves on long branches and 8.2% of the leaves on dwarf shoots both had anastomoses. Ginkgo leaves are thought to be the most primordial of all seed plant leaves.

According to Sakisaka (1958), both the pollen sacs and the ovules on the flower are marginal sporophyll forms. He further reasoned that the capellary leaf borders were connected to the conspicuous boundaries of the stony seed component. Gunckel and Wetmore (1946a) analyzed long and dwarf shoots and discovered that the lower, quickly expanding internodes in the long shoots release a substantial amount of auxin (Fig. 9.4).

9.5.2 G. biloba *Short- and Long-Shoot Leaves*

G. biloba is a heterophyllous plant with two distinct leaf shapes. Short-shoot leaves are consistently abundant and seem undivided or moderately bilobed, but long-shoot leaves are smaller and deeply split into two or more lobes. To more clearly comprehend how the two leaf types in male and female plants differ. The hydrology, morphology, and economics of *G. biloba* leaves displayed sex-related heterogeneities and significant shoot-based, according to Liu and Cheng (2023). The enhanced gas exchange rates and leaf hydraulic efficiency of male ginkgo trees may also improve performance.

9.5.3 Pollen Morphology of G. biloba

Boat-shaped grains feature a single longitudinal furrow that, in accordance with Yamazaki and Takeoka (1962), has a propensity to open when the grain is damp. They are all the same size and shape. The enonnous furrow encircles the whole length of the grain and more than half of its surface. He claimed that the size and shape of the germinal opening in *G. biloba* vary. Monocolpate has been replaced with 1-sulcate or anacolpate as the word for the furrow (aperture) (Fig. 9.5a–f).

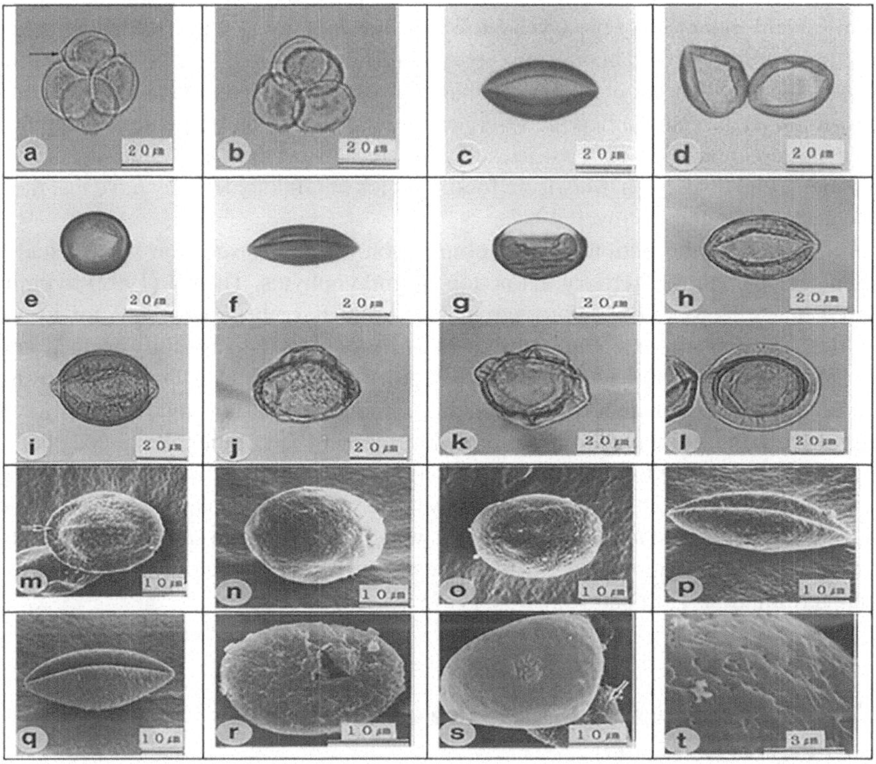

Fig. 9.5 a–t Scannable electron and light microscopy images of *G. biloba* pollen grains. **a, b** Tetrad. A cross-tetrad late tetrad stage. **b** a tetrahedral tetrad or a cross tetrad, depending on the context grains with **c–g** analyzed by LM. **c** Polar view: The grain is elliptical and extremely shrunken. **d** Polar view: subhemispherical furrow and inflated grains. Polar view: a grain that has completely expanded and a nearly circular furrow. Equatorial view: drastically reduced

9.6 Biology of Germination and Phenology

The typical belief regarding Ginkgo biloba is that it is a "prephanerogame" (a living or fossil seed plant whose ovules do not develop into embryos until they are liberated from the parent sporophyte (Stein and Beck 1987). The term “oviparous” was coined by Favre-Duchartre (1958) to define the fundamental components of G. biloba’s reproductive biology. Initiation of the ovule and seed germination take about 12 to 14 months, and fertilization occurs in September directly prior to or shortly after cutting off from the maternal tree. After the embryo has fully developed on the ground, germination usually happens the next springtime or early summer. Apart from several members of the Cycadales, this developmental scenario distinguishes among all other live seed plants, ginkgo. According to research by Gifford and Foster (1989), seeds with a sarcotesta present had a substantially lower overall germination percentage than seeds having the sarcotesta removed. Sarcotesta induces dormancy prior to decomposition and, as a result, inhibits fall germination. Due to animal-mediated seed dispersion, Mesozoic fossil species of Ginkgoales may have lost their sarcotesta (Holt and Rothwell 1997).

In addition to the notion that the elimination of these dispersion vectors had a significant part in the Tertiary extinction of ginkgophytes, Tiffney (1992) hypothesized that ginkgo’s odoriferous sarcotesta would have been excellent for luring reptiles (as herbivorous). The embryos in *G. biloba* seeds were fully formed and not dormant. *G. biloba* seeds benefited from germination at a high temperature (30 °C). Endosperm metabolism and changes in endogenous hormone concentration are particularly significant during embryo development. Like other living seed plants, some Ginkgo tree seeds have fully formed embryos when they absciss. Furthermore, germination can start in as soon as 8 weeks if the sarcotesta is taken out. Even while these Ginkgo trees are not ‘oviparous’ as is widely believed, the sarcotesta appear to function as a germination delaying mechanism and their presence leads in decreased total germination (Holt and Rothwell 1997).

9.7 Phytochemistry (A Treasure of Functional Phytochemicals)

The primary substances identified in the leaf extract by GC-MS included catechol, 1-hydroxy-2-propanone, 2,4-dimethyl-3-hexanol, dihydro-4-hydroxy-2(3H)-furanone, 3-methylmannoside, acetic acid, and 4-methylphenyl ester. (Fig. 9.6).

The rich flavonoids found in *G. biloba* leaves are utilized as an industrial raw material to create *G. biloba* extract (GbE), which is used to remedy cerebrovascular and cardiovascular illnesses in humans (Toghueo 2020). Terpenoids and flavonoids, as the primary pharmacological elements and have anticancer and antioxidant activities, are found in ginkgo leaves. Both flavonoids and terpenoids, which make up many

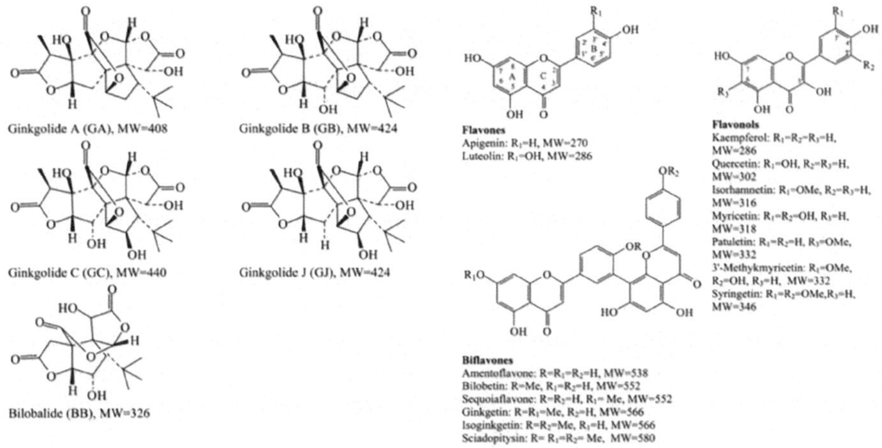

Fig. 9.6 *G. biloba* leaf terpene lactone and phenolic component structures. Lin et al.

of the secondary metabolites in plants, are made up mostly of quercetin, kaempferol, and isorhamnetin and ginkgolides and bilobalides (Wu et al. 2020a, b).

9.8 Economic Importance, Uses and Health Benefits

9.8.1 Economic Importance

Ginkgo raw kernels are poisonous; they are consumed roasted or cooked. In addition to being used as an expectorant and sedative, seeds are used to make detergent. Fruit has flammable oil. In China and Japan, toys and chessboards are made from wood. In China, the nut-like gametophytes are eaten as food. In recent years, ginkgo nuts, also known as ginkgo fruits or ginkgoes, have been used to make a variety of goods. Examples include shortbread (Liu et al. 2014), vinegar (Wang et al. 2014), drinks (Qian 2015), and wines (Yang 2022). Earlier ginkgo powder made from ginkgo nuts was employed in some recipes (Xu and Zhao 2022).

A very popular raw material, *G. biloba,* is employed not only in industrial technology but also in medicine. Its vast applicability is made possible by the presence of bilobalides, ginkgolides, flavonoids, and another bioactive component. A natural dietary supplement (EGb761) extracted from ginkgo is presently marketed, an alternative medicine ingredient, and a cosmetic component (Matsui et al. 1991).

The amounts of bioactive compounds in *G. biloba* leaves and fruit extracts, as well as the extracts' effectiveness against six distinct types of phytopathogenic bacteria that harm horticulture crops, were examined by Sánchez-Hernández et al. in 2023. The only extract that exhibited bacteriostatic activity was the Ginkgo biloba leaf extract, which mostly included 3,4-dimethyl-3-hexanol and 3-O-methyl-D-fructose.

According to Oh et al. (2015), *G. biloba* possesses antibacterial action against *Staphylococcus aureus Rosenbach, Clostridium perfringens, Bacillus thuringiensis Berliner, Listeria monocytogenes*, and *Haemophilus influenza.*

Ražná et al. (2020) mention antifungal activity of *G. biloba* extract of leaf against *Candida albicans.*

To clarify the connection between proteins and the medicinal activities of the nuts, Hatano et al. (2011) identified an antifungal protein from ginkgo nuts. Seeds of Ginkgo and their coats have poisonous chemicals such as ginkgolic acids and ginkgotoxin.

Cydia pomonella L. (the cold moth) is a widespread apple pest that can cause significant harm to the fruit. *G. biloba* and its synthetic metabolites have the capacity to suppress apple eating and infestation by *C. pomonella* neonate larvae. Fruit infestation behavior is influenced by ginkgo components, which may have uses in fruit protection. (Pszczolkowski et al. 2011).

G. biloba leaf extract (GBE) is a popular feed ingredient in aquaculture and is crucial for boosting the resistance of aquatic animals. The effects of incorporating GBE into *Penaeus vannamei* (P. vannamei) diet in intensive aquaculture were assessed by Liao et al. in 2023. *P. vannamei*'s ability to adapt to its environment was enhanced by the addition of 2 g/kg GBE by enhancing its defense mechanisms, lipid metabolism, and detoxification.

Tan and colleagues (2018) investigated the effects of GBE on the high-fat diet-fed hybrid grouper's growth performance, plasma biochemical markers, fish composition, immunological responses, and liver histology. They discovered that hybrid grouper fed diets enriched with GBE experienced hypolipidemic effects, maintained normal liver histology and function, enhanced hepatic antioxidant status, increased immune-related gene expression, and decreased apoptosis-related gene expression in the hybrid grouper's head kidney.

The LTR-RTs that made up most of the ginkgo genome were produced by two WGD events and ancient slow accumulation. The numerous defense mechanisms increasing the ginkgo's exceptional resilience are supported by a considerable enrichment in ancient duplicated and ginkgo-specific gene clusters. The work by Guan et al. (2016) gave information on large genome sequencing and opened the door for next genetic and evolutionary studies. The endosperm of ginkgo seeds, which grow from female gametophytes without fertilization and possess haploid genomes devoid of genetic recombination, was harvested to get genomic DNA.

Numerous benefits of *G. biloba* include its use in landscaping and medicine. Ukraine has a vast landmass, an enormous region of black soil, a sophisticated and varied geography, and forestry development capacities and innovative agricultural. Besides these organic circumstances, the *G. biloba* forest of Ukrainian is less dispersed and only contains one type. In botanical gardens, *G. biloba* is only occasionally employed and is only managed and grown for ornamental purposes. The Sumy region and all of Ukraine have favorable climates for reintroducing this extinct species, and studies on the cost- and efficiency-effectiveness of growing *G. biloba* naturally are being done. *G. biloba* may be planted in areas with significant soil erosion, used to beautify chemical plants and cities, and could enhance soil because it is resistant to radiation and toxic gases (Kovalenko et al. 2021). The woods of these extinct species may potentially be desirable for the building of new tourist itineraries for leisure or educational purposes.

9.8.2 Health Benefits of Ginkgo Supplements Are Associated with Several:

9.8.2.1 Therapeutic Outcomes

Due to its neuroprotective, immunomodulatory, and anti-inflammatory qualities, *G. biloba* has been studied and found to be clinically effective as a dietary supplement and medication for the enhancement of memory, treatment or prevention of Alzheimer's disease and other neurological disorders, and treatment of cardiovascular disorders. Traditional Chinese medicine has prolonged used of *G. biloba* to treat ailments like asthma, bronchitis, TB, cognitive impairment, stomachache, etc.

9.8.2.2 Neuroprotective Effect

G. biloba extract, which also guards against neurological damage, is one of the most well-known dietary supplements for enhancing memory (Ramassamy 2006). It has been demonstrated that the flavonoids and bilobalide in GBE have antioxidant and anti-aging activities. Catalase, superoxide dismutase, and glutathione peroxidase are anti-tissue-damaging enzymes. By scavenging free radicals and activating these enzymes, these drugs had an effect.

9.8.2.3 Anti-inflammatory Effect

Nitric oxide (NO), PGE2, and the mRNA expression of the proinflammatory cytokines IL-1, IL-6, and TNF- have been discovered to be decreased by GBE's anti-inflammatory characteristics (Mir and Albaradie 2015). According to studies on how the extract influenced the chronic inflammatory state detected in the colons of mice, GBE efficiently inhibits macrophage activation and downregulates inflammatory stress indicators.

9.8.2.4 Cardioprotective Effect

Prostaglandins and NO produced by GBE have been shown to enhance blood rheology, lower capillary permeability, and prevent platelet and aggregation hypoxia (DeFeudis 1998). The cardioprotective properties of GBE, a terpene-free extract and terpenoids (ginkgolide A and B), were investigated in rat hearts that had undergone isolated ischemia and reperfusion. According to the results (Liebgott et al. 2000), isolated terpenoids and GBE both aided in postponing the onset of contracture and accelerating functional recovery during ischemia and postischemia.

9.8.2.5 Anticancer Activity

According to studies on the antitumorigenic impact of GBE 761 utilizing in vivo xenograft models and in vitro cell models (Park et al. 2016), the extract roles as an anticancer factor by preventing the action of the enzyme aromatase in MCF-7 cells. In the treated cell model, there was also a reduction in the expression of CYP19 promoter 1.3, CYP19 mRNA, and PII. In an in vivo experiment, BALB/c nude mice were implanted with MCF-7 cells that overexpressed aromatase, and the animals were then treated orally with GBE 761 for three weeks. The size of their tumors was discovered to have greatly shrunk, and CYP19 mRNA expression had been downregulated.

9.8.2.6 Other Effects

In addition to treating hearing loss, vertigo, and tinnitus, Insufficient brain function has also been treated with ginkgo (Hilton and Stuart 2004). One of the first diabetes consequences that might potentially impair eyesight is diabetic cataracts. The pharmacological effects of *G. biloba* extract (EGb 761) for the avoidance of diabetic-induced cataract conditions have been demonstrated in rats with high-glucose settings (Lu et al. 2014) The anti-pathological effects of high glucose on lens epithelial cells were prevented by EGb761, which also improved lens opacity and reduced aldose reductase activation and oxidative stress.

9.8.2.7 Potential Side Effects

Before incorporating ginkgo into your routine, it's vital to consult a doctor. Although there are some circumstances in which ginkgo could be seriously harmful, for many adults, it comes with quite a low danger of taking it. Ginkgo should not be taken if you have an allergy to alkylphenol-containing plants or are on any drugs. The following adverse effects are possible: rash/allergic response, stomach pain, headaches, dizziness, nausea, and diarrhea. Ginkgo may have negative drug interactions with some prescription drugs. A few interactions could make bleeding more likely. Two examples of potentially dangerous medication interactions include blood thinners (Warfarin, aspirin) and SSRI/ MAOI/ antidepressants (Prozac, Zoloft) (Bent et al. 2005).

9.8.2.8 GSPs: Or Ginkgo Seed Proteins

GSPs have significant nutritional and medical value since they display a variety of biological actions, such as anti-aging, anti-tumor, anti-oxidation, and anti-bacterial ones (Huang et al. 2004). Conversely, some people who consume GSPs may experience allergic responses, particularly small children or newborns who are more sensitive to GSPs. Convulsions, exanimation, dyspnea, mydriasis, nausea, vomiting, diarrhea, stomach pain, and even death are clinical manifestations of these responses (Yang and Wu 2009).

9.9 Nutritional Studies

It has attracted a lot of attention that ginkgo leaves, seeds, and contemporary ginkgo leaf extract can delay age-related cognitive deterioration. Omega-3 fatty acids seem to be essential for maintaining neuronal structure and function and slowing cognitive decline, whereas antioxidant vitamins C and E seem to be particularly helpful for preventing age-related oxidative stress when given concurrently. Even though most research to yet has used rats, fruit and vegetable polyphenols also have a lot of potential. The B vitamins cobalamin (B12) and folate (B9), which are being researched for their ability to reverse age-related degeneration (Brown et al. 2010), have inconsistent results.

9.10 Current Agricultural Challenges

The current species Ginkgo biloba is morphologically highly like congenerics from the Mesozoic and Cenozoic, is phylogenetically isolated, and has a remnant distribution. What adaptations may have enabled this lineage to endure for more than 100 Myr with little to no morphological change was the subject of Royer et al. (2003) study. They examined information on 51 locations in North America and northern Europe's floral and sedimentological composition. That include fossilized Ginkgo trees and range from the most recent Cretaceous to the middle Miocene. The gathered data showed that Ginkgo was mostly restricted to degraded streamside and levee conditions during the late Cretaceous and Cenozoic, when it co-occurred with a recognized assortment of other plants. The ginkgo's life history traits, such as its slow growth rate, prolonged reproductive cycle, late reproductive maturity, complex and large seeds, and slowly and large developing embryos, are the exact opposite of those that are thought to be advantageous in today's disturbed habitats, so it's unexpected that these habitats are anticipated. Thus, palaeoecological findings from sedimentary settings and morphology are consistent with the idea that angiosperms with greater resilience to frequent disturbance replaced Ginkgo in riparian habitats.

According to Nakayama et al. (2017), heterophylly is an adaptation mechanism that enables plants to react to microenvironmental variability in the best way possible. The emergence of heteromorphic leaves may be a result of a plant's ecological adaptability. This phenomenon may be a way for plants to adjust to changing light conditions and maintain their ability to photosynthesize. Another example of heterophylly is when some plants generate distinctive leaf shapes from several kinds of shoots, such as long and short shoots. Long ones help branches grow quickly and undertake photosynthesis at high rates, whereas short shoots are associated with reproductive organs (Leigh et al. 2011).

The guidelines are set to seek for good variants with higher concentrations of these active ingredients and higher leaf yields than ordinary ones to obtain more Ginkgo biloba leaves rich in flavonoids and terpenlactones. To cultivate ginkgo trees for leaf use, many fine varieties were chosen in accordance with these criteria. Ginkgo kinds, sex, age, origin, environment, and harvest time, as well as placement in the tree's canopy and branches, branch types, and other variables that affect the active ingredient concentration and leaf count. The yields of leaves and medicinal components will increase due to proper fertilizer application, cutting, pruning, and planting density. concentrating on two more areas: enhanced directed cultivation of leaf-used Ginkgo and the collection, conservation, analysis, and selection of genetic resources. The yields of leaves and medicinal components will increase with proper fertilizer application, trimming, pruning, and planting density. Two further areas of emphasis include the directionally cultivating of leaf-used Ginkgo utilizing better varieties and standardized cultivation, as well as the collection, conservation, study, and selection of genetic resources (Gu et al. 2017).

According to Xu et al. (2020), *G. biloba* seedlings can withstand modest levels of soil salinity stress by altering a variety of genes and transcriptome variables, such as flavonoid production, which is elevated in response to environmental pressure. It is appropriate and relevant to plant the *G. biloba* plant in urban locations due to its strong resilience to environmental pressures, other pests, microbiological diseases (viral, fungal, and bacterial), and gaseous pollutants SO_2 and ozone. They are wise selections for individuals who landscape near the seaside because they are salt-tolerant plants.

Different geological eras (the Triassic, Jurassic, Cretaceous, and Paleocene) are known to have produced several extinct species and genera. Live *G. biloba* leaves have stomatal characteristics that are sensitive to variations in CO_2 concentration, according to CO_2-enrichment experiments. Paleo-CO_2 levels from the Tertiary and even earlier in the Mesozoic have been reconstructed from fossil Gingko leaves. The findings show that (i) gingko leaves' gas exchange diffusional limitation has most likely not changed much over the previous 200 million years and (ii) their physiology has been well conserved, exactly like their leaf structure (Sun et al. 2003).

However, greater temperatures or protracted droughts may cancel out the advantages of forest expansion (Fettig et al. 2019). Plants can be protected from environmental challenges by producing more defense metabolites, such as flavonoids, phenols, terpenoids, and aromatic chemicals. Phenotypic features, or growth and metabolic responses to the environment, are thought to be able to predict how various species would respond to ecological niches, climate change, and growth methods. In order to forecast the potential height growth for a particular planting location under a specific future climate, a climatic response function for *G. biloba* was constructed. China is home to more than 90% of the world's ginkgo sources, and to fulfil the growing demand for ginkgo-related goods, it has also created more than 200,000 ha of commercial plantations (Ahmad et al. 2020).

The growth and metabolism of genetically homogenous (containing several sexual categories) ginkgo clones cultivated in China over environmental (climatic) gradients were examined by Guo et al. in 2021. They also explored how climate change affects ginkgo leaf quality and production and created climate response models for phenotypic variables.

Among the most severe anthropogenic stressors placed on forest ecosystems is air pollution. The world's forests have been losing a lot of their greenery. According to Raven et al. (1986), ginkgo trees are the ones that are most resilient to air pollution. According to studies on the relative susceptibility of ginkgo trees to various air pollutants, SO_2, nitrogen oxides, and ozone are all tolerated by ginkgo trees; however, fluorides are only accepted to a lower amount. According to this, ginkgo trees are usually resistant to gaseous air contaminants.

Ginkgo trees in high-traffic sites had 5 times more sulfur than those in low-traffic sites, according to an examination of the leaves. The increased concentration of sulfur in the samples from the traffic region may have been caused by other particles and automotive exhaust emissions. Scanning electron microscopy (SEM) was used by Oh et al. (1983) to show that Ginkgo leaves from busy areas did not significantly tolerate cuticular wax degeneration other than from the deposition of unidentified particles. At higher magnifications, some degradation of the equally dispersed granular wax crystalloids on the leaf surface was seen. On the ginkgo, the majority of the granular wax crystalloids were still whole.

Acid rain has been demonstrated to be able to harm ginkgo trees. Acid rain simulation influenced seed germination, seedling growth, and chlorophyll content. After being exposed to simulated acid rain, Ginkgo leaves significantly increased in contact angles and stomatal resistance, and when pH levels dropped, leaf damage increased (Lee et al. 1993).

9.11 Cultivation Practices

There are several different seasonal conditions where ginkgo may be produced, from Mediterranean to cold temperate, with wintertime low temperatures ranging from 22 to 8 °C. In the Southern Hemisphere, it completely vanished once its geographic range shrunk around 40o 8″N (Ziegler et al. 1996). The guidelines are designed to look for fine variants with greater concentrations of those active components and higher leaf yields than regular ones to obtain more *G. biloba* leaves rich in flavonoids and terpenlactones. To cultivate ginkgo trees for leaf usage, many exquisite kinds were chosen and used. Ginkgo varieties' active component concentrations and leaf count are influenced by a variety of variables, including placement in the tree's canopy and branches, age, origin, environment, and harvest time. The yields of leaves and medicinal components will increase with proper fertilizer application, trimming, pruning, and planting density. Future research should concentrate on two areas: first, the collection, conservation, study, and selection of genetic resources; second, the directionally culture of Ginkgo utilizing better varieties and standardized farming (Gu et al. 2017).

Crown type variation and its connection to the main bioactive metabolite's concentrations were looked at to increase the flavonoid and lactone contents of Ginkgo by phenotypic selection. According to the findings, there were substantial variations between the various crown types in terms of the leaf area (LA), tree height (HT), branching angle (BA), diameter at breast height (DBH), and leaf density. Additionally, the results demonstrated that the concentration of isorhamnetin in ginkgo leaves varied meaningfully based on the type of crown and tended to climb as the crown size rose, demonstrating the value of the crown type classification and suggesting its potential for use in selective breeding. Wide-crown medium-type trees had the quickest rates of HT and DBH development as well as the highest leaf yield. Wide-crown medium type individuals would have much higher active component levels to fully take use of the variation among Ginkgo individuals and improve breeding efficiency. The study of Wu et al. (2020a, b) outlineed the breeding and industrial production strategy.

9.11.1 Cultivation Practices

Modern ginkgo is frequently shade-intolerant and performs best in open locations, according to He et al. (1997). *G. biloba* seeds need an open canopy to grow and flourish on a "semi-wild" stand. In disturbed microsites including stream banks, rocky cliff edges, and steep slopes, several individual trees were found to be sprouting. However, he did assume that Ginkgo may function as a gap opportunist and wait around in the shaded understory of the forest until a gap appears that will allow it to grow into the canopy. Although one of the most widely cultivated and productive commercial species in temperate regions of the world, G. biloba's slow growth is what most limits its application. Ginkgo grows best on "sandy loam soils… along dams… or at the foot of hillslopes" and "well-watered and well-drained conditions" (He et al. 1997), at least when it is farmed.

Despite being generally referred to be a slow-growing tree, ginkgo may grow up to 30 cm per year throughout the first 30 years or so of its existence in favorable settings with warm summers (Del Tredici 1989). At this "bolting" stage of their development, seedlings and saplings have a straight main axis and a few sparse, excurrent branches. When the young tree's crown starts to fill up, growth suddenly abruptly comes to a stop. Trees that are 100 years old may have barely reached two-thirds of their initial height. Individual trees can continue to give fruit for more than 1000 years, whereas plants need 20 to 30 years to develop viable seeds. (He et al. 1997).

Light: It is best to grow *G. biloba* in maximum sun to moderate shadow.

The ginkgo is tolerant of most soil types and climates. It can endure compacted soil, acidic soil, and soil that is alkaline. It loves loam or sandy soil that drains well and has a pH between 5.0 and 8.0. Compared to many other tree species, they can tolerate compacted soil better. Ginkgo trees grow well as street trees and are good for coastal locations due to their resistance to air pollution.

Water: If the location is well drained, it is necessary to maintain the soil's moisture. When a tree is young, moisture is very crucial because as it matures, it is generally drought tolerant.

Temperature and Humidity: Urban areas in several regions often host the cultivation of *G. bilobas*, demonstrating the plant's adaptability to a variety of moisture levels and temperatures. In hot, dry conditions, they can, however, suffer.

Fertilizer: Urban settings across several climates frequently grow *G. bilobas*, demonstrating the plant's resilience to a broad variety of moisture levels and temperatures. In hot, dry regions, though, they might suffer (Beaulieu and Hughes 2022).

9.11.2 *Reproductive Behavior of* **Ginkgo biloba** *Tree Species*

– Male cultivar "Autumn Gold" (zones 3 to 8) is well-liked. It has a canopy that is widely distributed, grows between 40 and 50 feet tall and 25 and 30 feet wide.
– 'Saratoga' (zones 4 to 8) resembles 'Autumn Gold' in terms of size and form. This cultivar stands out due to the V-shaped leaves, which deviate from the typical fan form.
Choose the well-known male cultivar "Fastigiata" (zones 3 to 8) if you desire a slender, pillar-shaped tree. It is 30 to 50 feet tall and 10 to 15 feet wide.
– "Princeton Sentry" (zones 3 to 8) is columnar as well, but it develops at a height and breadth of 40 to 50 feet and 20 to 30 feet, respectively.
A minimum height of 50 feet is reached by the tall, slender Fairmont tree (zones 5 to 8), although it only reaches a width of 15 to 18 feet.
For a tiny area, "Pendula" (zones 5 to 8) are a fantastic option. It is a slow-growing male cultivar with a stocky structure that matures at just 8 feet tall and the greatest stretch of 10 feet. The tree here has a more umbrella-like form, despite the cultivar name "Pendula" usually connoting a weeping aspect. (Hughes and Beaulieu 2022).

9.11.3 *What Are Lignotubers (Basal Chichi)*

In addition to sexual reproduction, lignotubers, also known as basal chichi, which appear close to the base of the trunk allow Ginkgo to reproduce clonally. Positive geotropic development is encouraged within these lignotubers if the nearby soil is disturbed, as by severe erosion. The rhizomatous tissue that forms at the tip of a subterranean lignotuber can result in both aerial shoots and adventitious roots. In older persons, these aerial chichi buds also grow inside secondary stems. Like lignotubers, aerial chichi responds to disturbances by growing (for example, major crown damage) and has the capacity to form viable clones.

9.12 Genetic Improvement

Conventional and unconventional plant breeding are the two primary categories of plant breeding, which are distinguished by the instruments and procedures used. One definition of conventional breeding is the practice of creating new types using natural techniques. Conventional breeding involves combining desired features from various gene pools that are closely related through a process called cross-pollination. As a result, the result of conventional breeding is a crop with pre-existing traits that have been combined and matched to produce desirable crops (Acquaah 2015).

A certain set of procedures is followed in conventional plant breeding. The process starts with establishing a specific goal, then moving on to variant creation and assembly, assessment, selection, release, multiplication, and dissemination of the new plant.

Megagametophytes were used by Gui (2004) to create a linkage map for the ginkgo tree, however it only contained a small number of unidentified markers. He has produced a 12-LG high-density genomic map for the ginkgo in this instance. According to Murray (2013), the quantity of LGs is equal to the amount of Ginkgo's haploid chromosomes. Seedlings of *G. biloba* can withstand low-level soil salinity stress by regulating several genes and transcriptome variables, such as flavonoid synthesis, which is increased in response to environmental stress, according to study by Xu et al. (2020). The primary physiologically active components of *G. biloba* are flavonoids, which are crucial in neutralizing free radicals and defending the development and growth of plants. Numerous genes, including 4CL, FLS, PAL, and F30H genes, as well as DFR gene, have shown a considerable increase in the pathways that control flavonoid production. These highly expressed, considerably elevated genes may aid in the flavonoid's accumulation (Li et al. 2018).

Plant-specific transcription factors called lateral organ boundaries domain (LBD) proteins control various transcriptional regulatory processes. They discovered a family of 37 GbLBD genes in ginkgo, which are divided into classes I (33, with the maximum number of Id genes (16) and II (4) based on gene structure and phylogenetic analysis. Additionally, the promoters, gene duplications, chromosomal location, and introns/exons of the ginkgo LBD gene were examined. Abiotic stress treatment, of which class II genes played a significant role in stress treatment, may significantly enhance the expression of the GbLBD gene, according to the expression pattern analysis of cold/drought treatment and IAA/ABA hormone therapy. Plant-specific transcription factors called lateral organ boundaries domain (LBD) proteins control several transcriptional regulatory processes. They discovered a family of 37 GbLBD genes in ginkgo, which are divided into classes I (33, with the maximum number of Id genes (16) and II (4) based on gene structure and phylogenetic analysis. Additionally, the promoters, gene duplications, chromosomal location, and introns/exons of the ginkgo LBD gene were examined. Abiotic stress treatment, of which class II genes played a significant role in stress treatment, may significantly enhance the expression of the GbLBD gene, according to the expression pattern analysis of cold/drought treatment and IAA/ABA hormone therapy (Tian et al. 2022).

Shen et al. (2005) employed chloroplast DNA (cpDNA) with eight size variations in the trnK1-trnK2 section to examine the genetic makeup of eight alleged refugial populations and plantations and make inferences about the glacial refugia that sustained *G. biloba.* Most of cpDNA haplotypes are only found in small areas of the geographic range (in southwest China). It's possible that throughout the Pleistocene, this region was shielded from the extremes of climate variations because it is currently a biodiversity hotspot of worldwide significance. The regulations are made to look for good variants with greater contents of flavonoids and terpenlactones, larger leaf yields than regular ones to obtain more Ginkgo biloba leaves rich in terpenoids and flavonoids. To cultivate ginkgo trees for leaf usage, many exquisite kinds were chosen and used. Ginkgo varieties' active component concentrations and leaf count are influenced by a variety of variables, including placement in the tree's canopy and branches, age, origin, environment, and harvest time. The production of leaves and medicinal components will increase thanks to the proper application of fertilizer, trimming, pruning, and planting density. Future research should concentrate on two areas: first, the collection, conservation, study, and selection of genetic resources; second, the directed culture of Ginkgo species utilized for leaf production utilizing better cultivars and standardized cultivation (Gu et al. 2017).

Due to the medicinal advantages of its plentiful flavonoids, Ginkgo has drawn a lot of attention. Guo et al. (2023) cloned the 1314 bp full-length gingko GbFLSa gene. It encodes a 363 amino acid protein and possesses a typical 2-oxoglutarate (2OG)-Fe (II) oxygenase domain. The *E. coli* BL21(DE3) strain was used to produce the 41 kDa recombinant GbFLSa protein. The cytoplasm contained this type of protein. Additionally, transgenic poplar plants had considerably lower levels of proanthocyanins, such as epicatechin, catechin, gallocatechin, and epigallocatechin, than nontransgenic (CK) ones. Thus, GbFLSa encodes a functional protein that may inhibit the formation of proanthocyanins. Both a proposed molecular mechanism for flavonoid synthesis and the role of GbFLSa in plant metabolism were made clear.

Zumajo-Cardona et al. (2021) discovered new genes that are probably involved in ovule development by thorough comparative transcriptome investigations of growing pollen cones, megagametophytes, and ovules in *G. biloba.* Comparative molecular studies of the evolution of the integument in angiosperms and *G. biloba* give crucial information for understanding the beginnings and development of the seed. It's a gymnosperm Ginkgo, which is regarded as a living fossil, is one of the oldest extant plant lineages where seeds evolved and has undergone over 300 million years (mya) of development without morphological change. Flavonoids are produced by the flavonoid $3'$, $5'$-hydroxylase (F3′5′H) enzyme in *G. biloba.* A new GbF3′5′H1 gene with differential expression that is similar to the F3′5′H enzyme involved in the route of flavonoid synthesis was found by Wu et al. (2020a, b) using transcriptome sequencing. They discovered that GbF3′5′H1 is considerably expressed in April and is broadly dispersed in the leaf. Additionally, four metabolites that are closely related to flavonoid synthesis were found. Importantly, transgenic plants had much larger levels of the antioxidants 4′,5-dihydroxy-7-glucosyloxyflavanone, gallocatechin, and epicatechin than nontransgenic plants did. They discovered that this novel gene contributes to the metabolites manufacture which is linked to flavonoids, indicating that GbF3′5′H1 is a strong applicant to maximize the production of ginkgo flavonoids.

9.13 Germplasm Diversity and Conservation

9.13.1 Conservation and Management

Except for Antarctica, ginkgo trees have been planted everywhere across the world, and as of right now, they are spread throughout a latitudinal range of around 103° (Fig. 9.7). At a maximum latitude of 60° 03′ N in Europe, it occurs in the Northern Hemisphere. Eastern to both of Asia, [specifically China, Korea, and Japan (the CJK region)], North America, and Europe, are home to most live ginkgo trees. Its lovely golden leaves in the autumn have led to its planting in urban places like parks and along roadside embankments. This tree has an elongated botanical history in the CJK area of eastern Asia, where ancient trees are frequently found adjacent to temples and rural towns. Ginkgo occurrence in China spans a latitudinal range of around 25°, with the southernmost site being at 20° 42′ N and the northernmost one at 45° 32′ N. In 2019, Zhao et al.

The IUCN Red List of Vulnerable Species classifies *G. biloba* as an endangered plant. The Red Book of Chinese Plants, the China Species Red mention, and the List of Wild Plants in China with Priority Conservation (first-class status) are also among the publications that mention it Zhang et al. (2015) confirm this. Whether Ginkgo should be a top priority for conservation has long been up for discussion. However, natural populations were discovered in four regions of China using population genetics and field research. Additional lines of evidence that disproved the idea of the extinction vortex included the resilience of population size and range to climatic oscillation as well as the adaptive development of defense/response genes (Zhao et al. 2019a, b). Less than 100 individuals are dispersed individually in the rare natural populations that may be discovered in south-western China, according to Lin et al. (2022).

A startling finding was the dearth of young Ginkgo trees or seedlings in most natural populations (Xie 2014). Thus, conservation efforts should be prioritized for Ginkgo's natural populations. Their team and the reserve management have been conducting long-term field assessments for the forests of ginkgo in the Mt. Tianmu National Nature Reserve for the past 20 years.

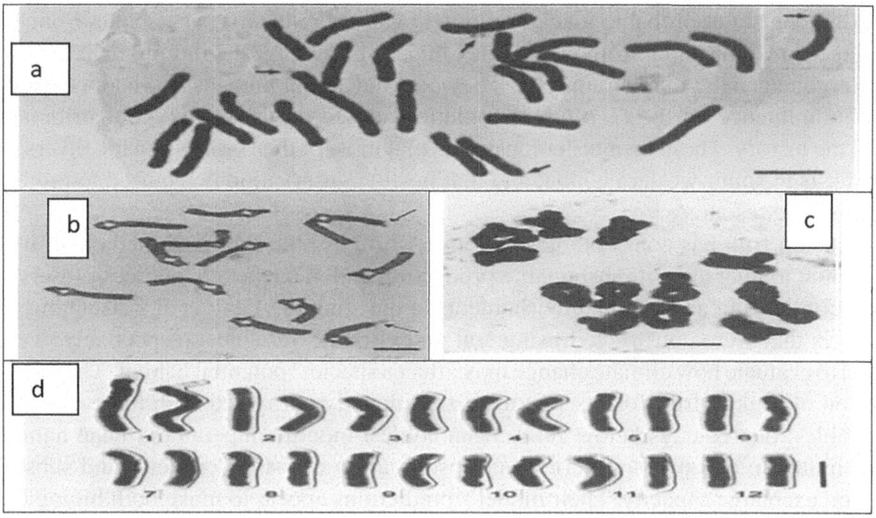

Fig. 9.7 a-d. Ginkgo biloba chromosomes. **a** Young Leaf in a Male Tree with a Somatic Diploid Chromosome (2n = 24). **b** A single ovule has 12 haploid chromosomes (after Lee). **c** During the first division's metaphase, meiotic chromosomes are found in a PMC. **d** the common karyotype. Arrows point to satellites or secondary constrictions

9.13.2 Plant Germplasm Conservation

In two succeeding years, juvenile half-sib lines ginkgo seedlings were examined for a variety of phenotypic traits and morphological traits. Ocokolji et al. (2011) state that this research established the foundation for future planting stock breeding and production for the construction of target plantations of this species as well as for urban coenoses. Half-sib families are seedlings that are descended from the same mother tree. To develop new ginkgo types, further breeding programmes might be used to identify planting stock using comparative analysis. This demonstrated the requirement for their proliferation as well as the need for the conservation of the investigated ginkgo genotypes. It has been claimed that the existence of wild populations is in danger because to the rising commercial demand for farmed ginkgo since the supply cannot keep up with it. Such worries might be partially allayed by the development of large-scale artificial ginkgo propagation. For many natural Ginkgo populations, logging remains a serious concern. Ginkgo has a very broad spectrum of flexibility, as demonstrated by experience. The suitable habitat for ginkgo would shift to higher latitudes and elevations, according to several climate predictions (Guo et al. 2019). Despite the Ginkgo's weak natural capacity to disperse, the changing environment could harm efforts to preserve it.

The genetic makeup of the plant *Ginkgo biloba* L. was ascertained by Gong et al. (2008a, b) using polymerase chain reaction-restriction fragment length polymorphism to examine changes in the trnK region of chloroplast DNA and make predictions about probable glacial refugia in China. With 220 participants from 12 villages in China and 3 from outside of China, it is the largest cohort to date to have undergone genetic marker analysis. They concluded that humans had had a considerable influence on the *G. biloba* populations based on the haplotype distribution and the history. These results demonstrated that most of the Ginkgo genetic diversity center is in south-western China, and that in situ conservation is required to protect the genetic resources.

Ginkgo fruit has considerable economic worth as both food and medicinal. It is essential to guarantee its sustainable production and the genetic resources preservation considering a changing environment. In the study by Feng et al. (2021), niche models that were constructed using soil and climatic variables, respectively, were used to evaluate how climate change may affect a species' potential habitat. The distribution of ginkgo fruit forests was most significantly correlated with three climatic variables: degree-days below zero, mean coldest month temperature, mean annual precipitation, and two soil variables: topsoil cation exchange capacity and subsoil cation exchange capacity. Their model's predictions appear to make both biological and financial sense based on the amount of expected habitat appropriateness and the variations in fruit attributes that were observed. This study identified suitable geographic locations for the production and protection of this species and improved understanding of how climate change affects ginkgo fruit forests.

9.13.3 In Situ Conservation

To help in situ conservation, 290 Ginkgo trees were discovered, identified, counted, and monitored. Additionally, 27 long-term Ginkgo plots of 20 m by 20 m were made (Lin et al. 2022). To monitor the ginkgo microhabitats, DBH growth rate, seed predators, and seedling survival on Mt. Tianmu, they began several monitoring studies (Gu et al.2021). Several elements or processes that were thought to be connected to the regeneration restriction were tested in the field and using manipulation studies.

9.13.4 Conservation Ex Situ

Researchers at Zhejiang University and the Zhejiang Forestry Academy have been using ex situ conservation to preserve seeds since 2017 to protect seed germination and seedlings. Ginkgo trees may have survived in China due to cultural and religious beliefs. According to severe cultural taboos, neither natives nor immigrants are allowed to add to or remove Ginkgo trees from Mt. Dalou, protecting the area's native Ginkgo trees in the process (Huang et al. 2020). Near Buddhist and Taoist temples, where monks have assisted the trees in surviving, are where you may find the oldest planted ginkgo trees (Isah 2015).

9.13.5 Cryopreservation

According to Nakanishi, it is well-known throughout the world as a landscape tree and as a significant source of medicinal chemicals (ginkgolides and bilobalide). *G. biloba* cell cultures, according to Park et al. (2004), are a helpful tool for targeted multiplication of this plant as well as a way for manufacturing ginkgolides A and B, bilobalide, and flavanol aglycones. Since samples in liquid nitrogen don't need to be subcultured on a regular basis and may be preserved for decades with the least amount of space, cryopreservation presents an alluring alternative to the time- and resource-intensive maintenance of in vitro cell cultures. *G. biloba* calli were successfully cryopreserved by Popova et al. using the desiccation technique, however post-freeze regeneration was only 23% or less. Comparing calli preculture in the presence of ABA to untreated controls and calli pretreatment with sucrose alone, a decreased desiccation rate was observed. Regardless of the amount of sugar in the medium, post-thaw regrowth was sporadic in calli precultured with sucrose alone. However, calli precultured with ABA and sucrose following cryopreservation ensured reliable regeneration. The effects of preculture with sucrose and sucrose + ABA on the tolerance of G. biloba cell cultures to desiccation and cryopreservation were investigated by Lu et al. (2009).

9.13.6 Genetic Studies (Cytogenetic Studies)

G. biloba has distinct male and female individuals and is dioecious. It is hypothesized that the monoecious progenitors of *G. biloba*, which contained ovulation and pollen cones on various regions of the plant, gave rise to dioecy (Nagataa et al. 2016). Chromosome number, chromosome banding, karyotype, genome size, and physical mapping are among the cytogenetic investigations performed on *G. biloba.*

9.13.6.1 Chromosomes of Ginkgo Biloba

The genome of G. biloba is approximately 10 Gb in size, although it has already been sequenced and assembled twice (Liu et al. 2021). The most current assembly, based on long-read sequencing, is "nearly complete," with 12 big scaffolds that match to this species' 12 chromosomes. Previous cytological studies revealed that G. biloba has $2n = 24$ somatic diploid chromosomes. According to Hizume, the karyotype consists of ten pairs of short subtelocentric chromosomes, one pair of long submetacentric chromosomes, and one pair of short metacentric chromosomes.

9.13.6.2 Karyotype

Figure 9.7 d shows the chromosomal pair composition of the organism.

No. 1: long, submetacentric chromosome with a small satellite at the end of a short arm.

No. 2: The most asymmetrical chromosome is a short subtelocentric one.

No. 3: Metacentric and short chromosome.

Nos. 4–10: Short, subtelocentric chromosomes that are identical to one another.

No. 11: chromosome with a little satellite at the end of its long arm, is short and subtelocentric.

No. 12: the subtelocentric, shortest chromosome.

9.13.6.3 Chromosome Banding

The chromosomes of G. biloba have been banded using methods such fluorescence banding, fluorescence staining, silver staining, and C-banding. Two of the satellite and secondary constriction of four chromosomes (numbers 1 and 11) are the C-bands and chromomycin A3 (CMA) bands. Near the secondary constrictions, there are three or four silver deposits, also known as active nucleolar organizing regions (Ag-NORs). Young leaves and root tips are examined for somatic chromosomes. Early in April, when haploid meiotic chromosomes are released from immature pollen in ovules and anthers, the female gametophyte enters its free nuclear stage.

9.13.6.4 In Situ Hybridization for Physical rDNA Mapping

Using in situ hybridization (ISH) with a rDNA probe, they looked at how rDNA contributed to the NOR variation found in this species. Following ISH using the 18S rDNA probe and the tips of the short arm of the long chromosomes (number 1) and the long arm of the short chromosomes, respectively, four hybridization signals (number 11) were seen in the mitotic chromosome. The C- and CMA bands, secondary constrictions, and minor satellites are all located nearby the loci. In PMCs split from the same plant, the main bivalent and one of the lesser bivalents each produced four signals on each homologous chromosome that were of comparable strength. Each nucleus in the tetrads held two signals of similar sizes. The major bivalent and one of the smaller bivalents each produced four signals on each homologous chromosome that were of equal intensity in PMCs separated from the same plant. Each nucleus in the tetrads held two signals of similar sizes. These ISH results demonstrate that there were variations in plant rDNA content.

Researchers recreated male- and female-specific genomic maps using transcriptome sequencing data from four genetic crosses, and they discovered the SDR in the middle of chromosome 2. The Ginkgo SDR is thought to be at least Cretaceous in age based on synonymous divergence between homologous X- and Y-linked genes in that area. Very little Y-degeneration is found when gene expression and substitution rates are analyzed. These results are consistent with studies on other homomorphic sex chromosomal dioecious plants. The sex chromosomes of Ginkgo are at least of Cretaceous origin, and the species' slow evolution is apparent in the relatively low rate of synonymous X: Y divergence, which is a result of the prolonged generation period and (presumably) low mutation rate. These findings are in accordance with research conducted in 2022 by Gong and Filatov.

In their 2019 study, Yalçin et al. investigated how well G. biloba leaf extract (GbE) shielded Swiss albino mice against hydrogen peroxide (H_2O_2) toxicity. They examined the effects of abnormal mitotic index (MI), abnormal micronucleus (MN), and abnormal chromosomal abnormalities (CAs) on cytogenetics. As markers of biochemical impairment, serum concentrations of the enzymes aspartate aminotransferase (AST), alanine aminotransferase (ALT), alkaline phosphatase (ALP), blood urea nitrogen (BUN), creatinine, glutathione (GSH), and malondialdehyde (MDA) were assessed in the liver and kidney tissues. It was discovered that the supply of H_2O_2 greatly decreased CAs, MN, AMn, and MI. The most frequent kind of CAs brought on by H_2O_2 was chromatin break. According to studies, GbE therapy boosts the number of mitotic cells, lowers the frequency of MN and CAs, and mitigates the clastogenic effects of H_2O_2. MDA levels, a marker of oxidative damage, increased while GSH levels decreased in the liver and renal tissues, in contrast. Following GbE treatment, the amount of oxidative damage caused by H_2O_2 in the liver and kidney tissues reduced, according to all of the biochemical markers examined. To counter the negative effects of elements like H_2O_2 in the environment, GbE can be taken as a dietary supplement.

9.14 Traditional Breeding of Ginkgo

(Breeding) Phenotypic characteristics and active ingredient concentrations are connected. High yields of the active ingredients have become a key goal for breeding. It is essential to comprehend the relationships between active component contents and morphological criteria to support breeders in their selection and improve breeding efficiency (Khadivi-Khub et al. 2014). Observations show that the Ginkgo tree's crown type varies greatly and was likely influenced by branching patterns, biomass, aesthetic value, etc. The crown type, however, cannot yet be determined by variations. It has the potential to be employed in selective breeding without regard to genetic origin if paired with an active chemical composition (Wu et al. 2020a, b).

9.14.1 Inbreeding Depression and Selection

IBMPC and SPQG30 software were used to assess the data and breeds. According to the findings of variance research, there are significant differences between clones in terms of the genetic features of heritability (h2) and genetic variability coefficient

(Gcv), which are determined by clone > sex > provenance to flavone glycosides in ginkgo leaves. On flavone glycosides, male tree clone leaves showed greater sigma g2, h2, Gcv, and delta G' values than female tree clone leaves. According to Xing et al. (2002), there is a maximum terpene clone in females and a maximum flavone content clone in men. Twenty higher terpenes clones' R-factor analyses and Q-cluster analyses both produced reliable findings. Ginkgo cultivars utilized for their leaves are more adapted to direct or index selection.

9.14.2 Heritability

Chloroplast DNA (cpDNA) with eight size variations in the trnK1-trnK2 region was utilized by Shen et al. (2005) to examine the genetic makeup of eight possible refugial populations and plantations and make inferences about the glacial refugia that sustained *G. biloba*. Haplotypes of 158 trees were taken from eight different locales make up the data. A small section of the geographical area is home to the bulk of the cpDNA haplotypes. Their findings imply that *G. biloba* refugia existed in southwest China. It's possible that throughout the Pleistocene, this region was shielded from the extremes of climate variations because it is currently a biodiversity hotspot of worldwide significance. Many academics believed the ginkgo trees on West Tianmu Mountain to be natural, however it's possible that they were planted there by Buddhist monks.

9.14.3 Genotypic and Phenotypic Correlation

Genome-wide association studies (GWASs) based on populations have been carried out on numerous tree species to better understand the links between characteristics and genetic variations (Ingvarsson et al. 2008). In plant genetic research, one of the RRGS techniques used is genotyping by sequencing (GBS). To conduct a GWAS for leaf traits across the widely dispersed ancient populations of *G. biloba*, Zhou et al. (2022) examined the leaf phenotypes. Identifying phenotypic differences between historical populations, locating excellent sources of leaf-producing germplasm, and examining how leaf traits are regulated were the study's main goals. The investigation of leaf phenotypes by Zhou et al. (2022) provided the foundation for a GWAS for leaf characteristics in the geographically dispersed ancient populations of *G. biloba*.

9.15 Genome Selection is Used to Forest Genetic Breeding

Meuwissen et al. (2001) originally suggested the novel breeding selection technique known as genome-wide selection (GS), which makes use of high-density molecular markers. Many genetic markers found throughout the entire genome are examined to determine the genomic estimated breeding value (GEBV). GS will next integrate the effect values throughout the whole genome after evaluating the impact values of various chromosomal regions or single markers. To calculate the genomic estimated breeding value (GEBV), many genetic markers are assessed across the genome. The impact values of various chromosomal regions or individual markers may then be assessed by GS, who will integrate the effect values throughout the whole genome. Then, the candidate population's members' known genotype data and SNP effect estimations may be used to produce the GEBV. Based on the GEBV rating, the reserved individuals can subsequently be selected from the application categories. The effects of each QTL may be replicated by an SNP since high-density SNP markers are dispersed across the genome and at least one SNP may be in linkage disequilibrium (LD) with the QTL controlling the target feature.

9.15.1 G. biloba *Reproductive Process: Secretory Fluids' Secret Properties*

The seed plant *G. biloba* possesses one of the oldest and most fundamental reproductive systems among all seed plants, and it also has a significant phylogenetic position. The G. biloba archegoniate has a unique pollen chamber, archegonial chamber, and flagellated gametes (spermatozoids), in which an egg cell develops inside an archegonium. The secretions involved in these processes are essential. In 2022, Mao et al. studied the development and functionality of pollination drops (PDs) and fertilization fluid in *G. biloba.* The endogenous and intracellular transport mechanisms of PDs, as well as their interactions with pollen, are discussed. They explore fresh understandings of the PD transport short RNAs, the processes for precisely controlling pollen tube development in *G. biloba*, and the structural patterns of pollination, fertilization, and the distinctiveness of sexual reproduction in ancient plants.

On the PacBio sequencing platform, full-length transcriptome sequencing might considerably improve the annotation of gene structures. Terpenoids found in abundance in *G. biloba* L. have therapeutic use. Using single-molecule real-time (SMRT) technology and Illumina sequencing, six reproductive organs essential to ginkgo's growth and development were sequenced to provide a high-quality full-length transcription library. These results predicted 4019 novel genes and 22,845 novel isoforms; 52.32% of the novel genes had annotations, and three novel isoforms had annotations for pathways linked to terpene production. The enrichment analysis of differentially expressed genes (DEGs) showed that 47 DEGs were enriched in the skeletal route of terpene synthesis, and 95 genes were classified into 21 categories crucial for seed development. Members of the phosphosynthase family, which generates terpene precursors, show a range of complicated expression patterns throughout seed development when combined with real-time quantitative reverse transcription PCR (qRT-PCR). The use of SMRT simplified a number of tasks, including the annotation of the ginkgo genome, the increase of the transcript collection for research on the development of ginkgo seeds, and a better understanding of how terpene synthesis is controlled in ginkgo seeds (Han et al. 2021).

To understand the regulation network of ginkgo flavonoids, many regulatory genes and transcription factors (TFs) have been discovered (Ying et al. 2020). However, diploid ginkgo was used in every single one of these investigations. Smarda et al. (2016) used flow cytometry to identify natural ginkgo variations between diploid and haploid ginkgo, which is crucial for identification the consequences of haploids. They also examined their cytological characteristics (Figs. 9.8 and 9.9).

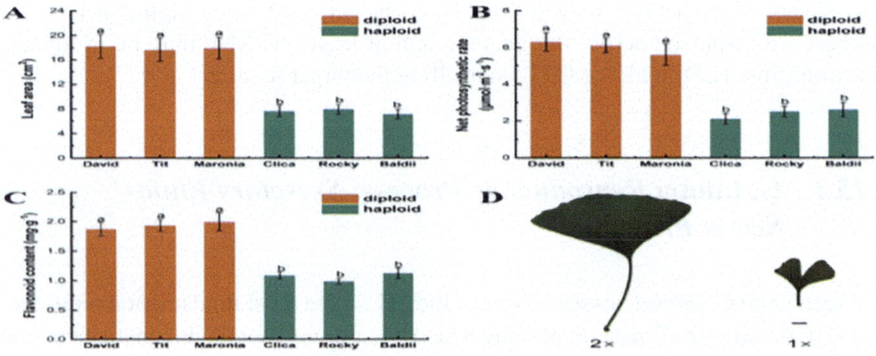

Fig. 9.8 Comparison of growth, morphology, flavonoid content, and photosynthesis of diploids and haploids. Leaf area (**a**). **b** The leaves' net photosynthetic rate. Leaf flavonoid concentration (**c**). **d** The morphology of haploid and diploid leaves. According to Hu et al. (2022)

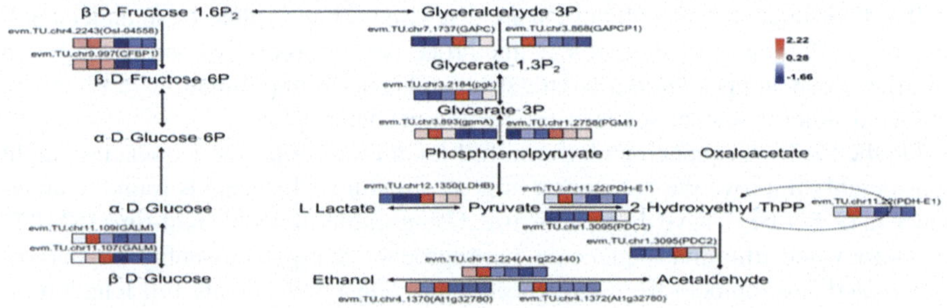

Fig. 9.9 Diagrammatic representation of DEGs in the haploid ginkgo's glycolysis/gluconeogenesis pathway. Ginkgo haploids are represented by the three squares on the left of the gene expression heatmap, whereas diploid ginkgo is represented by the three squares on the right. Hu et al. (2022)

Investigating the variations between diploid and haploid ginkgo trees is necessary to comprehend the effect of gene dosage on gene expressions (Hu et al. 2022). To examine the variations in biochemistry, physiology and growth between haploid and diploid ginkgo, Hu et al. (2022) performed a homogenous garden experiment. Additionally, a comprehensive transcriptome database for both was created to show the variations in their transcription. Their research showed that haploid ginkgo has less photosynthesis, less capacity for development, and less capacity for flavonoid accumulation. Despite the up-regulated expression of DEGs in haploid ginkgo reaching 46.7% of the total DEGs in the whole transcriptome data, the gene sets of photosynthesis metabolic, glycolysis/gluconeogenesis, and flavonoid biosynthesis pathways, which were significantly related to these differences, were found to show a significant down-regulated expression trend by gene set enrichment analysis (GSEA). Compared to diploid ginkgos, haploid ginkgos exhibit a significantly lower gene dosage.

9.15.2 SDS-PAGE Electrophoresis

According to Song et al. (2013), EGb's (leaves extracts of *G. biloba*) well-known antioxidant and anti-inflammatory characteristics may combine to have renoprotective benefits. *G. biloba* protein that agglutinated blood cells was isolated using a Sepharose CL-6B affinity column. These proteins are lectins with a clotting action, although galactose can reduce or decrease their clotting activity. By using simple affinity chromatography in one step, ginkgo fruit lectin may be isolated. The lectin protein found in ginkgo is a single subunit, but those found in bitter gourd and castor bean are multi-subunit lectins, all of which have a molecular weight of around 33 kDa. subunits -galactose particularly recognized by B-chain. As a result, the B-chain that galactose particularly recognizes is an old protein, supporting the notion that *G. biloba* is an old plant. Additionally, the study of ginkgo lectin suggests that the protein structure of extinct plants may be rather straightforward. The structure of functional proteins changed from being simple to complex together with the evolution of species (Van Holle and Van Damme 2019). In the study of Zhou et al. (2016), a brand-new antioxidant protein called GBSP was discovered in the seeds of *G. biloba*. Using MALDI-TOF/MS and SDS-PAGE, the apparent molecular weight of GBSP was discovered to be 18 kDa. The MALDI-TOF/TOF MS study of the amino acid sequence revealed that GBSP was a new protein. Significant anti-free radical defenses were shown by the protein against DPPH, ABTS, and superoxide anion.

9.15.3 Molecular Marker-Assisted Breeding

9.15.3.1 Genomic Resources

G. biloba, according to Hori et al. (2012), serves as a priceless bridge between seed plants and non-flowering, non-primitive plants. According to Heesch et al. genetic mapping is a dependable technique for map-based cloning, QTL mapping, and sequencing scaffold allocation. For Ginkgo genetic investigations, several sorts of markers have been created in the past. These include the biomarkers RAPD (Fan et al. 2004), isozyme (Tsumura and Ohba 1997), ISSR (Ge et al. 2003), and AFLP (Wang et al. 2008). Lin et al. (2011) created 204 EST-SSR markers in G. biloba using Roche 454 sequencing, whereas Han et al. (2015) used a Ginkgo cDNA library to create more than 4000 SSRs. These marker resources' accessibility has made it easier to publish several ginkgo genome maps. Tao (2016) created a connection map with just a few non-anonymous markers. It is crucial to create a highly dense linkage map using markers that are based on sequences to make the genome sequencing of Ginkgo possible.

9.16 Ginkgo Biloba Studies Using Tissue Culture

9.16.1 Embryo Culture

According to Li and Shen's (1934a, b) study, embryos that were taken out of the female gametophyte nutritive tissue of the seed and placed on a nutritious media grew more quickly. One of the first studies on Ginkgo embryos was conducted in this manner. Li stressed the need of adequate aeration and highlighted the main root's quick development as well as how the cotyledons became green in the presence of light. Kinetin, which inhibited the root, and gibberellic acid, which has been demonstrated to limit growth even at low quantities, were utilized by Bulard in 1967 as growth inhibitors. Radforth (1936) grew very young Ginkgo embryos in the female gametophyte's archegonia on a complicated media with yeast extract. On the other side, asparagine prevented growth. According to study by Webb et al. (1986), the use of Parafilm closures prevented embryo growth and promoted callus formation, which raised the demand for ideal aeration.

9.16.2 Gametophyte Female of G. Biloba

Rohr (1980) grew the female gametophyte of the Ginkgo and reported on the cytological changes over a two-month period using a basic Knops solution with additional glucose and thiamin. Rohr (1989) used female gametophytes that were 4–6 months post-pollination. Rohr found the aberrant cell development around the archegonia and the apical end of the gametophyte generating a whitish tissue. He also saw the zygotic embryos grow into fully grown plants.

The female ginkgo gametophyte's green haploid tissue cultures were created by taking the tissue from ovules that had undergone surface sterilization 11–13 weeks after pollination. The gametophyte was developed on White's medium containing 2,4-D and coconut water, which contains a range of amino acids, salts, and growth stimulants (Tulecke 1964). Additional investigation and comparisons of the free amino acid pools in the green tissue were also conducted, along with comparisons to albino, white pollen, and green tissue cultures created from Ginkgo stem and leaf tissue. Subculturing was employed to continuously develop all these tissues over a period of two to fifteen years (Tulecke 1967).

9.16.3 The Male Gametophyte

Friedman (1987), Friedman and Gifford (1988), and other researchers' observations and experiments clearly show the development of the male gametophyte inside the ginkgo plant's growing ovule. They asserted that the generative cell underwent an anticlinal annular ringlike division to form the stalk and the spermatogenous cell.

Comprehensive in vitro investigations of the male gametophyte were published by Friedman in 1987. He confirmed that gametophyte growth and the early starch buildup in pollen germination were both influenced by sucrose levels; lower sucrose promoted gametophyte branching. According to his statement, "Only a small portion of gametophytes in culture start the generative cell's division".

Rohr (1980) cultured the male gametophyte in a Bourgin and Nitsch medium devoid of adenine. He asserted that the reproductive (generative) cell was the ancestor of both the basal cell and the gamete mother cell (the spermatogenous cell). Two of the earliest discoveries about the development of the male gametophyte in culture were the building of a tissue from pollen and the synthesis of sperm cells in more than 50 gametophytes. Favre-Duchartre (1956) claimed that astral rays were also present in the sperm cells that developed from the large nucleated spermatogenous cell, also referred to as the body cell.

Friedman and Gifford's research has helped to improve the nomenclature used to describe the various Ginkgo male gametophyte forms (1988). These investigations utilized a complicated medium that included yeast extract or coconut milk to facilitate the growth of sperm cells. They were immature sperm cells since fully developed sperm cells produced in vitro would be visible swimming within the pollen tube or in a suitable media. On a medium with yeast extract and inoculations of fresh pollen, this development had taken place. Observations on 56 pairs of sperm cells showed that the orientation of the two daughter cells to the stalk cells, prothallial cells, and the exine of the pollen grain was exactly what occurred in vivo (Fig. 9.10).

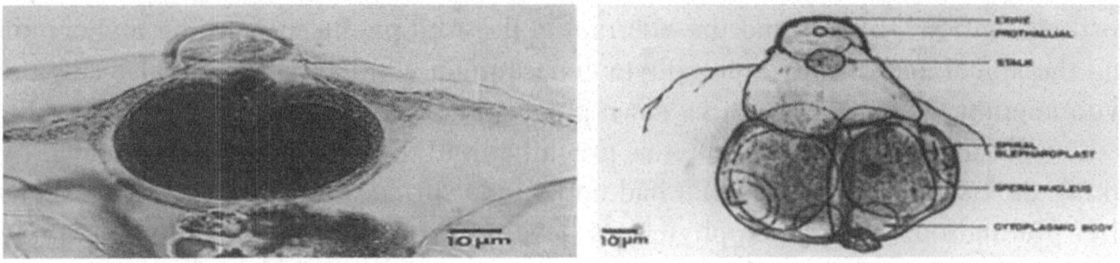

(a) (b)

Fig. 9.10 **a** A pair of developing sperm cells with the pollen grain's initial exine visible above them. **b** the spermatogenous cell's two sperm cells' spiral blepharoplasts. This was the male gametophyte's most advanced stage of in vitro development

9.16.4 Ginkgo Tissue Cultures

Ginkgo pollen was used to create the first pollen tissue culture. A complex medium containing White's and tomato juice was used to disseminate out mature strobili after they had been collected and surface sterilized. The tissues were easily created in subsequent tests using fresh and conserved pollen cultured on different media. The coenocytic gametophyte tube cells or septate pollen tubes, each of which can have 20–40 nuclei, are frequently produced when the tube cell nucleus divides (Tulecke 1965).

9.16.5 Embryogenesis

Numerous haploid or diploid tissues, cells, or protoplasts can be produced by woody perennials and develop into developing embryos (Tulecke 1987). From the haploid protoplasts of the female Ginkgo gametophyte produced at the incredibly early stage of septation, Laurain et al. (1993) were successful in directing embryogenesis. After three months in culture, the frequency of embryo development peaked at 1.5–1.9%. It was essential to have the right cell densities, proper culture medium, and a favorable uninucleate stage of microspore formation. Early embryo development in some cultures seems to be caused by electrostimulation. Ginkgo ovule tissues from trees in Chile, Argentina, and Australia were used to promote embryogenesis (Fontanel et al. 1994). Most of the imported Chilean ovules used in the embryogenic cultures; these cultures were carried out within the "critical period" for culture, which is 8–11 weeks after pollination. The apical and marginal cells of the female gametophyte underwent meristematic activity, which led to the development of an embryo (Figs. 9.11 and 9.12). Other attempts to promote embryogenesis using zygotic embryos or leaf tissue only yielded callus or green excrescences.

Ginkgo zygotic embryos were encouraged to form embryogenic callus, which led to the formation of embryos on the cotyledons. The embryos were in the torpedo or

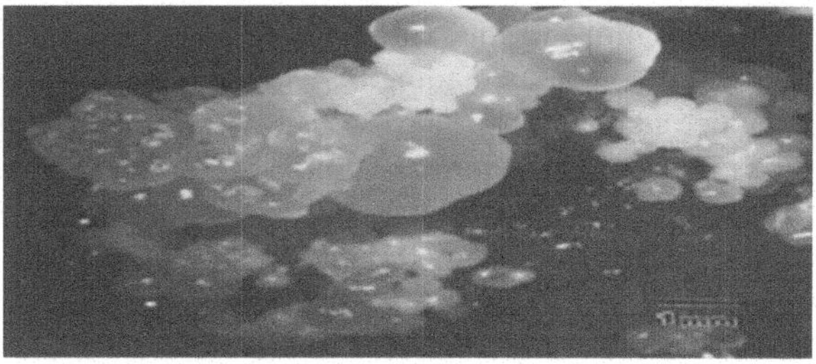

Fig. 9.11 Nine weeks after pollination, haploid Ginkgo megagametophytes were harvested and their embryos were cultured (Fontanel et al. 1994)

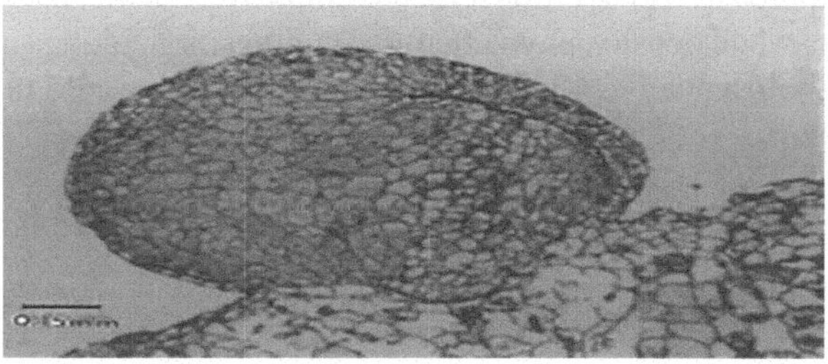

Fig. 9.12 After 3 weeks in the presence of 2,4-D and 2 months on a hormone-free medium, the L. S. of a globular embryo generated on a megagametophyte is obtained

early cotyledonary stages and were small (1–3.6 mm). A two-step process involving media for induction and development can be used to harvest the embryos from the cultures and develop them to the cotyledonary stage.

9.17 Genetic Engineering and Gene Editing

Ginkgo trees are significant both physiologically and commercially, therefore much work has gone into sequencing their DNA. Using markers based on sequences, this new map will provide a foundational framework for the future anchoring of sequence assemblies along Ginkgo chromosomes. Using Megagametophytes of 94 seeds from a single Ginkgo tree, a high-density genetic map was created using the specific locus amplified fragment (SLAF) sequencing method. With an average sequencing depth of 11.20×, 538,031 high-quality SLAFs were produced. 204,361 of these SLAFs were heterozygous in the mother tree and separated in the offspring. 12,263 SLAFs that were allocated to 12 linkage groups (LGs) were included in the created map. The number of chromosomes in Ginkgo was the same as the number of LGs on this map (Liu et al. 2017a, b).

The physiology, phenology, and growth of trees are significantly impacted by the multiple threats to forest ecosystems and the deterioration of these biological systems. The introduction of climate-adaptive genetically modified trees into damaged forests might help lower ecological hazards and create new opportunities. The fear of transgenic dissemination in forests should prevent any resistance to this. To adapt to change and survive, it is crucial to have a firm understanding of the genetic composition and phenotypic limits of the reproductive capacity of tree stands and populations. For a more thorough understanding of genes and their functional activities, the capacity to recognize adaptation progressions utilizing gene editing (GeEd) technology is essential. GeEd could therefore be applied to resolve allelic interactions for the expansion of gene diversity, slowing the spread of transgenic organisms among populations or species in various bioclimatic regions of the planet. This research focuses on the possibility of using CRISPR/Cas tools to alter genes in trees using a range of programmable changes based on genomic, transcriptomic, and epigenomic data that may be able to restore the gene function that is specific to a given phenotype. The application of several GeEd techniques to genetically increase several variables, such as wood density, phytochemical composition, biotic and abiotic stress tolerance, and photosynthetic efficiency in trees, was also covered in the 2022 article by Thapliyal et al.

9.18 Hybridization Events

All the close relatives of Ginkgo L. are extinct, making Ginkgo the only remaining species (Zhou et al. 2020). Therefore, no instances of hybridization between Ginkgo and any other species have been documented.

9.19 Conclusion and Future Prospects

The pyramidal broadleaf, deciduous ginkgo tree, often called maidenhair, has an artistic form has distribution throughout the temperate areas of Argentina, India, and New Zealand, also is grown throughout most of Europe and America. The tree is native to China, Japan, and Korea. Thanks to a variety of chemical defense mechanisms, including the repellence and anti-feedance against insect herbivores exerted by terpenic lactones (ginkgolides and bilobalides), the ginkgo is noted for its broad spectrum of resistance and tolerance to illnesses and insects. Chemical toxicity caused by terpene-based volatile organic compounds (VOCs), flavonoids, and/or lactones that attract predatory herbivorous insects. Due to dosage effects, gene duplication may enhance the effectiveness of resistance-related responses. It may also allow for the adaptive evolution of duplicated copies. gathering, conservation, analysis, and selection of genetic resources, as well as the cultivation of Ginkgo species under controlled conditions to produce their leaves, which are a crucial component of the trees and have a variety of purposes. *G. biloba* is appropriate and pertinent for urban environments due to its great resilience to environmental pressures, microbiological diseases (bacterial, viral, and fungal), various pests, and gaseous pollutants ozone and SO_2. They are salt-tolerant plants, making them suitable options for individuals who landscape close to the seaside.

Due to its economic importance in many areas mentioned in this chapter of the book and the possibility of cultivating it using modern technologies and due to its tolerance of climatic conditions such as salinity, drought and temperatures, especially the phenomenon of global warming, therefore we recommend paying attention to its cultivation along the coastal strip of the Mediterranean Sea in Egypt so that we can benefit from it in research environmentally and in various industries.

References

Ahmad KS, Wazarat A, Mehmood A, Ahmad MSA, Tahir MM, Nawaz F, Ahmed H, Zafar M, Ulfat A (2020) Adaptations in *Imperata cylindrica* (L.) Raeusch. and *Cenchrus ciliaris* L. for altitude tolerance. Biologia 75:183–198. https://doi.org/10.2478/s11756-019-00380-2

Archangelsky S, Cuneo R (1990) *Polysper mophyllum*, a new Permian gymnosperm from Argentina, with considerations about the Dicranophyllales. Rev Palaeo Bot Palynol 63:117–135

Acquaah G (2015) Conventional plant breeding principles and techniques. In: Al-Khayri JM, Jain SM, Johnson DV (eds) Advances in plant breeding strategies: breeding, biotechnology, and molecular tools. Springer International Publishing, Cham, pp 115–158. https://doi.org/10.1007/978-3-319-22521-0

Beaulieu D, Hughes A (2022) How to grow and care for Ginkgo Biloba trees. The Spruce https://www.thespruce.com. Plants & Flowers Trees

Belwal T, Giri L, Bahukhandi A, Tariq M, Kewlani P, Bhatt ID, Rawal RS (2018) Ginkgo biloba. In: Nonvitamin and nonmineral nutritional supplements. Elsevier Inc

Bent S, Goldberg H, Padula A, Avins AL (2005) Spontaneous bleeding associated with Ginkgo biloba. J Gen Intern Med 20:657–661. https://doi.org/10.1111/j.1525-1497.2005.0121.x

Brown LA, Riby LM, Reay JL (2010) Supplementing cognitive aging: a selective review of the effects of Ginkgo Biloba and a number of everyday nutritional substances. Exp Aging Res 36:105–122. https://doi.org/10.1080/03610730903417960

Carr CJ (1998) Patterns of vegetation along the Omo River in southwest Ethiopia. Plant Ecol 135:135–163

Chen Y, Fu C, Wu Z, Xu H, Liu H, Schneider H, Lin J (2021) Ginkgo Biloba. Trends Genet 37:488–489

DeFeudis FV (1998) *Ginkgo biloba* extract (EGb 761): from chemistry to the clinic. Ullstein Medical, Wiesbaden, p 400

Del Tredici P (1989) Ginkgos and multituberculates: evolutionary interactions in the tertiary. BioSystems 22:327–339

Fan XX, Shen L, Zhang X, Chen XY, Fu CX (2004) Assessing genetic diversity of *Ginkgo biloba* L. (Ginkgoaceae) populations from China by RAPD markers. Biochem Genet 42:269–278. https://doi.org/10.1023/B:BIGI.0000034431.15308.57

Favre-Duchartre M (1958) *Ginkgo,* an oviparous plant. Phytomorphology 8:377–390

Feng L, Sun J, Wang T, Tian X, Wang W, Guo J, Feng H, Guo H, Deng H, Wang G (2021) Predicting suitable habitats of *ginkgo biloba* L. fruit forests in China. Climate Risk Manage 34:100364

Fettig CJ, Mortenson LA, Bulaon BM, Foulk PB (2019) Tree mortality following drought in the central and Southern Sierra Nevada, California, U.S. For Ecol Manage 432:164–178

Fontanel AI, Serraf I, Petiard V (1994) Regeneration in *Ginkgo biloba* L.: induction of embryogenesis from megagametophyte according to the development state. C R Acad Sci 317:149–155

Friedman WE, Gifford EM (1988) Division of the generative cell and late development in the male gametophyte of *Ginkgo biloba.* Amer 1 Bot 75:1434–1442

Ge YQ, Qiu YX, Fu CX (2003) Genetic diversity of the "living fossil" *Ginkgo biloba* based on ISSR marker. Biodivers Sci 11:276–287

Gertz HJ, Kiefer M (2004) Review about *Ginkgo biloba* special extract EGb 761 (Ginkgo). Curr Pharm des 10:261–264

Gifford EM, Foster AS (1989) Comparative morphology of vascular plants, 3rd edn. WH Freeman, San Francisco

Gong W, Chen C, Dobes C, Fu CX, Koch MA (2008a) Phylogeography of a living fossil: pleistocene glaciations forced *Ginkgo biloba* L. (Ginkgoaceae) into two refuge areas in China with limited subsequent postglacial expansion. Molec Phylog Evol 48:1094–1105

Gong W, Zeng Z, Chen Y-Y, Chen C, Qiu Y-X, Fu C-X (2008b) Glacial Refugia of *Ginkgo biloba* and human impact on its genetic diversity: evidence from chloroplast DNA. J Integr Plant Biol 50(3):368–374. https://doi.org/10.1111/j.1744-7909.2007.00375.x

Gu K-J, Lin C-F, Wu J-J, Zhao Y-P (2021) Ginkgo DB: a comprehensive database for empowering living fossil conservation. https://ginkgo.zju.edu.cn

Gu X-R, Jiang Y-N, Ni Y-L, Jia H (2017) Research progress on breeding and cultivation of *Ginkgo biloba* for leaf use in China. Chin Traditional Herbal Drugs 48(15):3218–3227. https://doi.org/10.7501/j.issn.0253-2670.2017.15.030

Guan R, Zhao Y, Zhang H, Fan G, Liu X, Zhou W, Shi C, Wang J, Liu W, Liang X, Fu Y, Ma K, Zhao L, Zhang F, Lu Z, Lee SM-Y, Xu X, Wang J, Yang H, Fu C, Ge S, Chen W (2016) Draft genome of the living fossil *Ginkgo biloba*. GigaScience 5:49. https://doi.org/10.1186/s13742-016-0154-1

Gui RY (2004) The construction of fingerprint and genetic linkage map of several cultivars of *Ginkgo biloba*. Master's thesis, Nanjing Forestry University, Nanjing

Gunckel JE, Wetmore RH (1946) Studies of development in long shoots and short shoots of *Ginkgo biloba* L. I. The origin and pattern of development of the cortex, pith and procambium. Amer J Bot 33:285–295

Guo Y, Lu Y, El-Kassaby YA, Feng L, Wang G, Wang T (2019b) Predicting growth and habitat responses of *Ginkgo biloba* L. to climate change. Ann For Sci 76:101. https://doi.org/10.1007/s13595-019-0885-0

Guo J, Wu Y, Wang T, Xin Y, Wang G, Zhou Q, Xu L-A (2023) *GbFLSa* overexpression negatively regulates proanthocyanin biosynthesis. Front Plant Sci (Sec Plant Metabolism Chemodiversity) 14. https://doi.org/10.3389/fpls.2023.1093656

Han X, He B, Xin Y, Xu M, Xu L-A (2021) Full-length sequencing of *Ginkgo biloba* L. reveals the synthesis of terpenoids during seed development. Ind Crops Products 170:113714. https://doi.org/10.1016/j.indcrop.2021.113714

Han Y, Zhao X, Cao G, Wang Y, Li Y, Liu D, Teng W, Zhang Z, Li D, Qiu L, Zheng H, Li W (2015) Genetic characteristics of soybean resistance to HG type 0 and HG type 1.2.3.5.7 of the cyst nematode analyzed by genome-wide association mapping. BMC Genomics 16:598. https://doi.org/10.1186/s12864-015-1800-1

Hara N (1997) Morphology and anatomy of vegetative organs in *Ginkgo biloba*. Chapter 1, pp 3–15

Hatano K, Miyakawa T, Sawano Y, Tanokura M (2011) Antifungal and lipid transfer proteins from Ginkgo (*Ginkgo biloba*) seeds. In: Nuts and seeds in health and disease prevention, pp 527–534. https://doi.org/10.1016/b978-0-12-375688-6.10063-5

He S-A, Yin G, Pang Z-J (1997) Resources and prospects of *Ginkgo biloba* in China, pp. 373–383 *in* Hori et al. 1997

Hilton MP, Stuart EL (2004) *Ginkgo biloba* for Tinnitus. Cochrane Database Syst Rev 2:CD003852

Holt B, Rothwell GW (1997) Is *Ginkgo biloba* really an oviparous plant? Am J Bot 84(6):870–872

Hori T, Ridge RW, Tulecke W, Tredici PD, Tremouillaux-Guiller J, Tobe H (2012) *Ginkgo biloba*, A global treasure: from biology to medicine. Springer Science & Business Media, Berlin

Hu Y, Šmarda P, Liu G, Wang B, Gao X, Guo Q (2022) High-depth transcriptome reveals differences in natural haploid *Ginkgo biloba* L. Due to the effect of reduced gene dosage. Int J Mol Sci 23:8958. https://doi.org/10.3390/ijms23168958

Huang L, Tian L, Zhou L, Jin C, Qian S, Jim CY, Lin D, Zhao L, Minor J, Coggins C, Yang Y (2020) Local cultural beliefs and practices promote conservation of large old trees in an ethnic minority region in southwestern China. Urban Forestry Urban Greening 49:126584. https://doi.org/10.1016/j.ufug.2020.126584

Huang W, Xie B, Wang Y, Yang EN, Luo R (2004) Study on separation and purification of protein from ginkgo seed and its antioxidant activity. Scientia Agricultura Sinica 37:1537–1543

Ingvarsson PK, Garcia MV, Luquez V, Hall D, Jansson S (2008) Nucleotide polymorphism and phenotypic associations within and around the phytochrome B2 locus in European aspen (Populus tremula, Salicaceae). Genetics 178:2217–2226

Isah T (2015) Rethinking *Ginkgo biloba* L.: Medicinal uses and conservation. Pharmacognosy Rev 9(18):140–148. https://doi.org/10.4103/0973-7847.162137

Khadivi-Khub A, Salehi-Arjmand H, Hadian J (2014) Morphological and phytochemical variation of Satureja bachtiarica populations from Iran. Ind Crops Prod 54:257–265. https://doi.org/10.1016/j.indcrop.2014.01.039

Kovalenko IM, Klymenko GO, Melnychuk SD, Skliar IuL, Melnyk OS, Kyrylchuk KS, Bondarieva LM, Zubtsova IV, Yaroshchuk RA, Zherdetska SV (2021) Potential adaptation of *Ginkgo biloba*–comparative analysis of plants from China and Ukraine. Ukrainian J Ecol 11(1):329–337. https://doi.org/10.15421/2020_301

Laurain D, Chenieux JC, Tremouillaux-Guiller J (1993) Direct embryogenesis from female haploid protoplasts of *Ginkgo biloba* L., a medicinal woody species. Plant Cell Rep 12:656–660

Lee KH, Chung GC, Lee JS (1993) Effects of simulated acid rain on stomatal resistance, wettability and anatomical changes in *Quercus acutissima* and *Ginkgo biloba* seedlings. J Korean for Soc 82:328–336

Leigh A, Zwieniecki MA, Rockwell FE, Boyce CK, Nicotra AB, Holbrook NM (2011) Structural and hydraulic correlates of heterophylly in *Ginkgo biloba*. New Phytol 189:459–470

Li HL (1956) A horticultural and botanical history of *Ginkgo*. Bull Morris Arboretum 7:3–12

Li TT, Shen T (1934) The effect of "pantothenic acid" on the growth of the yeast and on the growth of the radicle of *Ginkgo* embryo in artificial media. Sci Rep Nat! Tsinghua Univ Ser 2:53–60

Li TT, Shen T (1934b) The effect of "pantothenic acid" on the growth of the yeast and on the growth of the radicle of *Ginkgo* embryo in artificial media. Sci Rep Nat Tsinghua Univ Ser 2:53–60

Li W, Yang S, Lu Z, He Z, Ye Y, Zhao B, Wang L, Jin B (2018) Cytological, physiological, and transcriptomic analyses of golden leaf coloration in *Ginkgo biloba* L. Hortic Res 5:12

Liebgott T, Miollan M, Berchadsky Y, Drieu K, Culcasi M, Pietri S (2000) Complementary cardioprotective effects of flavonoid metabolites and terpenoid constituents of *Ginkgo biloba* extract (EGb 761) during ischemia and reperfusion. Basic Res Cardiol 95(5):368–377

Lin H-Y, Li W-H, Lin C-F, Wu G-R, Zhao Y-P (2022) International biological flora: *Ginkgo biloba*. J Ecol 110:951–982. https://doi.org/10.1111/1365-2745.13856

Lin X, Zhang J, Li Y, Luo H, Wu Q, Sun C, Song J, Li X, Wei J, Lu A, Qian Z, Khan IA, Chen S (2011) Functional genomics of a living fossil tree, *Ginkgo*, based on next-generation sequencing technology. Physiol Plant 143:207–218. https://doi.org/10.1111/j.1399-3054.2011.01500.x

Liu E (2014) Ginkgo and Tartary Buckwheat Vinegar. CN103981077A, 13

Liu H, Wang X, Wang G, Cui P, Wu S, Ai C, Hu N, Li A, He B, Shao X, Wu Z, Feng H, Chang Y, Mu D, Hou J, Dai X, Yin T, Ruan J, Cao F (2021) The nearly complete genome of *Ginkgo biloba* illuminate's gymnosperm evolution. Nat Plants 7(6):748–756. https://doi.org/10.1038/s41477-021-00933-x

Liu H, Ye M, Guo H (2020) An updated review of randomized clinical trials testing the improvement of cognitive function of *Ginkgo biloba* extract in healthy people and Alzheimer's patients. Front Pharmacol 10:1688

Liu H, Zhang C, Meng Y, Zhang F, Huang N, Wang J, Li Y (2023) Hydraulic and economical traits in short- and long-shoot leaves of *Ginkgo biloba* males and females. Forests 14:535. https://doi.org/10.3390/f14030535

Liu X, Yu W, Zhang X, Wang G, Cao F, Cheng H (2017a) Identification and expression analysis under abiotic stress of the R2R3-MYB genes in *Ginkgo biloba* L. Physiol Mol Biol Plants 23(3):503–516. https://doi.org/10.1007/s12298-017-0436-9

Liu H, Cao F, Yin T, Chen Y (2017b) A Highly dense genetic map for *Ginkgo biloba* constructed using sequence-based markers. Front Plant Sci 8:1041. https://doi.org/10.3389/fpls.2017.01041

Lu Q, Yang T, Zhang M, Du L, Liu L, Zhang N, Guo H, Zhang F, Hu G, Yin X (2014) Preventative effects of *Ginkgo biloba* extract (EGb 761) on high glucose-cultured opacity of rat lens. Phytother Res 28(5):767–773

Lu Z-W, Popova EV, Wu C-H, Lee E-J, Hahn E-J, Paek K-Y (2009) Cryopreservation of *Ginkgo biloba* cell culture: effect of pretreatment with sucrose and ABA

Matsui K, Shinkawa Y, Tsuboi M, Kojima H, Ando Y (1991) Simple production of extract with high content of flavonoid from Ginkgo Leaf. JPH03227985A

McElwain JC (2013) Ginkgo the tree that time forgot. Science 340:812–813

Meuwissen THE, Hayes BJ, Goddard ME (2001) Prediction of total genetic value using genome-wide dense marker maps. Genetics 157:1819–1829

Meyen SV (1987) Fundamentals of palaeobotany. Chapman and Hall, London

Mir MA, Albaradie RS (2015) Immunomodulation of inflammatory markers in activated macrophages by leaf extracts of *Gingko biloba*. Adv Neuroimmune Biol 6(1):9–17

Murray BG (2013) Karyotype variation and evolution in gymnosperms. In: Greilhuber J (ed) Plant genome diversity, vol 2. Springer, Vienna, pp 231–243. https://doi.org/10.1007/978-3-7091-1160-4_14

Nagataa T, Hasebeb M, Toribab T, Tanedac H, Crane PR (2016) Sex conversion in *Ginkgo biloba* (Ginkgoaceae). J Jpn Bot 91:120–127

Nakayama H, Sinha NR, Kimura S (2017) How do plants and phytohormones accomplish heterophylly, leaf phenotypic plasticity, in response to environmental cues. Front Plant Sci 8:1717

Norstog KJ, Gifford EM, Stevenson DW (2004) Comparative development of the spermatozoids of cycads and *Ginkgo biloba*. Bot Rev 70(1):5–15. https://doi.org/10.1663/0006-8101(2004)070[0005:cdotso]

Ocokoljić M, Grbić M, Anastasijević N (2011) Characteristics and variability of half-sib progeny as the base for ginkgo breeding. Afr J Biotech 10(85):19752–19757. https://doi.org/10.5897/AJB11.2058

Oh JW, Chae JS, Yi CK (1983) Effects of atmospheric sulfur dioxide concentration on growth of some tree species (in Korean, abstract in English) Ministry of Science and Technology, Seoul, pp 49–69

Oh T-S, Koo H-M, Yoon H-R, Jeong N-S, Kim Y-J, Kim C-H (2015) Antifungal action of *Ginkgo biloba* outer seedcoat on rice sheath blight. Plant Pathol J 31:61–66

Park YG, Kim SJ, Kang YM, Jung HY, Prasad DT, Kim SW, Chung YG, Choi MS (2004) Production of ginkgolides and bilobalide optimized the *Ginkgo biloba* cell culture. Biotechnol Bioprocess Eng 9:41–46

Park YJ, Ahn HY, Kim HR, Chung KH, Oh SM (2016) *Ginkgo biloba* extract EGb 761-mediated inhibition of aromatase for the treatment of hormone-dependent breast cancer. Food Chem Toxicol 87:157–165

Pszczolkowski MA Durden K, Sellars S, Cowell B, Brown JJ (2011) Effects of *Ginkgo biloba* constituents on fruit-infesting behavior of codling Moth (*Cydia pomonella*) in Apples. J Agric Food Chem 59(20):10879–10886. https://doi.org/10.1021/jf202386c

Qian L (2015) Preparation method of Ginkgo Drink. CN104770812A

Radforth W (1936) The development in vitro of the proembryo of *Ginkgo*. R Can Inst Trans 21:87–94

Ramassamy C (2006) Emerging role of polyphenolic compounds in the treatment of neurodegenerative diseases: a review of their intracellular targets. Eur J Pharmacol 545(1):51–64

Raven PH, Evert RF, Eichhorn SL (1986) Biology of plants, 4th edn. Worth, New Youk

Ražná K, Sawinska Z, Ivanišová E, Vukovic N, Terentjeva M, Strićík M, Kowalczewski PŁ, Hlavačková L, Rovná K, Žiarovská J et al (2020) Properties of *Ginkgo biloba* L.: antioxidant characterization, antimicrobial activities, and genomic MicroRNA based marker fingerprints. Int J Mol Sci 21:3087

Renzaglia KS, Garbary DJ (2001) Motile gametes of land plants: diversity, development, and evolution. Crit Rev Plant Sci 20:107–213

Rohr R (1980) De differentiation en culture in vitro des cellules a reserves dans l'endosperme de *Ginkgo biloba.* Z Pflanzenphysiol 96:423–434

Rohr R (1989) Maidenhair tree *(Ginkgo biloba* L.). In: Bajaj YPS (ed) Biotechnology in agriculture and forestry, vol 5. Trees. II. Springer, Heidelberg, pp 574–590

Rothwell GW, Serbet R (1994) Lignophyte phylogeny and the evolution of spermatophytes: a numerical cladistic analysis. Syst Bot 19:443–482

Royer DL, Hickey LJ, Wing SL (2003) Ecological conservatism in the "living fossil" *Ginkgo.* Paleobiology 29(1):84–104

Sakisaka M (1958) Study on *Ginkgo* (in Japanese). Kazama Shobou, Tokyo

Seiwa K, Kikuzwa K (1996) Importance of seed size for the establishment of seedlings of five deciduous broad-leaved tree species. Vegetatio 123:51–64

Shen L, Chen X-Y, Zhang X, Li Y-Y, Fu C-X, Qiu Y-X (2005) Genetic variation of *Ginkgo biloba* L. (Ginkgoaceae) based on cpDNA PCR-RFLPs: inference of glacial refugia. Heredity 94(4):396–401. https://doi.org/10.1038/sj.hdy.6800616

Singh B, Kaur P, Singh RD, Ahuja PS (2008) Biology and chemistry of *Ginkgo biloba.* Fitoterapia 79(6):401–418

Smarda P, Vesely P, Smerda J, Bures P, Knapek O, Chytra M (2016) Polyploidy in a "living fossil' *Ginkgo biloba.* New Phytol 212:11–14

Song J, Liu D, Feng L, Zhang Z, Jia X, Xiao W (2013) Protective effect of standardized extract of *Ginkgo biloba* against Cisplatin-induced nephrotoxicity. In: Evidence-based complementary and alternative medicine, vol 2013, Article ID 846126, 11 pages. https://doi.org/10.1155/2013/846126

Stein WE, Beck CB (1987) Paraphyletic groups in phylogenetic analysis: progymnospermopsida and prephanerogames in alternative views of seed plant relationships. Bull Soc Bot France 134 Actualites Botaniques 2:107–119

Sun B, Dilcher DL, Beerling DJ, Zhang C, Yan D, Kowalski E (2003) Variation in *Ginkgo biloba L.* leaf characters across a climatic gradient in China. Proc Natl Acad Sci 100(12):7141–7146

Tan X, Sun Z, Liu Q, Ye H, Zou C, Ye C, Wang A, Lin H (2018) Effects of dietary ginkgo biloba leaf extract on growth performance, plasma biochemical parameters, fish composition, immune responses, liver histology, and immune and apoptosis-related genes expression of hybrid grouper (*Epinephelus lanceolatus*♂ × *Epinephelus fuscoguttatus*♀) fed high lipid diets. Fish Shellfish Immunol 72:399–409

Tang CQ, Ohsawa M (1997) Zonal transition of evergreen, deciduous, and coniferous forests along the altitudinal gradient on a humid subtropical mountain, Mt. Emei, Sichuan, China. Plant Ecol 133:63–78

Tao R, Wang C, Ye J, Zhou H, Chen H (2016) Polyprenols of Ginkgo biloba enhance antibacterial activity of five classes of antibiotics. Biomed Res Int 2016(1):4191938. https://doi.org/10.1155/2016/4191938

Tian Y, Han X, Qu Y, Zhang Y, Rong H, Wu K, Xu L (2022) Genome-wide identification of the Ginkgo (*Ginkgo biloba* L.) LBD transcription factor gene and characterization of its expression. Int J Mol Sci 23:5474. https://doi.org/10.3390/ijms23105474

Tiffney BH (1992) The role of vertebrate herbivory in the evolution of land plants. Paleobotanist 41:87–97

Toghueo RMK (2020) Endophytes from *Gingko biloba*: the current status. Phytochem Rev 19:743–759

Tsumura Y, Ohba K (1997) The genetic diversity of isozymes and the possible dissemination of *Ginkgo biloba* in ancient times in Japan. In: Hori T, Ridge RW, Tulecke W, Del Tredici P, Tremouillaux-Guiller J, Tobe H (eds) *Ginkgo biloba* A global treasure. Springer, Tokyo, pp 159–172. https://doi.org/10.1007/978-4-431-68416-9_13

Tulecke W (1964) A haploid tissue culture from the female gametophyte of *Ginkgo biloba* L. Nature 203:94–95

Tulecke W (1967) Studies on tissue cultures derived from *Ginkgo biloba* L. Phytomorphology 17:381–386

Tulecke W (1987) Somatic embryogenesis in woody perennials. In: Bonga JM, Durzan DJ (eds) Cell and tissue culture in Forestry, vol 2. Specific principles and methods: growth and development. Nijhoff, Dordrecht, pp 61–69

Tulecke W, Taggart R, Colavito L (1965) Continuous cultures of higher plant cells in liquid media. Contrib Boyce Thompson Inst 23:33–46

Vakhrameev VA (1991) Jurassic and cretaceous floras and climates of the earth (Translated by Ju. V. Litvinov, edited by Norman F. Hughes). Cambridge University Press, Cambridge

Van Holle S, Van Damme EJM (2019) Messages from the past: new insights in plant lectin evolution. Front Plant Sci. www.frontiersin.org. Vol 10, Article 36

Wang F, Cui X, Liu G, Lan X, He Y (2014) Ginkgo beverage and preparation method thereof. CN104473273B

Wang L, Xing SY, Wang F, Han KJ (2008) AFLP analysis on genetic relationship of *Ginkgo biloba* female germplasms. Sci Silvae Sin 47:91–96

Webb DT, Arias W, Hostos E (1986) Callus formation by *Ginkgo biloba* embryos on hormone-free media controlled by closures and media components. Phytomorphology 36:121–128

Wu D, Feng J, Lai M, Ouyang D, Yu W, Wang G, Cao F, Jacobs DF, Zeng S (2020a) Combined application of bud and leaf growth fertilizer improves leaf flavonoids yield of *Ginkgo biloba.* Ind Crops Prod 150:112379

Wu Y, Ma X, Zhou Q, Xu L-A, Wang T (2020b) Selection of crown type provides a potential to improve the content of isorhamnetin in *Ginkgo biloba.* Ind Crops Prod 143:111943. https://doi.org/10.1016/j.indcrop.2019.111943

Xie H, Wang J-R, Yau L-F, Liu Y, Liu L, Han Q-B, Zhao Z, Jiang Z-H (2014) Quantitative analysis of the flavonoid glycosides and terpene trilactones in the extract of Ginkgo biloba and evaluation of their inhibitory activity towards fibril formation of β-amyloid peptide. Molecules 19:4466–4478. https://doi.org/10.3390/molecules19044466

Xing SY, Wu DJ, Xing LF, You XL, Zhang YP, Sun X, Liu YQ (2002) Quantitative genetic analysis and multiple trait selection of pharmaceutical composition on *Ginkgo biloba* L. leaves. Yi Chuan Xue Bao 29:928–935

Xu N, Liu S, Lu Z, Pang S, Wang L, Wang L, Li W (2020) Gene expression profiles and flavonoid accumulation during salt stress in *Ginkgo biloba* seedlings. Plants 9:1162. https://doi.org/10.3390/plants9091162

Xu W, Zhao Q (2022) Method for preparing Gingko powder from fresh Gingko fruits. CN114209048A

Yamazaki T, Takeoka M (1962) Electron-microscope investigations of the fine details of the pollen grain surface in Japanese gymnosperms. Grana PalynoI 3:3–12

Yang JT, Wu CE (2009) Advance in allergic composition and mechanism of allergy of *Ginkgo biloba* seeds. Food Sci Technol 6:282–286

Yang X (2022) Process for preparing Gingko Wine. CN114292718A

Ying G, Cga B, Mwa B, Fffa B, Ek C, Tw C, Gwa B (2020) Metabolome and transcriptome analyses reveal flavonoids biosynthesis differences in *Ginkgo biloba* associated with environmental conditions—ScienceDirect. Ind Crops Prod 158:112963

Zhang C, Fan Q, Chen SL, Ma H (2015) Reversal of P-glycoprotein overexpression by Ginkgo biloba extract in the brains of pentylenetetrazole-kindled and phenytoin-treated mice. Kaohsiung J Med Sci 31(8):398–404. https://doi.org/10.1016/j.kjms.2015.05.007

Zhao J, Yang J, Xie Y (2019a) Improvement strategies for the oral bioavailability of poorly water-soluble flavonoids: an overview. Int J Pharm 570:118642

Zhao Y-P, Fan G, Yin P-P, Sun S, Li N, Hong X, Hu G, Zhang H, Zhang F-M, Han J-D, Hao Y-J, Xu Q, Yang X, Xia W, Chen W, Lin H-Y, Zhang R, Chen J, Zheng X-M, Lee SM-Y, Lee J, Uehara K, Wang J, Yang H, Fu C-X, Liu X, Xu X, Ge S (2019b) Resequencing 545 ginkgo genomes across the world reveals the evolutionary history of the living fossil. Nat Commun 10:4201. https://doi.org/10.1038/s41467-019-12133-5

Zhou Q, Mu K, Ni Z, Liu X, Li Y, Xu LA (2020) Analysis of genetic diversity of ancient Ginkgo populations using SSR markers. Ind Crops Prod 145:111942. https://doi.org/10.1016/j.indcrop
Zhou Q, Shen X, Li Y (2022) Variation pattern and genome-wide association study of leaf phenotypic traits among ancient *Ginkgo biloba* L. populations. Forests 13:1764. https://doi.org/10.3390/f13111764
Zhou H, Wang C, Ye J, Chen H, Tao R, Cao F (2016) Effects of high hydrostatic pressure treatment on structural, allergenicity, and functional properties of proteins from ginkgo seeds. Innov Food Sci Emerging Technol (IFSET) 34:187–195
Ziegler AM, Rees PM, Rowley DB, Bekker A, Qing L, Hulver ML (1996) Mesozoic assembly of Asia: constraints from fossil floras, tectonics, and paleomagnetism. In: Yin A, Harrison M (eds) The tectonic evolution of Asia. Cambridge University Press, Cambridge, pp 371–400
Zumajo-Cardona C, Little DP, Stevenson D, Ambrose BA (2021) Expression analyse s in *Ginkgo biloba* provide new insights into the evolution and development of the seed. Sci Rep 11:21995. https://doi.org/10.1038/s41598-021-01483-0

Chapter 10
Recapitulating the Significance of Breeding Strategies in *Gymnema sylvestre* for Gymnemic Acid Enhancement

Saurabha Bhimrao Zimare, Atif Khurshid Wani, and Prachi Saurabha Zimare

Abstract *Gymnema sylvestre* (Retz.) R.Br. ex Sm., is a perennial woody climber of Apocynaceae family. It is native to Asia, Africa, Australia, and some parts of Arabian Peninsula. This is an important medicinal herb commonly used in various traditional medicinal systems to treat diabetes mellitus (DM). Plant parts of *G. sylvestre* have wide array of phytochemicals, however scientific community has paid more attention to its triterpenoid saponin, 'gymnemic acid'. This compound is considered to be the principle bioactive compound responsible for alleviating the blood sugar level. Considering the diverse medicinal importance of *G. sylvestre*, it has been haphazardly exploited from its natural habitat for commercial use leading in the depletion. Thus, strategies to conserve and propagate this commercially important plant are required. The content of gymnemic acid in *G. sylvestre* is more prone to seasonal and geographical variations. Hence, more emphasis has been given to the in vitro culture techniques. Biotechnological approaches such as plant tissue culture, shoot and cell culture, somatic embryogenesis, elicitors and precursors are the alternatives to enhance the gymnemic acid content up to commercial level. Along with these in vitro approaches use of arbuscular mycorrhizal fungi has proved to be a useful strategy for gymnemic acid production. This chapter gives an insight to the wide range of applications associated with *G. sylvestre* besides highlighting the role of different approaches for its sustainability.

S. B. Zimare
Department of Botany, Sir Parashurambhau College (Autonomous), Pune 411030, Maharashtra, India

A. K. Wani
Department of Biotechnology, School of Bioengineering and Biosciences, Lovely Professional University, Jalandhar 144411, Punjab, India

P. S. Zimare (✉)
Department of Botany, Comrade Godavari Shamrao Parulekar College of Arts, Commerce and Science, Talasari, Palghar 401606, India
e-mail: prachik16@gmail.com

J. M. Al-Khayri et al. (eds.), *Biodiversity and Genetic Improvement of Medicinal and Aromatic Plants I*, Advances in Plant Breeding Strategies 9,
https://doi.org/10.1007/978-3-031-81846-2_10

Keywords Callus and cell culture · Diabetes mellitus · Gymnema saponins · Gymnemic acids · Micropropagation

10.1 Introduction

Diabetes mellitus (DM) is a major endocrine disorder that progresses into integrative metabolic disorders such as hyperglycemia and carbohydrate, lipid and protein metabolic disorders, eventually leading into discrepancy in the insulin secretion and its efficacy (Solis-Herrera et al. 2000). The two major types of DM are: Type I diabetes or insulin-dependent DM (IDDM), which is caused due to failure in releasing insulin from the β-cells of the islets of Langerhans in the pancreas; and Type II diabetes or non-insulin-dependent DM (NIDDM) is caused due to insulin resistance possibly because of very few insulin receptors (King 2012). Nearly 13% population has been affected globally due to DM which has become major concern, as its leads to multi-organ complications in its serious conditions (Burke et al. 2003). If left untreated, can result in the damage of vital organs like heart and kidney, congestive kidney failure, cardiovascular complications, diabetic retinopathy, cataract, diabetic foot infection, and bone and joint disorders (Pálsson and Patel 2014). According to WHO (2022), at present about 422 million people worldwide have diabetes, and each year around 1.5 million deaths are directly attributed to diabetes. The incidence of diabetes is gradually increasing since few decades (https://www.who.int/health-topics/diabetes).

Various approaches such as non-pharmacological interventions (diet and physical activity), synthetic drugs, and immuno-suppressive compounds has been employed for the management of DM. Although these approaches are effective, they have certain disadvantages like harmful side-effects, uneconomical, and unreachable to many communities (Kibiti and Afolayan 2015). Taking into account all the above facts, researchers are exploring alternative methods for the management of this disease. Herbal therapy is a new avenue where discovering novel herbal preparations which do not have side-effects and are non-toxic in nature, affordable, and easily available to people, can be that alternative approach that can aid in the disease management. The inclination of people towards naturopathic treatment for chronic disease like DM is gaining momentum, and the traditional approach has the potential to offer a solution to challenge to the diabetic complications. As per World ethnobotanical data, approximately 800 herbs are used to control DM (Alarcon-Aguilara et al. 1998). However, very few herbs have been explored for their anti-diabetic property. Compilation of most notable medicinal plants exhibiting antidiabetic potential have been reviewed (Alarcon-Aguilara et al. 1998; Mentreddy 2007; Kibiti and Afolayan 2015; Alam et al. 2022).

One of the extensively explored plant is *Gymnema sylvestre* (Retz.) R.Br. ex Sm. (Apocynaceae), known for its antidiabetic activity. *G. sylvestre* has been commonly used in the traditional medicine system (Ayurveda, Homeopathy, Siddha, and Unani) for almost 2000 years. Sushruta (sixth century B.C.) and Charaka (first century

A.D.) recommended the use of leaves of '*Sālasārādi*' group of medicinal plants to control the maturity onset of diabetes, and *G. sylvestre* is one the plant species of this group. Because of its ability to abolish the sweet taste of sugar, it has been named as '*Madhunashini*' or '*Gurmar*', which literally means 'destroyer of sugar'. According to Sushruta chewing of *G. sylvestre* leaves destroyed the ability to identify the sweet taste in case of glycosuria (*madhumeha* = 'honey urine') and other urinary diseases (Laha and Paul 2019). Moreover, *G. sylvestre* is effective against malaria, glycosuria, dysuria, biliousness, jaundice, constipation, water retention, dyspepsia, furunculosis, bronchitis, asthma, cough, fever, hemorrhoids, renal and vesicle calculi, conjunctivitis, ophthalmia, amenorrhea, leukoderma, epilepsy, adenopathy, and snakebites (Tiwari et al. 2014; Khan et al. 2019). Based on the traditional knowledge considerable development has been brought about in the pharmacological and clinical research of *G. sylvestre* to ascertain its hypoglycemic activity, and unravel its bioactive hypoglycemic constituents involved in the treatment of diabetes. Besides these, investigations leading to reveal medicinal importance of *G. sylvestre* exhibited its diverse therapeutic and pharmacological activities like antidiabetic, anti-inflammatory, anticancer, cytotoxic, antihyperlipidemic, antioxidant, antimicrobial, antibiotic, anti-allergic, hypolipidemic, hypercholesterolemia, immunomodulatory, immunosuppressive, gastro-protective, hepatoprotective, larvicidal activities and demonstrated promising effects on arthritis, obesity, wound healing, and Parkinsonism (Yadav et al. 2019; Khan et al. 2019).

Different phytochemicals have been reported by several researchers in *G. sylvestre*. Major active compounds include gymnemic acid, saponins and oleanane-type of triterpenoid, and stigmasterol, and quercitol (Kapoor 1990). Gymnemic acid I–VII, VIII–XVIII, gymnemasaponins I–V, and gymnemagenin (Yoshikawa et al. 1989a, 1989b, 1991, 1993, 1997; Liu et al. 1992) were isolated and characterized from *G. sylvestre*.

10.1.1 Botanical Classification

Kingdom	Plantae
Subkingdom	Tracheobionta
Super-division	Spermatophyta
Division	Magnoliophyta
Class	Magnoliopsida
Sub-class	Asteridae
Order	Gentianales
Family	Apocynaceae
Sub-family	Asclepiadaceae
Genus	*Gymnema*

(continued)

(continued)

Kingdom	Plantae
Species	*G. sylvestre* (Retz.) R.Br. ex Sm

Source Kirtikar and Basu (1987)

10.1.2 Geographical Distribution

G. sylvestre is a perennial woody climber, found in the tropical and subtropical regions of the world (Fig. 10.1). It is widely spread in Asia (India, Sri Lanka, Japan, Malaysia, Philippines, Indonesia, South China, Vietnam, and Saudi Arabia), Africa (Ethiopia, East, and Tropical Africa), Australia, and some parts of Germany and USA. In India, it is reported from Western and Northern India, parts of Central and Southern India, Deccan peninsula, Western ghats, and Konkan (Tiwari et al. 2014; Kishore and Singh 2015; Khan et al. 2019; Laha and Paul 2019; Jamadagni et al. 2021).

10.1.3 Description and Vernacular Names

G. sylvestre is a vulnerable slow growing, perennial, large woody climber. Stem is cylindrical, twining, branched, whereas leaves are opposite decussate with a few variations for shape, margin, and pubescence. Petioles 5–10 mm long and pubescent

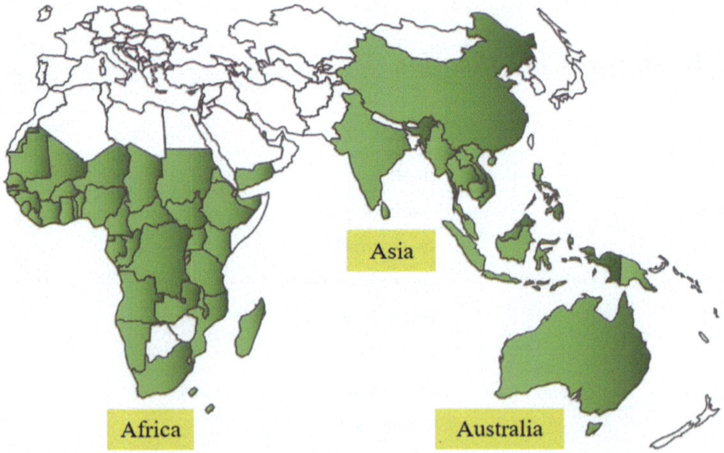

Fig. 10.1 Worldwide distribution of *Gymnema sylvestre*

Fig. 10.2 *Gymnema sylvestre* **a** habit, **b** climbing twig, **c** leaves, **d** flowering twig, **e** fruiting, **f** roots

with milky latex. Flowers of Apocynaceae/ Asclepiadaceae show the presence of gynostegium. Flowers are nearly sessile, small, and yellowish in colour with 5 lobed calyx and corolla. Fruits are of follicle types which are lanceolate in shape and 5.0–7.5 cm long. Seed 10–12, glabrous and brown, flat with thin marginal wing (Singh et al. 2008; Nikalje et al. 2013; El Shafey et al. 2013; Pramanick 2016).

Vernacular names of *G. sylvestre* in India are *Meshashringi* (Sanskrit); *Bedkicha pala, Kavali, Kalikardori* (Marathi); *Gurmar* (Hindi); *Kadhasige, Sannagerasehambu* (Kannada); *Cakkarakkolli, Madhunasini* (Malayalam); *Sirukurunkay, Sakkaraikkolli, Adigam, Cherukurinja* (Tamil); *Podapatra* (Telugu); *Dhuleti, mardashingi* (Gujarati) (Kanetkar et al. 2006; Shrinet et al. 2021) (Fig. 10.2).

10.2 Cultivation Practices

10.2.1 Current Cultivation Practices

In spite of growing in the wild, *G. sylvestre* is now categorized as vulnerable, and has been recognized and prioritized for promotion of cultivation by the National Medicinal Plant Board (NMPB), Government of India (https://nmpb.nic.in/sites/default/files/publications/gudmar.pdf). The vulnerability of *G. sylvestre* and commercial exploitation of this valuable medicinal plant lays emphasis on the need of exploration of alternative propagation techniques for conservation, methods that would accelerate the large-scale production, and improvement of the plant species. Conventional methods of cultivation comprise of vegetative reproduction through stem cuttings, while sexual reproduction through seeds. However, seeds show poor viability and low germination rate, while vegetative stem-cuttings show poor rooting ability (Lee et al. 2006). Although *G. sylvestre* can grow in any type of soil, it flourishes in medium black soil or well-drained red loamy soil, prefers growing in dry areas and cannot stand water-logging. The tropical and sub-tropical regions of the world with an average to high rainfall are amiable for its abundant growth. Highest germination (92%) was observed when propagated on coir-dust. In a nursery, seeds are usually sown from November to February, transplanted on field after 3–4 months, and harvested after 8–9 months. Fruiting takes place around November and December. There existed no substitute method of propagation until recently for propagating and conservation of genetic stock (Singh et al. 2008; Laha and Paul 2019). Pandey (2012) performed field experiments to improve certain aspects in the cultivation techniques of *G. sylvestre.* The findings endorsed that the pre-treatment of seeds with cold water for 24 h helps to improve germination. To vegetatively propagate, treatment of the hard-wood cuttings (10–15 mm diameter) with IBA (500 ppm) for 30 min. produces maximum rooting. March and July were the appropriate time to obtain higher vegetative propagation frequency. For improved growth and production 4000 kg ha^{-1} FYM was essential, along with regular weeding and hoeing. Harvesting was recommended after two years of plantation, two time a year. Karthic and Seshadri (2009) opted hydroponics as a potential alternative for large scale production of *G. sylvestre.* Stem cuttings of twenty months-old tissue cultured plantlets were used as explants. Maximum rooting of 66% was obtained on 1/10-Murashige and Skoog (MS) supplemented with IBA (0.5 mg L^{-1}) with 96% survival rate.

10.2.2 Cultivation Challenges

Maturation post transplantation and harvesting period after plantation of *G. sylvestre* is comparatively longer. Propagation through seeds is difficult due to poor seed viability. There has been no report on alternative mode of propagation available for propagating and conserving genetic stock until recently. Ghosh and Chatterjee (2016) conducted a ten-year study on the incidence of anti-diabetic plants from selected regions of eastern India. An extensive survey included research on agro-technology, cultivation and domestication of few medicinal and aromatic plants (MAPs) which included *G. sylvestre.* Domestication of *G. sylvestre* under ex situ conditions showed improved growth and yield. Their findings showed that many factors are responsible for the depletion of these medicinal plants, which mainly include climatic change, biotic pressures, destructive harvesting, loss of habitat due to shifting cultivation by tribals, urbanization, and agriculture. There is a need for improved agro-techniques to encounter these challenges. Application of biotechnology-based breeding methods (BBBMs) can be an option to overcome these conventional cultivation barriers.

10.2.3 Genetic Improvement Objectives

The unscientific collection and over-exploitation of *G. sylvestre* due to its economic importance is leading towards the depletion of this species. This situation demands not only the domestication of *G. sylvestre*, but needs involvement of modern biotechnological techniques, high through-put screening methods, genomics and proteomics in the improvement of production system.

10.3 Germplasm Diversity

Plant germplasm existence has been assisted by the developments in the in vitro culture technology, molecular markers, and cryopreservation that offer alternative to diversity analyses and genetic resource management and conservation. Germplasm conservation ensure secure preservation of germplasm of beneficial medicinal plants by accumulating and maintaining plant parts or tissues.

10.3.1 Morphological and Phytochemical Diversity

The wide distribution of *G. sylvestre*, its potential to adapt to diverse environmental conditions, and its ability to yield broad range of phytoconstituents of pharmacological importance, impels innovative approaches of metabolomics to procure chemical profile data and correlate with the ecological variables. Environmental factors not only impact the production of phytochemicals, but also influences the organoleptic and phenotypic variation within and between populations (Naik and Al-Khayri 2016; Mankar et al. 2021). Investigating elite chemotypes of *G. sylvestre* accessions having higher potential for phytochemicals is imperative. Characterization of morphological and biochemical variations were accessed for 93 accessions of *G. sylvestre* collected from different ecoclimatic regions of Kerala, India. Morphological characterization was performed by selecting seven vegetative traits, and biochemical characterization was done using TLC-Densitometry for estimation of total saponin concentration for all the accessions. The findings showed significant variations in morphological and biochemical characters (Nair and Keshavachandran 2007a). Manohar et al. (2009) investigated the concentrations of gymnemic acid in different plant parts of *G. sylvestre* using HPLC. Shoot tips (54.29 mg g^{-1} DW) of *G. sylvestre* has higher gymnemic acid, while other plant parts have detectable amount of gymnemic acid. Various studies have validated that the concentrations of saponin under in vivo conditions are mainly governed by the environmental factors (Zhang et al. 2019). In India, gymnemic acid content of *G. syslesvtre* has been reported from different regions ranging from Central and Northern India (Madhya Pradesh and Uttar Pradesh)

(Pandey and Yadav 2010; Verma et al. 2016), Western India (Gujarat) (Maheshwari and Nadgauda 2016) to the Southern India (Tamil Nadu) (Nalina and Rajamani 2020). The variation in the gymnemic acid content in *G. sylvestre* populations from Seoni, Mandla, Panna, Chhatarpur, Amarkantak, and Chitrakoot regions of Madhya Pradesh ranged in between 0.96 and 1.58% (Pandey and Yadav 2010). Chemical fingerprinting of fourteen accessions of *G. sylvestre* collected from Madhya Pradesh and Uttar Pradesh was performed for gymnemic acid content. The study indicated that this technique was capable in differentiating the studied accessions based on their geographical locations (Maheshwari and Nadgauda 2016). Thus, the observations of Pandey and Yadav (2010) and Verma et al. (2016) implied that gymnemic acid content in *G. sylvestre* is greatly influenced by geographical variations. Morphological and biochemical variations among the *G. sylvestre* from diverse geographical regions of Gujarat was assessed by Maheshwari and Nadgauda (2016). The authors have recorded wide variations in morphological and biochemical parameters and concluded that environmental factors arising from seasonal variations played a crucial role. Dhanani et al. (2015) investigated eighteen accessions from different ecotypes of *G. sylvestre* from India to assess their variability. Morphological (leaf-based) parameters and biochemical parameters (gymnemagenin content) exhibited considerable significant variability. Thus, their results recommended the use of these accessions for further breeding programs of cultivars having potential for production of gymnemagenin, and promoted the cultivation of this valuable medicinal plant. These comprehensive studies have drawn attention towards correlation between geographical regions and gymnemic acid and gymnemagenin content of *G. sylvestre*.

10.3.2 Genetic Diversity Analysis

Genetic diversity analysis using molecular markers plays a crucial role in understanding the effect of numerous factors on the genetic variability and its population structure, and in the implementation and regulation of effective conservation strategies (Haji et al. 2014). Molecular markers are a useful tool to study genetic variation within and between populations, for population genetic studies, and in differentiating morphologically correlated species giving an evaluation of genetic diversity. Isozymes and RAPD markers were used to score phenotypically and biochemically distinct *G. sylvestre* accessions from Kerala region of India (Nair and Keshavachandran 2007a, b). Significant genetic variation was detected in the studied accessions, which was in concurrence with their previous study based on the characterization of morphological and biochemical variations for same accessions. Genetic variation between twelve *G. sylvestre* progenies and the parent was detected using AFLP markers, which demonstrated sensitivity in the detection of polymorphism among the parent and seed raised progenies at molecular level (Osman et al. 2011). Morphological and genetic diversity among twenty-two populations of *G. sylvestre* from northern Western Ghats of India were investigated using RAPD and ISSR markers. Significant distinction was recorded in the morphological study based

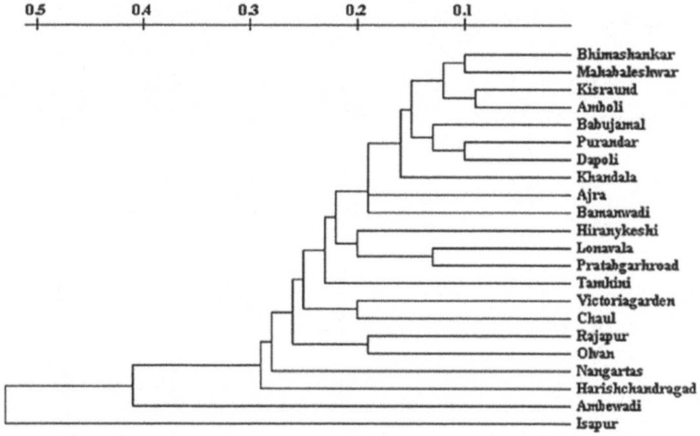

Fig. 10.3 RAPD analysis of *Gymnema sylvestre* populations collected from northern Western Ghats, India. *Source* Shahnawaz et al. (2012)

on leaf morphology, which could be due to the variability in the environmental and natural conditions (habitat fragmentation, geographical separation, type of pollination, wind seed dispersal, and out-crossing) and anthropogenic activities (Shahnawaz et al. 2012) (Fig. 10.3). Verma et al. (2016) assessed 14 accessions of *G. sylvestre* from Central India for their genetic and chemical diversity and evaluated correlation between them. Their investigation asserted that both the ISSR and HPLC fingerprinting methods were applicable in identifying high-yielding accessions and chemotypes. Although chemical fingerprinting proved to be a better preference as compared to genetic fingerprinting to differentiate accessions based on their site of collection, habitat and their agro-climatic conditions, their combined implementation had the potential to reveal the evolutionary development, chemotypic variation, and adaptation.

10.4 Application of Biotechnological Tools

The sources for naturally occurring saponins are inadequate due to its restricted availability, geographical, and seasonal variations. The complex structures of triterpenoid saponins make chemical synthesis difficult and uneconomical for large scale production (Palazón et al. 2008; Lambert et al. 2011). Nowadays, plant tissue culture is not only used for the conservation and mass multiplication of medicinal plants, but also for scaling-up of secondary metabolites using elicitors, precursors, genetic engineering, and synthetic biology (Lambert et al. 2011; Niazian 2019). The culturing of plants cell and tissue on artificial growth media is free from different environmental and climatic barriers and is able to produce various secondary metabolites similar to the whole plant (Sarin 2005). *G. sylvestre* is subjected to unsustainable use by the pharmaceutical industries because of its therapeutic and pharmacological importance. In order to fulfill the demand, *G. sylvestre* is being extensively collected and the natural strands are fast disappearing from their habitats. There arose a requirement to pursue alternative ways to meet the increasing demand and cut down the burden of over-harvesting from naturally grown wild populations. Large-scale production of *G. sylvestre* is the only way, and this could be done by employing biotechnological approach.

10.4.1 Micropropagation Using Shoot Cultures

Reddy et al. (1998) performed micropropagation in *G. sylvestre* using axillary buds as explants, however the multiplication frequency was very much low. Komalavalli and Rao (2000) established rapid multiplication protocol using various explants (axillary node, shoot tip and cotyledonary node, leaf, petiole, root and internodal segments) from in vitro generated seedlings, along with media manipulation. Axillary nodes showed significant response (57 shoots explant^{-1}) as compared to other explants used, and the developed plantlets were successfully hardened. Thus, offering an efficient protocol for mass propagation and conservation of *G. sylvestre.* Inclusion of antioxidants, vitamin B_2 along with the growth regulators in MS medium influenced shoot proliferation in *G. sylvestre.* Incorporation of antioxidants prevents phenolic leaching and increase the production of healthy normal shoots (Devi and Srinivasan 2008). MS medium along with 6-Benzylaminopurine (BAP, 1 mg L^{-1}), indole acetic acid (IAA, 0.5 mg L^{-1}), citric acid (100 mg L^{-1}) and vitamin B_2 (100 mg L^{-1}) have produced 28.3 shoots explant^{-1}. While, half-strength MS medium with IBA (3 mg L^{-1}) was suitable for rooting (Devi and Srinivasan 2008). Bharti (2010) developed a direct plant regeneration protocol using apical bud of *G. sylvestre* as explant. Combination of BAP and Kinetin (4.44 + 4.64 μM) produced multiple shoots (10–12 multiple shoots culture^{-1}). Higher rooting frequency was achieved on half-strength MS + IAA (2.85 μM), and successful hardening of rooted plantlets and acclimatization was observed. In another investigation, nodal segment of *G. sylvestre* produced 7 shoots explant^{-1} having 3.4 cm of shoot length after 1 week of inoculation on MS medium containing BAP (1 mg L^{-1}). Regenerated shoots were separated and transferred to MS medium with different concentrations of 1-Naphthalene acetic acid (NAA). Maximum rooting (2.7 roots explant^{-1}) was recorded on NAA (1 mg L^{-1}) having length 0.7 cm (Manonmani and Francisca 2012). Kaushalya and Senarath (2014) observed highest shoot elongation (14.8 mm) not only on MS + BAP (1 mg L^{-1}), but also on basal MS (14.2 mm), and half-strength MS + BAP (10.4 mm). Thus, when basal salt concentrations were reduced, the shoot elongation response also decreased, indicating that MS showed better response for shoot elongation than half-strength medium. Further, the separated shoots showed 100% rooting on MS + indole-3-butyric acid (IBA, 3.0 mg L^{-1}), however at higher concentrations of IBA basal callusing was noted. Survival rate was 62.5% when the plantlets were transferred from rooting medium to sand: soil (1: 2). After three weeks, the transplantation rate under normal garden conditions was 75%. Ashwani and Dipali (2013) induced shoot multiplication on MS and B5 medium using cytokinins and auxins alone. MS media supplemented with BAP (5.0 mg L^{-1}) produced maximum 2 shoots explant^{-1} having shoot length of 2.57 cm., while no response was observed on either of the auxins. About 80% rooting was observed on half strength MS medium. Shah et al. (2013) employed a number of parameters to developed an in vitro micropropagation protocol for *G. sylvestre*. In their investigation they used five nutrient media, five cytokinins, and six concentrations, all in various combinations. Nitsch and Nitsch medium supplemented with 5 μM BAP showed maximum response (1.82 shoots explant^{-1}) with 5.2 nodes explant^{-1} for one-month inoculation. Obtained shoots were cultured on all the five media with auxins (IAA, IBA, and NAA) and coumarin (non-auxin treatment) at different concentrations (0, 5, 10, 15 μM). Highest rooting was observed on Gamborg B5 medium supplemented with 5 μM NAA exhibiting 85% rooting response with 4.27 roots explant^{-1} after 35 days of inoculation. 60–70% survival rate was recorded for hardening. Multiple shoot induction protocol was established by supplementing MS media with BAP (4.44 μM) + KIN (2.32 μM) + NAA (2.69 μM) which produced 11 shoots culture^{-1} (Zimare and Malpathak 2017). While maximum rooting was recorded on half strength MS media having IBA (4.92 μM) + IAA (5.70 μM) (Fig. 10.4). After 8-week study, these shoots were further investigated for gymnemic acid accumulation using HPTLC which revealed 10.37 mg g^{-1} DW of gymnemic acid content (Zimare and Malpathak 2017).

Isah (2019) studied the sprouting of axillary bud from nodal segment and plant regeneration from shoot-tip generated callus cultures of *G. sylvestre*. Highest shoots were observed on MS media + BAP (2 mg L^{-1}) which produced 7.92 micro-shoots of length 2.38 cm on 0.5 g callus^{-1}. The obtained microshoots were transferred to half-strength MS amended with auxins like NAA, IAA and IBA. Half-strength MS + IBA (3 mg L^{-1}) produced highest rooting response with an acclimatization survival rate of 87%.

Fig. 10.4 Production of multiple shoots and rooting in *Gymnema sylvestre*. *Source* Zimare and Malpathak (2017)

10.4.2 Callus and Cell Suspension Cultures

A comparative account of callus production was studied in young and mature leaf segments of *G. sylvestre*. MS medium was supplemented with various concentrations of auxins and cytokinins in combination. Callusing was observed in all combinations, but the callus index was higher in mature leaf segments than that of young leaf segments (Nazeem et al. 1991). Inter-nodal explants of *G. sylvestre* were used for callus induction. Influence of PGRs like 2,4-D (2,4-Dichlorophenoxyacetic acid), kinetin, IAA, BAP was studied on the growth of callus. 2, 4-D (1 - 2 mg L^{-1}) alone showed callus induction but the growth rate was very slow. Higher callus induction was recorded for 2, 4-D and kinetin in combination (1–5 mg L^{-1}: 1 mg L^{-1}) with friable callus initiated in short period of time (Roy et al. 2008). Nan and WTPSK (2013) used leaf explants of *G. sylvestre* for the induction of callus. They used MS with lower concentrations of BAP, kinetin and 2,4-D to induce callus, but this is failed for callogenesis. Increased concentration of 2,4-D above 4 mg L^{-1} showed callus induction after 3 weeks incubation. Higher callusing response was seen on 5 mg L^{-1} 2,4-D. They also performed GS-MS analysis to compare the phytochemicals from fresh *G. sylvestre* leaves from naturally grown plants with callus. Methanolic leaf and callus extracts revealed total of nine compounds in each extract, while six phytochemicals were common in both samples. Thus, their findings showed that callus is also a source of important bioactive compounds and can be employed instead of leaves. Gopi and Vatsala (2006) elucidated the response of PGRs on cell growth and production of gymnemic acid in *G. sylvestre*. Nodal and leaf explants were used for callus initiation, where significant callusing was recorded for nodal explants as compared to leaf on MS medium containing 0.5 mg L^{-1} 2,4-D. High fresh biomass yielding medium containing 2,4-D and NAA concentrations were selected to study their effects on callus culture. Maximum biomass [FW (370 g L^{-1}) and DW (23.12 g L^{-1})] was recorded on MS medium supplemented with 2,4-D (0.5 mg L^{-1}) + NAA (2.5 mg L^{-1}) + BAP (0.5 mg L^{-1}). Gymnemic acid was quantified using HPLC analysis of aqueous and methanolic cell suspension which revealed that accumulation of active principles was higher in cell culture as compared to the leaf extracts of naturally growing *G. sylvestre*. *G. sylvestre* seeds were germinated on MS basal medium, and leaves of the developed seedlings (12 cm height) were used for callus induction. Maximum callus induction was recorded on MS medium with BAP (0.5 mg L^{-1}) and IAA (1.5 mg L^{-1}). For suspension culture 0.345 g initial inoculum was used, and on day 20 maximum gymnemic acid accumulation was noted as 98.63 mg L^{-1}. Further, effect of various external phytohormones, pH of medium and shaking speed was studied on the gymnemic acid yield. Maximum biomass yield (1.98 g L^{-1} DW) and gymnemic acid yield (109.71 mg L^{-1}) was obtained on BAP (1 mg L^{-1}) and IAA (0.5 mg L^{-1}). Further, pH 5.6 stimulated and yielded higher gymnemic acid (98.01 mg L^{-1}), and 84.68 mg L^{-1} gymnemic acid was produced at 120 rpm (Devi et al. 2006). Lee et al. (2006) studied the effects of chemical [medium salt strength—¼-, half-strength and full-MS; sucrose—(0, 1, 3, 5, 7 and 9%); and PGRs—IBA, NAA, and 2,4-D (1, 3, 5, 7 and 9 mg L^{-1})]

and physical [(inoculum size—20, 40, 60, 80, 100 and 120 mg L^{-1}; and aeration volume—0.05, 0.1, 0.2, 0.3, and 0.4 vvm)] environments on *G. sylvestre* cell cultures through bioreactor. Strengths of MS medium did not considerably influence the cell growth, while sucrose concentration affected cell growth, with maximum cell growth at 3% concentration. Inhibition of cell growth was observed at concentrations higher that 3% sucrose. Among the PGRs used, significant result was recorded for 5 mg L^{-1} NAA, while IBA hindered cell growth significantly. Highest growth rate of 0.2 day^{-1} was observed using 60 g L^{-1} of inoculum size, and was found optimum for employing in economical culture. Maximum cell growth (255.3 g L^{-1}) was attained by increasing aeration volume from 0.05 to 0.4 vvm throughout culture period. Praveen et al. (2011) studied the influence of various macro-elements and nitrogen source on cell suspension cultures of *G. sylvestre* and the accumulation of biomass and gymnemic acid. Maximum biomass accumulation [(165 g L^{-1} FW) and (15.42 g L^{-1} DW)] was on MS medium with NH_4NO_3 (0.5 × concentration), maximum gymnemic acid production (11.32 mg g^{-1} DW) was recorded in the MS medium fortified with KH_2PO_4 (2.0 ×), and maximum biomass growth [(159.72 g L^{-1} FW) and (14.95 g L^{-1} DW)] was reported for NH^{4+}/NO^{3-} ratio of 7.19/18.80, and this ratio also enhanced cell growth and gymnemic acid production. Zimare et al. (2017) developed a protocol for accumulation of higher biomass and gymnemic acid in cell culture of *G. sylvestre.* MS media along with different PGRs in combinations were used to attain higher biomass which was assessed using parameters like packed cell volume, cell count, fresh weight and dry weight of the cell suspension cultures of *G. sylvestre.* Maximum biomass and highest gymnemic acid production were recorded on MS medium supplemented with 2,4-D (4.52 μM) + BAP (4.44 μM) after 4 weeks' study.

10.4.3 Somatic Embryogenesis

A developmental pathway of plant regeneration is somatic embryogenesis, where embryos are formed from vegetative cells that are similar to zygotic embryos following embryological stages without fusion of gametes. Somatic embryogenesis has been identified as an ideal technique for clonal large-scale propagation and genetic improvement, and can be used for commercial applications. Ashok Kumar et al. (2002) reported regeneration of *G. sylvestre* from somatic embryogenesis. Seedling explants (hypocotyl, cotyledon and leaf) were used to induced embryogenic callus, and subsequently plantlets were regenerated from the somatic embryos. Induction of embryogenic callus was achieved on MS medium containing 0.5–5.0 μM of 2,4-D, 0.5–2.0 μM BAP, and 2% w/v sucrose within 6–8 weeks of culture. Two types of embryos, globular or heart were initiated on this medium. While torpedo and cotyledon stage of embryos were developed after sub-culturing them on embryo maturation medium containing MS salts, B5 vitamins, 0.5 μM BAP and 2% sucrose. Germination of embryo and plantlets formation was attained on same medium by sub-culturing, and 77% plantlets were successfully acclimatized in greenhouse. Ahmed et al. (2009) established a protocol for somatic embryogenesis for *G. sylvestre* via embryonic suspension cultures. MS medium with coconut water (10%) and 2,4 -D (0.5 mg L^{-1}) or NAA (1.0 mg L^{-1}) was used for callus induction. For attaining somatic embryogenesis, obtained callus was transferred to liquid MS medium with coconut water (10%), NAA (1.0 mg L^{-1}), BAP (1.0 mg L^{-1}), sucrose (3.0%), citric acid (1 mg L^{-1}) and glutamine (10 mg L^{-1}). After 8 weeks, four different stages of embryos (globular, heart, torpedo, and cotyledonary) were observed in the suspension cultures. The maturation of embryos was significantly influenced by PGRs and photo-period. About 5–7% embryos developed into plantlets on semi-solid MS medium containing vitamin B_5, sucrose (3.0%) and agar (0.8%).

10.4.4 Genetic Transformation

Different explants of *G. sylvestre* (cotyledons, leaves, roots, stems, hypocotyls, and cotyledonary nodal segments) were used for inducing hairy roots using *Agrobacterium rhizogenes* (KCTC 2703). Cotyledons and leaf explants induced hairy roots. Maximum biomass accumulation (97.63 g L^{-1} fresh mass and 10.92 g L^{-1} DW) on day 25, and maximum gymnemic acid production (11.30 mg g^{-1} DW) on day 20 was observed on MS liquid medium. Suspension cultures showed 9.4-fold increment of biomass at 25 days of culture, and 4.7-fold higher gymnemic acid content in hairy root biomass when with the untransformed control roots (Nagella et al. 2013).

10.4.5 Production of Secondary Metabolites

10.4.5.1 Elicitors

Elicitation is one of the biotechnological strategies that is well-known for improving the production of preferable secondary metabolites in plant cell and organ cultures. Veerashree et al. (2012) used biotic (yeast extract, chitin and pectin) and abiotic (methyl jasmonate) elicitors for the production of gymnemic acid in suspension cultures of *G. sylvestre*. Although all elicitors enhanced gymnemic acid production, highest response was reported for yeast extract treated cultures. Their work indicated that gymnemic acid production was dependent on elicitor concentration and treatment duration. Mahendran et al. (2021) evaluated the production of deacylgymnemic acid, gymnemagenin, gymnemic acid—IV and—XVII using abiotic elicitors like salicylic acid, methyl jasmonate and sodium nitroprusside on the cell suspension culture of *G. sylvestre*. The cell suspension cultures treated with 20 μM sodium nitroprusside showed maximum biomass, deacylgymnemic acid, and gymnemic acid XVII, while maximum accumulation of gymnemagenin and gymnemic acid IV was recorded in 10 μM salicylic acid treated cultures. Use of heavy metals as abiotic stress agents has increased because metals tend to alter the secondary metabolism properties leading to the variation in the production of phytocontituents. Various metal salts (chlorides of cadmium, calcium, cobalt, copper, mercury and silver nitrate) were used as abiotic chemical elicitors to enhance gymnemic acid in *G. sylvestre* suspension cultures. The cultures treated with $CdCl_2$ accumulated maximum gymnemic acid (59.97 mg g^{-1} DW) at 2 mM after 24 h. Praveen et al. (2014) elicited hairy root cultures of *G. sylvestre* using polyunsaturated fatty acids (PUFAs) namely oleic acid and linolenic acid to enhance biomass and gymnemic acid content. Cultures containing 1 and 5 μM linolenic acid enhanced the gymnemic acid content significantly showing 7.78-fold increment as compared to control. This enhanced gymnemic acid accumulation indicated that linolenic acid stimulated the metabolic pathway that led to the biosynthesis of gymnemic acid.

Biotic elicitors of fungal (*Aspergillus niger* and *Saccharomyces cerevisiae*) and bacterial (*Bacillus subtilis* and *Escherichia coli*) origin were used to increase gymnemic acid content in cell suspension cultures of *G. sylvestre*. All studied elicitors enhanced gymnemic acid production, however highest increment was reported for *A. niger* which was 11.2-fold higher than control (Chodisetti et al. 2013). Two endophytic fungal strains namely, *Polyancora globosa* and *Xylaria* sp. were isolated from *G. sylvestre* leaves. Dried powder of fungal mycelia (DPFM) and extracellular culture filtrate (ECF) of endophytic fungi were used as elicitors in the suspension cultures of *G. sylvestre*. The DPFM showed higher gymnemic acid accumulation as compared to ECF. Cell suspension cultures elicited using DPFM of *P. globosa* yielded higher gymnemic acid content (124.23 mg g^{-1} DW) as compared to those elicited with DPFM of *Xylaria* sp. (102.24 mg g^{-1} DW). The consortium of DPFM and ECF of *P. globosa* and *Xylaria* sp. showed highest gymnemic acid accumulation (139.98 and 94.86 mg g^{-1} DW, respectively) as compared to DPFM and ECF alone (Netala et al. 2016).

The use of elicitors in callus and cell suspension cultures is a common strategy, while little attention has been given to the elicitation in shoot culture. Zimare et al. (2018) treated shoot cultures of *G. sylvestre* grown on MS media supplemented with BAP (4.44 μM) + KIN (2.32 μM) + NAA (5.37 μM) with biotic elicitors such as *Alternaria alternata*, *Botrytis cinerea*, and *Rhizoctonia solani*. Maximum gymnemic acid content was recorded in the cultures treated with *A. alternata* followed by *B. cinerea* and *R. solani* (23.22, 17.36, 16.54 mg/g DW).

10.4.5.2 Precursor Feeding

Any compound that happens to be an intermediate, or is at the initial of a secondary metabolite biosynthetic pathway, stands a chance to enhance the production of the end product. Feeding of appropriate precursors to the cell culture medium can induce the accumulation of secondary products. Isopentenyl pyrophosphate (IPP), squalene, and 2,3-oxidosqualene regulate the biosynthetic pathway for the production of triterpenoid saponin. Shoots cultures of *G. sylvestre* grown on MS media with BAP (4.44 μM) + KIN (2.32 μM) + NAA (5.37 μM) were fed with IPP, squalene, and 2,3-oxidosqualene to study their effects on gymnemic acid production. The

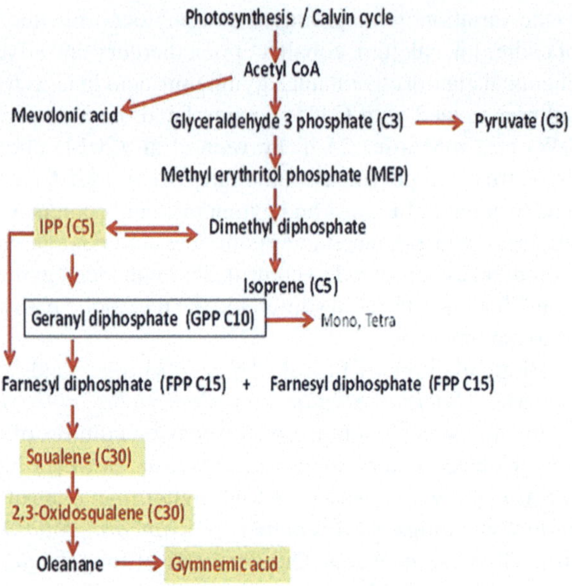

Fig. 10.5 Pathway for gymnemic acid production in *Gymnema sylvestre*. *Source* Zimare and Malpathak (2020)

cultures treated with 2,3-oxidosqualene showed higher accumulation of gymnemic acid (23.31 mg g^{-1} DW) followed by squalene and IPP. The results imply that these precursors may increase the activity of certain enzymes like IPP isomerase, squalene synthase, squalene epoxidase, and β-amyrin synthase (β-AS) which are important for the accumulation of terpenoid saponins (Zimare and Malpathak 2020). Based on these results Zimare and Malpathak (2020) proposed a pathway for gymnemic acid accumulation in *G. sylvestre* (Fig. 10.5) which was similar to Tiwari et al. (2014).

10.5 Mycorrhizal Treatment

Use of arbuscular mycorrhizal (AM) fungi has been employed to increase biomass of medicinal plants (Zubek et al. 2012) and phytochemical constituents (Thangavel et al. 2009). The use of AM fungi for medicinal plants is well documented (Ceccarelli et al. 2010; Zubek et al. 2012; Akhtar et al. 2022). Though it is an important strategy, few reports are available of its use in *G. sylvestre*. For the first time *Glomus fasciculatum* (*Gf*) and *G. mosseae* (*Gm*) was used for biomass and gymnemic acid enhancement in *G. sylvestre* under in vivo conditions. Plants treated with *Gf* had significantly higher biomass for shoot length (32.70 cm), root length (34.70 cm), fresh weight (3.33 g), and dry weight (0.92 g) and gymnemic acid content than that of *Gm*-treated plants and control (Fig. 10.6). These results were in concurrence with Maier et al. (1997) and Jurkiewicz et al. (2010) who stated that mycorrhizal colonization can induce quantitative and/or qualitative changes in biomass and terpenoid content.

Fig. 10.6 Increased biomass of *Gymnema sylvestre* **a** control, **b** *Glomus fasciculatum*, **c** *Glomus mosseae* after 90 days. *Source* Zimare et al. (2013)

10.6 Functional Genomics

Saponins are widely distributed in higher plants and glycosidic in nature. Saponins are amphiphilic due to the presence of a hydrophilic sugar moiety and a hydrophobic genin called 'sapogenin'. The structure of gymnemic acid is nothing but a sapogenin (triterpene aglycone moiety) attached to a sugar chain (Podolak et al. 2010). Glycosylation is a crucial course of action in the generation or modification of triterpene saponins. Investigations comprising functional genomics along with identification of genes, cloning, and their functional characterization can facilitate in decoding the operational role of these genes involved in the biochemical pathway responsible for the medicinal and pharmacological properties of the compounds in the plants. Tiwari et al. (2009) and Sabir et al. (2009) submitted two partial cds namely, accession no. GU191124 and GU181368 to NCBI database to comprehend the molecular mechanism of genes responsible for the therapeutic activities of *G. sylvestre*. The mechanism of gymnemic acid biosynthesis has not yet been reported due to lack of characterization of numerous proteins and enzymes at the biochemical level, and unavailability of the transcriptome data of *G. sylvestre*. The biosynthetic pathway of gymnemic acids and its related mechanism of pharmacological properties can be decoded by identifying and characterizing the genes responsible for the biosynthesis of triterpene glycosides and gymnemic acids. Although, the gymnemic acid biosynthesis pathway is unknown, a putative pathway for triterpene glycosides has been derived from the isoprenoid pathway by means of triterpene aglycone glycosylation at the terminal transformation of gymnemagenin (Fig. 10.7) (Tiwari et al.

2014). A putative steroidal pathway of polyoxypregnane glycosides and candidate genes of these pathways was explored by Kalariya et al. (2021) using de novo transcriptomic study in *G. sylvestre*. The study has made available data on key enzymes of pregnane glycoside biosynthesis, that has provided valuable resource for the upcoming functional genomic studies in *G. sylvestre*. Ayachit et al. (2019) investigated the transcripts involved in terpenoid biosynthesis, since the medicinal properties of *G. sylvestre* mainly influenced by its terpenoid profile. De novo transcriptome sequencing of *G. sylvestre* leaf was performed using RNA sequencing technology to identify the functional gene profile of *G. sylvestre*. For the biosynthesis of terpenoid the precursor molecules come from the cytosolic mevalonate (MVA) and plastidial methyl-erythritol phosphate (MEP) pathways. Their findings showed agreement with Tiwari et al. (2014) who demonstrated a putative pathway for biosynthesis of gymnemic acid. Squalene synthase (SQS) is an important enzyme for the production of squalene molecules which are responsible for the formation of secondary products such as saponins, triterpenoids, and polyoxypreganens in *G. sylvestre*. It is also a key enzyme involved in the regulation of triterpenoid biosynthesis. SQS gene was characterized in *G. sylvestre* (Kalariya et al. 2021). To provide a scope for gene cloning and over-expression of SQS to obtain enhanced quantity of compounds having squalene as its precursor molecule. Further, it was assumed that gymnemagenin (sapogenin) gave rise to gymnemic acids and derivatives through glycosylation mechanism by glycosyltransferases (Tiwari et al. 2014).

10.7 Conclusion and Future Prospects

10.7.1 An Overview of the Current Status

G. sylvestre is a rich source of various secondary metabolites and has attracted the attention of researchers to explore its bioactivities. Enormous research has been conducted on the pharmacological aspects and clinical trials to establish the hypoglycemic activity, and to discover novel constituents involved in the treatment of diabetes as well as the allied complications that are directly attributed to diabetes. Likewise, different phytochemicals the chemical synthesis of triterpenoid saponins is economically unfeasible mainly because of complexity of the structures, higher cost of chemicals, and required time. The difficulties in the synthesis and higher annual demand is creating a heavy pressure on the wild population making the plant vulnerable. This situation needs immediate management attention considering the utilization of wild populations and demands a need of developing cultivation practices to suffice the requirement of raw material for industrial use. There is a need to develop alternative sources for the production of pharmaceutically important saponins to not only fulfil the increasing demand, but to conserve the natural populations.

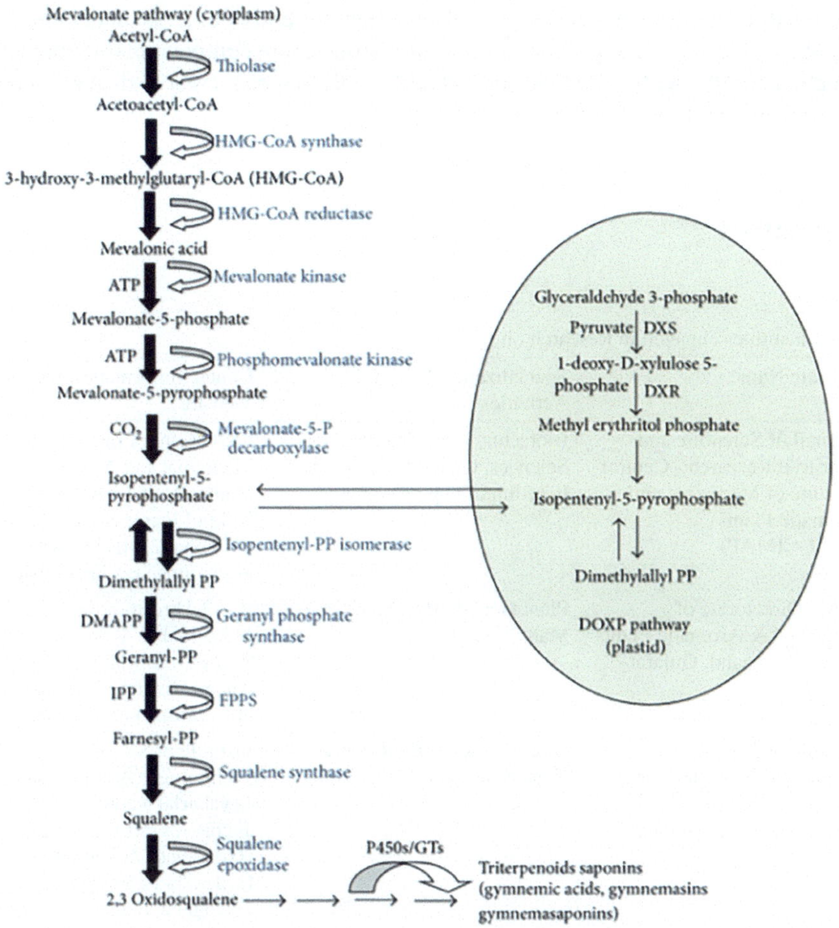

Fig. 10.7 Biosynthesis of gymnemic acid in *Gymneam sylvestre*. *Source* Tiwari et al. (2014)

10.7.2 Recommendations for Future Research

Extraction of saponins from plants is extremely tedious and expensive. Therefore, biotechnological approach/ plant tissue culture can be an alternative to obtain clonal, uniform, genetically stable plants that can yield higher secondary metabolites of pharmaceutical interest. Although the pharmacological importance of *G. sylvestre* has been revealed, very less is known about the saponin biosynthesis pathway, and the molecular biology of triterpene biosynthesis in *G. sylvestre*. Identification of genes and characterization shall cater novel opportunities to concoct structure–function relationships of medicinally valuable saponins. Modulation in production of saponins can be achieved by the genetic identification of genes and its functional analysis. Bio-informatic tools can aid to uncover bioactive compounds with potential anti-diabetic activity, improve the druggability of molecular targets, and offer a vast understanding of diabetes mechanisms. Databases of naturally occurring compounds showing anti-diabetic activity can be effective and valuable tools that can assist to discover novel anti-diabetic medicines or natural compounds.

Appendix I

List of institutes engaged in Research on *G. sylvestre*

Institute Name	Specialization Research Activities	Contact Information and Website
Council of Scientific and Industrial Research - Central Institute of Medicinal and Aromatic Plants (CSIR-CIMAP)	Biotechnology, Chemical Sciences, Genetics and Plant Breeding, Crop Protection	CSIR-Central Institute of Medicinal and Aromatic Plants, P.O-CIMAP, Near Kukrail Picnic Spot Lucknow-226015 https://www.cimap.res.in/english/
ICAR-Directorate of Medicinal & Aromatic Plants Research, Anand, Gujarat	Plant Genetic Resource (PGR) Management	ICAR-Directorate of Medicinal & Aromatic Plants Research, Boriavi - 387310, Anand District, Gujarat. https://dmapr.icar.gov.in/
Institute of Science & Technology (Autonomous), Jawaharlal Nehru Technological University Hyderabad	Biotechnology, Environment, Nano Science and Technology	Institute of Science & Technology (Autonomous), Jawaharlal Nehru Technological University Hyderabad, Kukatpally, Hyderabad - 500085, Telangana, India
Yeungnam University, Gyeongbuk, Republic of Korea	Medical Biotechnology	280 Daehak-Ro, Gyeongsan, Gyeongbuk 38541, Republic of Korea (38541 Korea) https://m.yu.ac.kr/english/index.do
Academy of Scientific and Innovative Research (AcSIR), Ghaziabad, Uttar Pradesh, India	Biotechnology	Academy of Scientific and Innovative Research (AcSIR) CSIR- Human Resource Development Centre, (CSIR-HRDC) Campus Postal Staff College Area, Sector 19, Kamla Nehru Nagar, Ghaziabad, Uttar Pradesh- 201 002 https://acsir.res.in/

(continued)

(continued)

Institute Name	Specialization Research Activities	Contact Information and Website
Gujarat Biotechnology Research Centre	Plant Biotechnology, Bioinformatics & Systems Biology	Gujarat Biotechnology Research Centre, Department of Science & Technology, Gandhinagar - 382011, India https://gbrc.gujarat.gov.in/
Jamia Hamdard University, New Delhi	Botany, Biotechnology, Biochemistry, Transgenic plant development	Department of Botany, School of Chemical and Life Sciences, Hamdard University, New Delhi - 110062, India http://jamiah amdard.edu/
Kanazawa University, Japan	Cultivation of medicinal plants, Quality evaluation of herbal and ayurvedic medicines, Biotechnology	Kanazawa University, Kakumamachi, Kanazawa, Ishikawa 920–1192, Japan https://www.kanazawa-u.ac.jp/e
Govt. of India, Dept. of Atomic Energy, Bhabha Atomic Research Centre	R&D-Health, Food and Agriculture: - Development of molecular markers, transgenics, biosensors, fertilizer formulations, Development of new elite crop varieties including oil seeds and pulses. Using radiation-induced mutagenesis, hybridization, and tissue culture techniques	Bhabha Atomic Research Centre, Trombay, Mumbai - 400085 India https://www.barc.gov.in/index.html
Botanical Survey of India, Central Botanical Laboratory	Genetic and Chemical Profiling	AJC Bose Indian Botanic Garden, Howrah, West Bengal, India. https://bsi.gov.in/bsi-units/en?rcu=140

References

Akhtar N, Wani AK, Dhanjal DS, Mukherjee S (2022) Insights into the beneficial roles of dark septate endophytes in plants under challenging environment: resilience to biotic and abiotic stresses. World J Microbiol Biotechnol 38(5):79. https://doi.org/10.1007/s11274-022-03264-x

Alam S, Sarker MMR, Sultana TN, Chowdhury MNR, Rashid MA, Chaity NI, Zhao C, Xiao J, Hafez EE, Khan SA, Mohamed IN (2022) Antidiabetic phytochemicals from medicinal plants: prospective candidates for new drug discovery and development. Front Endocrinol (Lausanne) 13:800714. https://doi.org/10.3389/fendo.2022.800714

Alarcon-Aguilara FJ, Roman-Ramos R, Perez-Gutierrez S, Aguilar-Contreras A, Contreras-Weber CC, Flores-Saenz JL (1998) Study of the anti-hyperglycemic effect of plants used as antidiabetics. J Ethnopharmacol 61(2):101–110. https://doi.org/10.1016/s0378-8741(98)00020-8

Ashok Kumar H, Murthy H, Paek K (2002) Somatic embryogenesis and plant regeneration in *Gymnema sylvestre*. Plant Cell Tissue Organ Cult 71(1):85–88. https://doi.org/10.1023/A:1016529322046

Ashwani S, Dipali G (2013) In vitro shoot multiplication of *Gymnema sylvestre* R. Br. Adv Plant Sci 26(2):321–323

Ayachit G, Shaikh I, Sharma P, Jani B, Shukla L, Sharma P, Bhairappanavar SB, Joshi C, Das J (2019) De novo transcriptome of *Gymnema sylvestre* identified putative lncRNA and genes regulating terpenoid biosynthesis pathway. Sci Rep 9(1):1–13. https://doi.org/10.1038/s41598-019-51355-x

Bharti S (2010) *In vitro* propagation of *Gymnema sylvestre* Retz. R. Br through apical bud culture. J Med Plants Res 4(14):1473–1476

Burke JP, Williams K, Narayan KMV, Leibson C, Haffner SM, Stern MP (2003) A population perspective on diabetes prevention: whom should we target for preventing weight gain? Diabetes Care 26(7):1999–2004. https://doi.org/10.2337/diacare.26.7.1999

Ceccarelli N, Curadi M, Martelloni L, Sbrana C, Picciarelli P, Giovannetti M (2010) Mycorrhizal colonization impacts on phenolic content and antioxidant properties of artichoke leaves and flower heads two years after field transplant. Plant Soil 335(1):311–323. https://doi.org/10.1007/s11104-010-0417-z

Chodisetti B, Rao K, Gandi S, Giri A (2013) Improved gymnemic acid production in the suspension cultures of *Gymnema sylvestre* through biotic elicitation. Plant Biotechnol Rep 7(4):519–525. https://doi.org/10.1007/s11816-013-0290-3

Devi CS, Murugesh S, Srinivasan VM (2006) Gymnemic acid production in suspension cell cultures of *Gymnema sylvestre*. J Appl Sci 6(10):2263–2268

Devi CS, Srinivasan VM (2008) *In vitro* propagation of *Gymnema sylvestre*. Asian J Plant Sci 7(7):660–665

Dhanani T, Singh R, Waman A, Patel P, Manivel P, Kumar S (2015) Assessment of diversity amongst natural populations of *Gymnema sylvestre* from India and development of a validated HPLC protocol for identification and quantification of gymnemagenin. Ind Crops Prod 77:901–909. https://doi.org/10.1016/j.indcrop.2015.09.076

El Shafey AAM, El-Ezabi MM, Seliem MME, Ouda HHM, Ibrahim DS (2013) Effect of *Gymnema sylvestre* R. Br. leaves extract on certain physiological parameters of diabetic rats. J King Saud Univ Sci 25(2):135–141. https://doi.org/10.1016/j.jksus.2012.11.001

Ghosh M, Chatterjee SK (2016) Sustainable development of medicinal and aromatic plants as potential cash crops for the welfare of rural people in eastern India. Acta Hortic 1125. ISHS 2016. https://doi.org/10.17660/ActaHortic.2016.1125

Gopi C, Vatsala T (2006) In vitro studies on effects of plant growth regulators on callus and suspension culture biomass yield from *Gymnema sylvestre* R. Br. Afr J Biotechnol 5(12):1215–1219

Haji RFA, Bhargava M, Akhoon BA, Kumar A, Brindavanam NB, Verma V (2014) Correlation and functional differentiation between different markers to study the genetic diversity analysis in medicinally important plant *Plumbago zeylanica*. Ind Crops Prod 55:75–82. https://doi.org/10.1016/J.INDCROP.2014.01.045

Isah T (2019) *De novo in vitro* shoot morphogenesis from shoot tip-induced callus cultures of *Gymnema sylvestre* (Retz.) R. Br. ex Sm. Biol Res 52. https://doi.org/10.1186/s40659-019-0211-1

Jamadagni P, Pawar S, Jamadagni S, Gautam M, Gaidhani S, Prasad GP, Gurav A, Gurav A (2021) Recent updates in research on *Gymnema sylvestre*. Pharmacogn Rev 15(30):128–133. https://doi.org/10.5530/phrev.2021.15.15

Jurkiewicz A, Ryszka P, Anielska T, Waligórski P, Białońska D, Góralska K, Tsimilli-Michael M, Turnau K (2010) Optimization of culture conditions of *Arnica montana* L.: effects of mycorrhizal fungi and competing plants. Mycorrhiza 20(5):293–306. https://doi.org/10.1007/s00572-009-0280-z

Kalariya KA, Meena RP, Poojara L, Shahi D, Patel S (2021) Characterization of squalene synthase gene from *Gymnema sylvestre* R. Br. Beni-Suef Univ J Basic Appl Sci 10(1):6. https://doi.org/10.1186/s43088-020-00094-4

Kanetkar PV, Singhal RS, Laddha KS, Kamat MY (2006) Extraction and quantification of gymnemic acids through gymnemagenin from callus cultures of *Gymnema sylvestre*. Phytochem Anal 17(6):409–413. https://doi.org/10.1002/pca.939

Kapoor LD (1990) CRC handbook of ayurvedic medicinal plants. CRC Press, Boca Raton

Karthic R, Seshadri S (2009) Cost effective mass multiplication of *Gymnema sylvestre* in hydroponic system. Nat Prec :1–1. https://doi.org/10.1038/npre.2009.3770.1

Kaushalya N, Senarath W (2014) Callus induction and *in vitro* plantlet regeneration of *Gymnema sylvestre* R. Br. (Retz.) and the phytochemical screening of natural plants and callus cultures. Plant Tissue Cult Biotech 23(2):201–210. https://doi.org/10.3329/ptcb.v23i2.17521

Khan F, Sarker MdMR, Ming LC, Mohamed IN, Zhao C, Sheikh BY, Tsong HF, Rashid MA (2019) Comprehensive review on phytochemicals, pharmacological and clinical potentials of *Gymnema sylvestre*. Front Pharmacol 10:1223. https://doi.org/10.3389/fphar.2019.01223

Kibiti CM, Afolayan AJ (2015) Herbal therapy: a review of emerging pharmacological tools in the management of diabetes mellitus in Africa. Pharmacogn Mag 11(Suppl 2):S258–S274. https://doi.org/10.4103/0973-1296.166046

King AJ (2012) The use of animal models in diabetes research. Br J Pharmacol 166(3):877–894. https://doi.org/10.1111/j.1476-5381.2012.01911.x

Kirtikar KR, Basu BD (1987) Indian medicinal plants. Lalit Mohan Basu, Allahabad. Jayyd Press, New Delhi, India, vol 2, p 146

Kishore L, Singh R (2015) Protective effect of *Gymnema sylvestre* L. against advanced glycation end-product, sorbitol accumulation and aldose reductase activity in Homoeopathic formulation. Indian J Res Homoeopathy 9(4):240–248. https://doi.org/10.4103/0974-7168.172866

Laha S, Paul S, Paul S, Paul S (2019) *Gymnema sylvestre* (Gurmar): a potent herb with anti-diabetic and antioxidant potential. Pharmacogn J 11(2):201–206. https://doi.org/10.5530/pj.2019.11.33

Lambert E, Faizal A, Geelen D (2011) Modulation of triterpene saponin production: in vitro cultures, elicitation, and metabolic engineering. Appl Biochem Biotechnol 164:220–237. https://doi.org/10.1007/s12010-010-9129-3

Lee EJ, Mobin M, Hahn EJ, Paek KY (2006) Effects of sucrose, inoculum density, auxins, and aeration volume on ceil growth of *Gymnema sylvestre*. J Plant Biol 49(6):427. https://doi.org/10.1007/BF03031122

Liu HM, Kiuchi F, Tsuda Y (1992) Isolation and structure elucidation of gymnemic acids, antisweet principles of *Gymnema sylvestre*. Chem Pharm Bull (Tokyo) 40(6):1366–1375. https://doi.org/10.1248/cpb.40.1366

Mahendran G, Iqbal Z, Kumar D, Verma SK, Rout PK, ur Rahman L (2021) Enhanced gymnemic acids production in cell suspension cultures of *Gymnema sylvestre* (Retz.) R. Br. ex Sm. through elicitation. Ind Crops Prod 162:113234. https://doi.org/10.1016/j.indcrop.2020.113234

Maheshwari NV, Nadgauda RS (2016) Variation in gymnemic acid content in Gymnema germplasm from different geographical regions of Gujarat. J Chem Pharm Res 8(3):37–41

Maier W, Hammer K, Dammann U, Schulz B, Strack D (1997) Accumulation of sesquiterpenoid cyclohexenone derivatives induced by an arbuscular mycorrhizal fungus in members of the Poaceae. Planta 202(1):36–42. https://doi.org/10.1007/s004250050100

Mankar GD, Gulave AB, Datkhile K, SB Zimare (2021) Altitudinal gradients influence the accumulation of pharmaceutically important phenolic compounds in the leaves of *Lobelia nicotianifolia* Roth. and regulates its antioxidant and anticancer property. Indian J Biochem Biophys 58(3):253–260

Manohar SH, Naik PM, Praveen N, Murthy HN (2009) Distribution of gymnemic acid in various organs of *Gymnema sylvestre*. J Forestry Res 20(3):268–270. https://doi.org/10.1007/s11676-009-0046-7

Manonmani R, Francisca P (2012) *In vitro* multiplication of *Gymnema sylvestre* R. Br. through nodal explants. Int J Pharma Bio Sci 3(2):49–53

Mentreddy SR (2007) Medicinal plant species with potential antidiabetic properties. J Sci Food Agric 87(5):743–750. https://doi.org/10.1002/jsfa.2811

Nagella P, Thiruvengadam M, Jung S, Murthy H, Chung I (2013) Establishment of *Gymnema sylvestre* hairy root cultures for the production of gymnemic acid. Acta Physiol Plant 35(10):3067–3073. https://doi.org/10.1007/s11738-013-1327-5

Naik PM, Al-Khayri JM (2016) Abiotic and biotic elicitors-role in secondary metabolites production through *in vitro* culture of medicinal plants. In: Abiotic and biotic stress in plants—recent advances and future perspectives. InTech, Rijeka, pp 247–277. https://doi.org/10.5772/61442

Nair S, Keshavachandran R (2007a) Genetic variability of chakkarakolli (*Gymnema sylvestre* R. BR.) in Kerala assessed using morphological and biochemical markers. J Trop Agri 44(1–2):64–67

Nair S, Keshavachandran R (2007b) Molecular diversity in chakkarakolli (*Gymnema sylvestre* R. Br.) assessed through isozyme and RAPD analysis. J Trop Agri 44(1–2):31–36

Nalina L, Rajamani K (2020) Morphologically elite genotype in *Gymnema* TNGSy-55-IC-0630558. Electron J Plant Breed 11(1):276–278

National medicinal plant board, India. https://nmpb.nic.in/sites/default/files/publications/gudmar.pdf. Accessed on 21st June 2022

Nazeem P, Sudhadevi P, Josepha L, Geetha C (1991) *In vitro* callus production from leaves of *Gymnema sylvestre* R. Br. Anc Sci Life 11(1–2):43

Netala VR, Kotakadi VS, Gaddam SA, Ghosh SB, Tartte V (2016) Elicitation of gymnemic acid production in cell suspension cultures of *Gymnema sylvestre* R. Br. through endophytic fungi. 3 Biotech 6(2):1–11. https://doi.org/10.1007/s13205-016-0555-y

Niazian M (2019) Application of genetics and biotechnology for improving medicinal plants. Planta 249(4):953–973. https://doi.org/10.1007/s00425-019-03099-1

Nikalje GC, Zimare SB, Malpathak NP (2013) A comparative pharmacognostic study on the leaf, stem and root components of *Gymnema sylvestre* (Retz) R. Br. ex. Sm. Proc Natl Acad Sci India Sect B Biol Sci 83:125–134. https://doi.org/10.1007/s40011-012-0088-z

Osman MA, Dhawan SS, Bahl JR, Darokar MP, Khanuja S (2011) AFLP marking and polymorphism among progenies of *Gymnema sylvestre*: an important medicinal plant of India. Nat Prod Commun 6(11):1679–1682

Palazón J, Navarro-Ocaña A, Hernandez-Vazquez L, Mirjalili MH (2008) Application of metabolic engineering to the production of scopolamine. Molecules 13(8):1722–1742. https://doi.org/10.3390/molecules13081722

Pálsson R, Patel UD (2014) Cardiovascular complications of diabetic kidney disease. Adv Chronic Kidney Dis 21(3):273–280. https://doi.org/10.1053/j.ackd.2014.03.003

Pandey A (2012) Cultivation technique of an important medicinal plant *Gymnema sylvestre* R. Br. (Gurmar). Acad J Plant Sci 5:84–89. https://doi.org/10.5829/idosi.ajps.2012.5.3.2021

Pandey AK, Yadav S (2010) Variation in gymnemic acid content and non-destructive harvesting of *Gymnema sylvestre* (Gudmar). Pharmacognosy Res 2(5):309–312. https://doi.org/10.4103/0974-8490.72330

Podolak I, Galanty A, Sobolewska D (2010) Saponins as cytotoxic agents: a review. Phytochem Rev 9(3):425–474. https://doi.org/10.1007/s11101-010-9183-z

Pramanick DD (2016) Anatomical studies on the leaf of *Gymnema sylvestre* (Retz.) R.Br. ex Schult. (Apocynaceae): a magical herbal medicine for diabetes. Int J Herb Med 4(1):70–72

Praveen N, Murthy H, Chung I (2011) Improvement of growth and gymnemic acid production by altering the macro elements concentration and nitrogen source supply in cell suspension cultures of Gymnema sylvestre R. Br. Ind Crops Prod 33(2):282–286. https://doi.org/10.1016/j.indcrop.2010.12.015

Praveen N, Thiruvengadam M, Yang Y, Kim S, Murthy H, Chung I (2014) Production of gymnemic acid from hairy root cultures of *Gymnema sylvestre* R. Br. as influenced by polyunsaturated fatty acids (PUFAs) and their antioxidant activity. Ind Crops Prod 54:54–61. https://doi.org/10.1016/j.indcrop.2014.01.008

Reddy SP, Gopal RG, Sita LG (1998) *In vitro* multiplication of *Gymnema sylvestre* R.Br. – An important medicinal plant. Curr Sci 75(8):843–845

Roy A, Ghosh S, Chaudhari M, Saha PK (2008) Effect of different plant hormones on callus induction in *Gymnema sylvestre* R.Br. (Asclepiadaceae), Afr J Biotechnol 7(13):2209–2211

Sabir F., P. Tiwari, Asha, and N. S. Sangwan, "*Gymnema sylvestre* UDP-glucosyltransferase mRNA, partial cds (accession no. GU191124)," NCBI Database, 2009.

Sarin R (2005) Useful metabolites from plant tissue cultures. Biotechnology 4(2):79–93

Shah SN, Amjad M, Ansari S (2013) Micropropagation of *Gymnema sylvestre* R. Br. Sky J Med Plant Res 2(3):18–28

Shahnawaz M, Zanan RL, Wakte KV, Mathure SV, Kad TD, Deokule SS, Nadaf AB (2012) Genetic diversity assessment of *Gymnema sylvestre* (Retz.) R. Br. ex Sm. populations from Western Ghats of Maharashtra, India. Genet Resour Crop Evol 59(1):125–134. https://doi.org/10.1007/s10722-011-9757-7

Shrinet K, Singh RK, Chaurasia AK, Tripathi A, Kumar A (2021) Bioactive compounds and their future therapeutic applications. In: Natural bioactive compounds, technological advancements. Elsevier Inc. pp 337–362. https://doi.org/10.1016/b978-0-12-820655-3.00017-3

Singh VK, Umar S, Ansari SA, Iqbal M (2008) *Gymnema sylvestre* for diabetics. J Herbs Spices Med Plants 14(1–2):88–106. https://doi.org/10.1080/10496470802341508

Solis-Herrera C, Triplitt C, Cersosimo E, DeFronzo RA (2000) Pathogenesis of Type 2 Diabetes Mellitus. In: Feingold KR, Anawalt B, Boyce A, Chrousos G, de Herder WW, Dhatariya K, Dungan K, Hershman JM, Hofland J, Kalra S, Kaltsas G, Koch C, Kopp P, Korbonits M, Kovacs CS, Kuohung W, Laferrère B, Levy M, McGee EA, McLachlan R, Morley JE, New M, Purnell J, Sahay R, Singer F, Sperling MA, Stratakis CA, Trence DL, Wilson DP (eds) Endotext. MDText.com, Inc, South Dartmouth (MA)

Thangavel S, Nisha MC, Sevanan R (2009) Effect of indigenous arbuscular mycorrhizal fungi on some growth parameters and phytochemical constituents of Pogostemon patchouli Pellet. Maejo Int J Sci Technol 3(1):222–223

Tiwari P, Mishra BN, Sangwan NS (2014) Phytochemical and pharmacological properties of *Gymnema sylvestre*: an important medicinal plant. Biomed Res Int 2014:830285. https://doi.org/10.1155/2014/830285

Tiwari PF, Sabir A, Sangwan NS (2009) *Gymnema sylvestre* glycosyltransferase mRNA, partial cds (accession no. GU181368), NCBI Database

Veerashree V, Anuradha C, Kumar V (2012) Elicitor-enhanced production of gymnemic acid in cell suspension cultures of *Gymnema sylvestre* R. Br. Plant Cell Tissue Organ Cult 108(1):27–35. https://doi.org/10.1007/s11240-011-0008-6

Verma AK, Dhawan SS, Singh S, Bharati KA, Null J (2016) Genetic and chemical profiling of *Gymnema sylvestre* accessions from entral India: its implication for quality control and therapeutic potential of plant. Pharmacogn Mag 12(Suppl 4):S407–S413. https://doi.org/10.4103/0973-1296.191443

World Health Organization (WHO) (2022). https://www.who.int/health-topics/diabetes. Accessed on 21st June 2022

Yadav D, Kwak M, Jin J-O (2019) Clinical applications of *Gymnema sylvestre* against type 2 diabetes mellitus and its associated abnormalities. Progr Nutr 21(2):258–269. https://doi.org/10.23751/pn.v21i2.7780

Yoshikawa K, Amimoto K, Arihara S, Matsuura K (1989a) Structure studies of new antisweet constituents from *Gymnema sylvestre*. Tetrahedron Lett 30(9):1103–1106. https://doi.org/10.1016/S0040-4039(01)80371-3

Yoshikawa K, Amimoto K, Arihara S, Matsuura K (1989b) Gymnemic acid V, VI and VII from gurmar, the leaves of *Gymnema sylvestre* R. Br. Chem Pharm Bull 37(3):852–854. https://doi.org/10.1248/cpb.37.852

Yoshikawa K, Arihara S, Matsuura K (1991) A new type of antisweet principles occurring in *Gymnema sylvestre*. Tetrahedron Lett 32(6):789–792. https://doi.org/10.1016/S0040-4039(00)74887-8

Yoshikawa K, Kondo Y, Arihara S, Matsuura K (1993) Antisweet natural products. IX. structures of gymnemic acids XV-XVIII from *Gymnema sylvestre* R. BR. V. Chem Pharm Bull 41(10):1730–1732. https://doi.org/10.1248/cpb.41.1730

Yoshikawa M, Murakami T, Kadoya M, Li Y, Murakami N, Yamahara J, Matsuda H (1997) Medicinal foodstuffs. IX. The inhibitors of glucose absorption from the leaves of *Gymnema sylvestre* R. BR. (Asclepiadaceae): structures of gymnemosides a and b. Chem Pharm Bull 45(10):1671–1676. https://doi.org/10.1248/cpb.45.1671

Zhang T, Han M, Yang L, Han Z, Cheng L, Sun Z, Yang L (2019) The effects of environmental factors on ginsenoside biosynthetic enzyme gene expression and saponin abundance. Molecules 24(1):14. https://doi.org/10.3390/molecules24010014

Zimare S, Kakade P, Malpathak N (2018) Biotic elicitation: a strategy for enhancement of antidiabetic gymnemic acid from *Gymnema sylvestre* 97:43–48

Zimare SB, Borde MY, Jite PK, Malpathak NP (2013) Effect of AM fungi (Gf, Gm) on biomass and gymnemic acid content of *Gymnema sylvestre* (Retz.) R. Br. ex Sm. Proc Natl Acad Sci India Sect B Biol Sci 83(3):439–445. https://doi.org/10.1007/s40011-013-0159-9

Zimare SB, Kakade PS, Malpathak NP (2017) Unorganized (cell suspension) cultures: a biotechnological tool for enhancement of biomass and gymnemic acid in *Gymnema sylvestre*. Plant Cell Biotechnol Mol Biol 18(5–6):281–289

Zimare SB, Malpathak NP (2017) *In vitro* multiple shoot and gymnemic acid production in Gymnema sylvestre (Retz.) R. Br. Ex. Sm. Indian J Biotechnol 16(4):635–640

Zimare SB, Malpathak NP (2020) Potential mevalonate pathway precursors for enhanced production of gymnemic acid. Indian J Exp Biol 58(09):651–655

Zubek S, Mielcarek S, Turnau K (2012) Hypericin and pseudohypericin concentrations of a valuable medicinal plant *Hypericum perforatum* L. are enhanced by arbuscular mycorrhizal fungi. Mycorrhiza 22(2):149–156. https://doi.org/10.1007/s00572-011-0391-1

Index

J. M. Al-Khayri et al. (eds.), *Biodiversity and Genetic Improvement of Medicinal and Aromatic Plants I*, Advances in Plant Breeding Strategies 9,
https://doi.org/10.1007/978-3-031-81846-2

C

D

E

F

G

H

I

The manufacturer's authorised representative in the EU is Springer
Nature Customer Service Centre GmbH, Europaplatz 3, 69115 Heidelberg,
Germany. If you have any concerns regarding our products, please
contact ProductSafety@springernature.com

Printed and bound by CPI Group (UK) Ltd, Croydon, CR0 4YY
30/09/2026
02225568-0010